Quantifying and Understanding Plant Nitrogen Uptake for Systems Modeling

Quantifying and Understanding Plant Nitrogen Uptake for Systems Modeling

Edited by
Liwang Ma
Lajpat R. Ahuja
Tom Bruulsema

CRC Press
Taylor & Francis Group
Boca Raton London New York

CRC Press is an imprint of the
Taylor & Francis Group, an **informa** business

Published 2009 by CRC Press
Taylor & Francis Group
6000 Broken Sound Parkway NW, Suite 300
Boca Raton, FL 33487-2742

© 2009 by Taylor & Francis Group, LLC
CRC Press is an imprint of Taylor & Francis Group, an Informa business

First issued in paperback 2019

No claim to original U.S. Government works

ISBN-13: 978-0-367-45259-9 (pbk)
ISBN-13: 978-1-4200-5295-4 (hbk)

**Visit the Taylor & Francis Web site at
http://www.taylorandfrancis.com**

**and the CRC Press Web site at
http://www.crcpress.com**

Library of Congress Cataloging-in-Publication Data

Quantifying and understanding plant nitrogen uptake systems modeling /
    editors, Liwang Ma, Lajpat R. Ahuja, Thomas W. Bruulsema.
        p. cm.
    Includes bibliographical references and index.
    ISBN 978-1-4200-5295-4 (hardback : alk. paper)
    1. Plants, Effect of nitrogen on. 2. Crops and nitrogen. 3. Nitrogen--Fixation. I.
Ma, Liwang. II. Ahuja, L. (Lajpat) III. Bruulsema, Thomas W. IV. Title.

QK753.N54Q36 2008
631.8'4--dc22   2008019573

Contents

Editors vii

Contributors ix

Chapter 1   Current Status and Future Needs in Modeling Plant Nitrogen
Uptake: A Preface 1

Liwang Ma, Lajpat R. Ahuja, and Thomas W. Bruulsema

Chapter 2   Modeling Nitrogen Fixation and Its Relationship to Nitrogen
Uptake in the CROPGRO Model 13

*Kenneth J. Boote, Gerrit Hoogenboom, James W. Jones, and
Keith T. Ingram*

Chapter 3   Modeling Nitrate Uptake and Nitrogen Dynamics in Winter
Oilseed Rape (*Brassica napus* L.) 47

*Philippe Malagoli, Frédéric Meuriot, Philippe Laine,
Erwan Le Deunff, and Alain Ourry*

Chapter 4   Control of Plant Nitrogen Uptake in Native Ecosystems by
Rhizospheric Processes 71

Hormoz BassiriRad, Vincent Gutschick, and Harbans L. Sehtiya

Chapter 5   Dissolved Organic Nitrogen and Mechanisms of Its Uptake by
Plants in Agricultural Systems 95

*David L. Jones, John F. Farrar, Andrew J. Macdonald,
Sarah J. Kemmitt, and Daniel V. Murphy*

Chapter 6   Water and Nitrogen Uptake and Responses in Models of Wheat,
Potatoes, and Maize 127

P. D. Jamieson, R. F. Zyskowski, F. Y. Li, and M. A. Semenov

Chapter 7   Modeling Grain Protein Formation in Relation to Nitrogen
Uptake and Remobilization in Rice 147

*Yan Zhu, Hongbao Ye, Gregory S. McMaster, Weiguo Li, and
Weixing Cao*

Chapter 8 Modeling Water and Nitrogen Uptake Using a Single-Root
Concept: Exemplified by the Use in the Daisy Model 169

Søren Hansen and Per Abrahamsen

Chapter 9 Modeling Plant Nitrogen Uptake Using Three-Dimensional and
One-Dimensional Root Architecture 197

*Lianhai Wu, Ian J. Bingham, John A. Baddeley, and
Christine A. Watson*

Chapter 10 Simulation of Nitrogen Demand and Uptake in Potato Using a
Carbon-Assimilation Approach 219

*Dennis Timlin, Mikhail Kouznetsov, David Fleisher,
Soo-Hyung Kim, and V. R. Reddy*

Chapter 11 Roots below One-Meter Depth Are Important for Uptake of
Nitrate by Annual Crops 245

Hanne L. Kristensen and Kristian Thorup-Kristensen

Chapter 12 Nitrogen-Uptake Effects on Nitrogen Loss in Tile Drainage as
Estimated by RZWQM 259

Robert W. Malone and Liwang Ma

Chapter 13 Simulated Soil Water Content Effect on Plant Nitrogen Uptake
and Export for Watershed Management... 277

*Ping Wang, Ali Sadeghi, Lewis Linker, Jeff Arnold,
Gary Shenk, and Jing Wu*

Index.. 305

Editors

Liwang Ma is a soil scientist with the USDA-ARS, Agricultural Systems Research Unit in Fort Collins, Colorado. Dr. Ma received his B.S. and M.S. in agricultural biophysics from Beijing Agricultural University (now China Agricultural University) in 1984 and 1987, respectively, and his Ph.D. in soil science from Louisiana State University in 1993. He has authored and coauthored 70 journal papers and 50 other published works (books, book chapters, and proceedings). His research interests center on agricultural systems modeling and include pesticide fates, plant growth, soil carbon/nitrogen dynamics, plant water and nitrogen uptake, and soil water and nutrient movement. He is the principal scientist responsible for developing, enhancing, and maintaining the USDA-ARS Root Zone Water Quality Model (RZWQM2). He is a Fellow of the American Society of Agronomy and has served as an associate editor of the *Journal of the Soil Science Society of America* (2001–2007). He is now serving as associate editor of the *Agronomy Journal* and the *Journal of Environmental Quality*.

Lajpat (Laj) R. Ahuja is a supervisory soil scientist and research leader of the USDA-ARS, Agricultural Systems Research Unit in Fort Collins, Colorado. He has made original and pioneering research contributions in several areas of agricultural systems, including infiltration and water flow in soils, estimation and scaling of hydraulic properties, transport of agrochemicals to runoff and to groundwater through soil matrix and macropores, quantification of the effects of tillage and other management practices on relevant properties and processes, and modeling of entire agricultural systems and application of system models in field research and technology transfer. Dr. Ahuja has authored or coauthored 270 publications and served as associate editor (1987–1992) and technical editor (1994–1996) of the *Journal of the Soil Science Society of America*. He is a Fellow of the Soil Science Society of America (SSSA), the American Society of Agronomy (ASA), and the American Association for the Advancement of Science (AAAS). Dr. Ahuja is a recipient of the USDA-ARS, Southern Plains Area, Scientist of the Year Award; USDA-ARS, Northern Plains Area, Scientist of the Year Award; USDA-ARS, Technology Transfer Award; Federal Laboratory Consortium (FLC), Excellence in Technology Transfer Award; SSSA Don Kirkham Soil Physics Award; and ASA Environmental Quality Award.

Thomas W. Bruulsema, a native of Ancaster, Ontario, is director of the northeast region of the North American Program of the International Plant Nutrition Institute (IPNI). His research focuses on the benefits of plant nutrition for the crops of the region, and his educational activities feature responsible, science-based use of fertilizer nutrients. Before joining the institute as regional director in 1994, Dr. Bruulsema held a research associate position at the University of Minnesota. He holds a Ph.D. in soil science from Cornell University. He also has experience in international agriculture, having served four years with the Mennonite Central Committee as a research agronomist in Bangladesh. He currently serves as president of the Canadian Society of Agronomy. He served as president of the Northeast Branch of the American Society of Agronomy and of the Soil Science Society of America from 1999 to 2002. He has also been active in the Certified Crop Adviser (CCA) program, having served as chair of the International (2001–2004) and Ontario (1999–2000) boards, and currently represents CCA on the board of directors of the American Society of Agronomy. Dr. Bruulsema is a Fellow of the American Society of Agronomy.

Contributors

Per Abrahamsen
Department of Agricultural Sciences,
Faculty of Life Science
University of Copenhagen
Taastrup, Denmark

Jeff Arnold
USDA-ARS
Grassland Soil and Water Research
Lab
Temple, Texas, USA

Lajpat R. Ahuja
USDA-ARS
Agricultural System Research Unit
Fort Collins, Colorado, USA

John A. Baddeley
Crop and Soil Systems Research, SAC
Craibstone Estate, Aberdeen
United Kingdom

Hormoz BassiriRad
Department of Biological Sciences
University of Illinois at Chicago
Chicago, Illinois, USA

Ian J. Bingham
Crop and Soil Systems Research, SAC
Craibstone Estate, Aberdeen
United Kingdom

Kenneth J. Boote
Agronomy Department
University of Florida
Gainesville, Florida, USA

Thomas W. Bruulsema
International Plant Nutrition Institute
Guelph, Ontario, Canada

Weixing Cao
Hi-Tech Key Lab of Information
Agriculture of Jiangsu Province
Nanjing Agricultural University
Nanjing, Jiangsu
People's Republic of China

John F. Farrar
School of Biological Sciences
University of Wales
Bangor, Gwynedd
United Kingdom

David Fleisher
USDA-ARS
BA, PSI, Crop Systems and Global
Change Lab
Beltsville, Maryland, USA

Vincent Gutschick
Department of Biology
New Mexico State University
Las Cruces, New Mexico, USA

Søren Hansen
Department of Agricultural Sciences,
Faculty of Life Science
University of Copenhagen
Taastrup, Denmark

Gerrit Hoogenboom
Department of Biological and
Agricultural Engineering
University of Georgia
Griffin, Georgia, USA

Keith T. Ingram
Agricultural and Biological Engineering
Department
University of Florida
Gainesville, Florida, USA

P. D. Jamieson
NZ Institute for Crop and Food
  Research
Christchurch, New Zealand

David L. Jones
School of the Environment and Natural
  Resources
University of Wales
Bangor, Gwynedd
United Kingdom

James W. Jones
Agricultural and Biological Engineering
  Department
University of Florida
Gainesville, Florida, USA

Sarah J. Kemmitt
School of the Environment and Natural
  Resources
University of Wales
Bangor, Gwynedd, United Kingdom
Agriculture and Environment Division,
  Rothamsted Research
Harpenden, Hertfordshire
United Kingdom

Soo-Hyung Kim
College of Forest Resources
University of Washington
Seattle, Washington, USA

Mikhail Kouznetsov
J. Blaustein Institutes for Desert
  Research
Ben-Gurion University of the Negev
Sede Boker Campus, Israel

Hanne L. Kristensen
Department of Horticulture
University of Aarhus
Aarslev, Denmark

Philippe Laine
UMR INRA-UCBN 950 EVA,
  Ecophysiologie Végétale, Agronomie
  and Nutritions N, C, S
Institut de Biologie Fondamentale et
  Appliquée
Université de Caen Basse Normandie
Caen Cedex, France

Erwan Le Deunff
UMR INRA-UCBN 950 EVA,
  Ecophysiologie Végétale, Agronomie
  and Nutritions N, C, S
Institut de Biologie Fondamentale et
  Appliquée
Université de Caen Basse Normandie
Caen Cedex, France

F. Y. Li
NZ Institute for Crop and Food
  Research
Christchurch, New Zealand

Weiguo Li
Hi-Tech Key Lab of Information
  Agriculture of Jiangsu Province
Nanjing Agricultural University
Nanjing, Jiangsu
People's Republic of China

Lewis Linker
U.S. Environmental Protection Agency,
  Chesapeake Bay Program
Annapolis, Maryland, USA

Liang Ma
USDA-ARS
Agricultural Systems Research Unit
Fort Collins, Colorado, USA

Andrew J. Macdonald
Agriculture and Environment Division,
  Rothamsted Research
Harpenden, Hertfordshire
United Kingdom

Philippe Malagoli
Department of Life Science
University of Toronto
Toronto, Ontario, Canada

R. W. Malone
USDA-ARS
National Soil Tilth Lab
Ames, Iowa, USA

Gregory S. McMaster
USDA-ARS
Agricultural Systems Research Unit
Fort Collins, Colorado, USA

Frédéric Meuriot
UMR INRA-UCBN 950 EVA,
  Ecophysiologie Végétale, Agronomie
  and Nutritions N, C, S
Institut de Biologie Fondamentale et
  Appliquée
Université de Caen Basse Normandie
Caen Cedex, France

Daniel V. Murphy
School of Earth and Geographical
  Sciences
The University of Western Australia
Crawley, Australia

Alain Ourry
UMR INRA-UCBN 950 EVA,
  Ecophysiologie Végétale, Agronomie
  and Nutritions N, C, S
Institut de Biologie Fondamentale et
  Appliquée
Université de Caen Basse Normandie
Caen Cedex, France

V. R. Reddy
USDA-ARS
BA, PSI, Crop Systems and Global
  Change Lab
Beltsville, Maryland, USA

Ali Sadeghi
USDA-ARS
Hydrology and Remote Sensing Lab
Beltsville, Maryland, USA

Harbans L. Sehtiya
Department of Biological Sciences
University of Illinois at Chicago
Chicago, Illinois, USA

M. A. Semenov
Biomathematics and Bioinformatics,
  Rothamsted Research
Harpenden, Hertfordshire
United Kingdom

Gary Shenk
U.S. Environmental Protection Agency,
  Chesapeake Bay Program
Annapolis, Maryland, USA

Kristian Thorup-Kristensen
Department of Horticulture
University of Aarhus
Aarslev, Denmark

Dennis Timlin
USDA-ARS
PSI, Crop Systems and Global Change
  Lab
Beltsville, Maryland, USA

Ping Wang
University of Maryland Center for
  Environmental Science
Annapolis, Maryland, USA

Christine A. Watson
Crop and Soil Systems Research,
  SAC
Craibstone Estate, Aberdeen
United Kingdom

Jing Wu
University of Maryland Center for
  Environmental Science
Annapolis, Maryland, USA

Lianhai Wu
Crop and Soil Systems Research,
  SAC
Craibstone Estate, Aberdeen
United Kingdom

Hongbao Ye
Hi-Tech Key Lab of Information
  Agriculture of Jiangsu Province
Nanjing Agricultural University
Nanjing, Jiangsu
People's Republic of China

Yan Zhu
Hi-Tech Key Lab of Information
  Agriculture of Jiangsu Province
Nanjing Agricultural University
Nanjing, Jiangsu
People's Republic of China

R. F. Zyskowski
New Zealand Institute for Crop and
  Food Research
Christchurch, New Zealand

1 Current Status and Future Needs in Modeling Plant Nitrogen Uptake: A Preface

Liwang Ma, Lajpat R. Ahuja, and
Thomas W. Bruulsema

CONTENTS

1.1   Introduction 1
1.2   Current Status 2
1.3   Future Research Needs 8
Acknowledgment 10
References 10

1.1   INTRODUCTION

Managing nitrogen (N) to increase crop production is one of the success stories of modern agriculture. However, with recent concerns about environmental quality, agricultural managers face new challenges and have had to shift from a single goal of increasing crop production to dual goals of increasing production and reducing environmental impacts. The long-term sustainability of both food security and the environment has set new goals for modern agriculture enterprises. Site-specific optimal management of N under different agroclimatic conditions is an important part of these goals. This optimal management involves synchronizing the timing of soil N availability (mineralization and fertilization) with crop N demand and appropriate water management under uncertain weather conditions. Better understanding and quantification of plant N uptake at different growth stages and under varying conditions is an essential requirement for this purpose.

This new challenge is beyond the many decades of field trials for measuring crop N uptake under specific experimental conditions. It requires a quantitative systems approach to integrate all the factors affecting plant N uptake together to develop integrated management practices (IMP) on a whole system level. Synchronizing plant N demand and soil N supply is the inevitable goal of 21st century agriculture and requires a complete understanding of the dynamics and interactions between N

1

demand and N supply. Since plant N demand is largely unknown and depends on plant growth (e.g., carbon assimilation, phenological stage) and plant stress conditions (e.g., N and water stresses), it is a real-world challenge to decide when and how much fertilizer should be applied to ensure the right amount of available soil N is in the root zone to minimize any potential N loss to the environment. There is also a lag period between the start of plant N stress and the observable stress phenomena in the field.

Agricultural system models are able to take up this challenge because they are developed to integrate all the factors affecting plant N uptake into a virtual management system. As shown in Figure 1.1, N uptake affects and is affected by all the processes in the soil–plant system directly or indirectly. There are many feedback mechanisms that are still unknown to agricultural scientists. Some of the known growth regulators are not considered in models, partially due to lack of quantification of their functionality. Nonetheless, agricultural system models present the potential of integrating all of the processes together based on decades of experimental observations. Although there are numerous system models of different types, N uptake is generally determined by soil N supply and plant N demand (Figure 1.1). The differences lie in how to estimate the supply and demand terms, and how to quantify various factors affecting them.

As shown in Figure 1.1, all the factors are intertwined together. For example, water uptake (transpiration) is one of the main processes affecting N uptake, although some authors have found that N uptake and transpiration can be decoupled (Hansen and Abrahamsen 2008; BassiriRad et al. 2008). Nitrogen cannot be transported in the soil and plant without water. Therefore, water movement in the soil would affect both plant N-uptake estimation and soil N supply. Generally, transpiration is estimated by soil water supply and energy-driven potential transpiration. Plant root growth/distribution and leaf area index are the two main biological parameters determining transpiration in models, and these are in turn affected by plant N uptake. Plant N demand is also determined from biomass production, photosynthate partitioning, and phenology development. Another important issue is that, although qualitative knowledge of the feedback mechanisms exists after decades of field experiments, quantitative representation of these mechanisms is empirical and is not fully understood, especially how the mechanisms dynamically respond to environmental conditions. As a result, plant N-uptake simulations in any given model are generally empirical in nature and are not transferable to other models without taking into account other associated assumptions (e.g., water uptake, carbon allocation). Another aspect that may be important in some situations is that most system models ignore uptake of dissolved organic N (DON) from soils (Wu et al. 2008).

1.2   CURRENT STATUS

The level of detail of plant N-uptake simulations varies among system models, and their accuracy depends on the quality and quantity of data used. Occasionally, model users also contribute to the quality of simulation results, because tremendous skills and knowledge are needed to correctly use a system model. User errors are inevitable, but can be reduced with training and experience. Often, models are calibrated

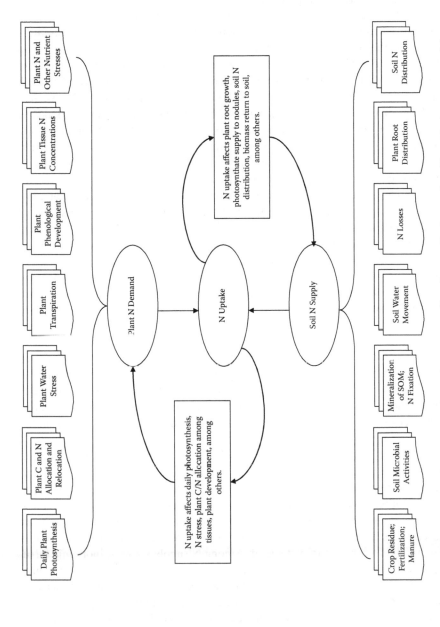

FIGURE 1.1 A schematic of the interactions among the main processes directly contributing to plant N uptake in system models.

and evaluated against only a limited number of sites and years, and using only a few data measurements on residual soil N, aboveground plant N, and grain N. Nitrogen in the root is largely ignored, as are the N contents of individual tissues (e.g., stem, leaf). Also, very few studies have looked into soil N distribution in the soil profile and rooting depth during a crop growing season.

One of the main differences among models of plant N uptake concerns the mechanisms whereby N enters the root system. There are several levels of sophistication. One is the semi-empirical approach, where N uptake is calculated from soil water content, root distribution, and asymptotic exponential functions (similar to the Michaelis–Menten equation) of ammonium (NH_4^+) and nitrate (NO_3^-) in each soil layer, as is the case in CERES and CROPGRO models (Boote et al. 2008). Second is the macromechanistic approach as convective (passive) and active uptake, as in the RZWQM model (Malone and Ma 2008). The third is the micromechanistic approach, where N uptake by single roots is simulated, as in the Daisy model (Hansen and Abrahamsen 2008). The fourth is to simulate N uptake by each of the N transporters on the root surface (Malagoli et al. 2008).

Most models assume no preferential uptake of ammonium and nitrate, but a few give higher priorities to ammonium over nitrate, as is the case in the Daisy model (Hansen and Abrahamsen 2008). Uptake of soil dissolved organic N (DON) is evident in the literature, but incorporation of DON in modeling has not been common due to its complexity and a lack of understanding (Wu et al. 2008; Jones et al. 2008).

With respect to N demand, in principle, crop modelers agree that crop N demand varies among tissues and with growth stages. To make plant response to N status flexible, three N concentrations (minimum, optimum, and maximum) are often assumed for each tissue. In addition, most models have a N storage pool that may be relocated to individual tissues when needed, and have the capability of remobilizing some N from lower-priority tissues to high-priority tissues (Jamieson et al. 2008). However, these relationships are qualitative and not dynamically responsive to environmental or plant growth conditions. Nitrogen supply to N demand ratios, as well as tissue N concentrations relative to the minimum and optimum targets, are also used to define N stress factors for plant growth and development, which again is not fully understood (Jamieson et al. 2008; Boote et al. 2008).

Soil N supply varies greatly among models due to differences in simulating soil organic C/N dynamics, N fixation, root growth, and soil water movement. These differences contribute to the great uncertainty in N supply simulations, especially for long-term studies (Smith et al. 1997). In most cases, model users find it difficult to identify exactly which processes contribute to errors in N supply when they compare measured and simulated soil N concentrations. Soil N is also affected by actual N uptake, which is a function of N demand. Among N supply sources, inorganic fertilizer application is the most precisely known input into the system (if volatilization of ammonia is not an issue). Manure application is subject to mineralization processes, the quantification of which is not universally agreed upon among modelers. On top of these, N fixation in legumes is not simulated in most models. Of course, the amount of soil N available to the plant depends on root distribution and soil water content. Management practices, such as tillage and irrigation, and weather also affect soil N availability (Kristensen and Thorup-Kristensen 2008).

In this book, 12 chapters cover a broad range of topics related to N uptake in agricultural systems from laboratory, to field, to watershed scales, and they outline the current status of research and modeling in plant N uptake.

Chapter 2 presents the state-of-the-science N-fixation submodel in the CROP-GRO model (Boote et al. 2008), whereas most other agricultural system models assume the difference between legume N demand and N uptake as N fixation. Dynamic nodule growth and nitrogenase activity are simulated in CROPGRO as functions of temperature and water deficit conditions. Boote et al. (2008) describe in detail the evolution of the CROPGRO model and the development of the N-fixation module. They also note the difference between CROPGRO and CERES in simulating N stresses and plant N demand. The use of sensitivity analysis and step-by-step reasoning to parameterize the N-fixation module should be very helpful to other model developers. The completeness of evaluating the N-fixation module for N-deficit-based triggering of nodule growth, priority for assimilation, water deficit, temperature response, sink-source relationships, N fertilization, and soil and climate conditions sets an example for model development.

In chapter 3, Malagoli et al. (2008) identify two different transporters in the winter oilseed rape (*Brassica napus* L.) in a ^{15}N study, each with a constitutive and an inducible component. One is the high-affinity transport system (HATS), which exhibited a Michaelis–Menten kinetics behavior. The other is the low-affinity transport system (LATS), which did not show a saturated kinetic behavior. The HATS transporter is shown to be responsible for N intake at low soil NO_3^- concentration, and LATS at high soil NO_3^- concentrations. In addition, functionality of the two transport systems was regulated from stage to stage throughout the growing season. Although the experiments were conducted under no water stress, the transporters were affected by environmental conditions such as root temperature, day/night cycle, photosynthetically active radiation (PAR), and soil N deprivation. Using ^{15}N-labeled fertilizer in the field, the authors also quantified N partitioning in various tissues, and the relative contributions of N taken up versus N remobilized in plant tissues. This study provided further evidence that the assumption of a constant maximum N-uptake rate in the Michaelis–Menten equation may not be adequate.

In chapter 4, BassiriRad et al. (2008) provide a thoughtful review of the interactions among various factors contributing to N uptake in native ecosystems, such as the root/shoot ratio, root longevity, root architecture, and mycorrhizal colonization. Mineralization and atmospheric N deposition are more important in supplying N in native ecosystems than in agricultural systems. They also touch the issue of DON uptake and its interaction with assimilation of nitrate. Similar to the findings of Malone and Ma (2008) described later, BassiriRad et al. (2008) found that the maximum N-uptake rate in the Michaelis–Menten equation was a strong controller of plant N uptake. The root/shoot ratio was determined more by plant water demand than by plant N demand. One interesting result of this chapter is the analysis of mass flow versus diffusive flow of N to plant roots. Their modeled nitrate concentrations and N-uptake rates suggest that diffusion can substitute for mass flow and that N uptake can be independent of transpiration.

In chapter 5, Jones et al. (2008) address the mechanisms and extent of N uptake in the form of soil dissolved organic N (DON) (e.g., amino acids, amino sugars, nucleic acids, chlorophyll, and phospholipids). Although the amount of DON may be as high as 10–50 kg N/ha in some soil–plant systems, studies on plant uptake of DON are sparse, especially with regard to modeling. Leaching of DON is usually not accounted for in soil N balance calculations. Uptake of DON (e.g., amino acids) by soil microorganisms is well recognized and may be preferred from the energy point of view. The transporters for DON absorption in microorganisms are similar to those in plant roots. Plant roots have leaky root surfaces that not only release DON, but also recapture DON via transporters. There are many different transporters specific to each type of DON in plant roots, and their functionality may change with plant age. Jones et al. (2008) reviewed a broad spectrum of literature on soil DON and provided unequivocal evidence on direct uptake of DON by the plants, as well as survival of plants on DON alone. However, the exact role of DON uptake by plants in terms of total plant N budget is largely unknown, which makes it difficult to directly simulate N uptake via DON without fully understanding soil microbial activities and release of DON by plants.

Chapter 6 by Jamieson et al. (2008) presents new approaches to calculate plant N demand by critically reviewing the differences between water and N stresses in crop models. After presenting the current empiricism in the N demand concept, they proposed a new approach to define N demand by separately simulating plant structural N, green tissue N, and storage N. They found that this new concept improved grain protein, green area index, yield, and biomass in wheat, potato, and maize models presented. Different from other studies, grain protein was simulated rather than grain N concentration in general.

Chapter 7 by Zhu et al. (2008) focuses on grain protein formation in rice, which has become increasingly important in recent years. The newly developed protein-formation module used simple approaches to pre- and post-anthesis N uptake, N remobilization, and grain protein accumulation, and that module was then integrated into the RiceGrow model. The integrated model showed promising results for different N fertilization rates, rice varieties, and irrigation levels in China, Japan, and Thailand. This module can potentially be used by other rice models such as CERES-Rice, RCSODS, SIMRIW, and ORYZA2000.

In chapter 8, Hansen and Abrahamsen (2008) detail the single-root approach to water and N uptake in the Daisy model, where water and N uptake can behave very much differently in the soil profile. Water uptake is regulated by the crown potential, whereas N uptake is assumed to be an active process regulated by the N concentration at the root surface. It is also assumed that both mass flow and diffusion contribute to movement of ammonium and nitrate to the root surface. The plant exhibits preference for taking up ammonium over nitrate. Daily N uptake is limited either by plant N demand or a maximum N absorption rate that can be obtained from hydroponic experiments. In this approach, two-thirds of N uptake can occur when water uptake (transpiration) is zero. This approach seems to provide satisfactory simulation of N uptake and dry-matter production without parameter fitting.

In chapter 9, active and passive N-uptake mechanisms are also used in the SPAC-SYS model by Wu et al. (2008), with no preferential absorption of ammonium over nitrate. Different from the RZWQM model, a user-specified factor is utilized to partition total N uptake to passive and active uptake. These N-uptake mechanisms were evaluated with both three-dimensional (3-D) and one-dimensional (1-D) root distributions. They found that the 3-D root structure improved simulations of N uptake and mineral N in the soil profile compared with the 1-D root growth. The 1-D root system is used by most agricultural system models, and since the 3-D root growth simulation requires more parameters, it may not be feasible for larger-scale applications where a wide range of soil types and weather conditions occurs. The simplified 1-D root growth model may be a better choice for practical purposes, unless the goal is to understand the relationship between soil N uptake and root structure.

In chapter 10, in addition to a concise review of N-uptake mechanisms, Timlin et al. (2008) present a unique study on estimating passive and active (diffusive) N uptake in a potato study by maintaining constant soil N concentrations throughout the laboratory experiments and measuring canopy-level carbon-assimilation rates. Passive N uptake was calculated by daily transpiration and soil N concentration, and active uptake was the difference between total N uptake and passive uptake. Total N uptake was calculated from daily carbon (C) assimilation and the C/N ratio of plant tissues. The study was conducted under both ambient (370 μmol mol air^{-1}) and elevated (700 μmol mol air^{-1}) CO_2. They found that total N uptake was higher under ambient CO_2 compared with elevated CO_2 due to a lower C/N ratio at elevated CO_2. Since transpiration was not different between the two CO_2 levels, passive N uptake was similar as well. Thus, active N uptake was higher under the ambient CO_2 level than under elevated CO_2. The results also showed that there was a decrease in contribution from active N uptake as soil N concentration increased. A simulation model with a diffusive N-uptake mechanism and 2D-SOIL was successful in modeling active N uptake and its responses to N treatments under ambient CO_2.

In chapter 11, Kristensen and Thorup-Kristensen (2008) present a ^{15}N study of root N uptake from below 1-m soil profile and confirmed that correct simulation of rooting depth was important for plant N uptake and soil residual N for deep-rooted crops. Using the EU_ROTATE_N model, they found that as much as 30% of soil residual N was overestimated when maximal rooting depth was preset to 0.9 m compared with unlimited rooting depth for winter wheat. They also found that methods of root measurements could affect plant N uptake considerably. In their study with curly kale, they estimated four-times-higher N inflow when root distributions were measured by the root extraction method than by the minirhizotron method.

In chapter 12, Malone and Ma (2008) use the Root Zone Water Quality Model (RZWQM) to investigate the importance of N-uptake simulation on N loss in tile drainage flow. In RZWQM, N uptake is simulated by both a passive uptake (via transpiration stream) and an active uptake (via the Michaelis–Menten approach). Active uptake only plays a role when passive uptake cannot meet plant N demand that is estimated by daily C assimilation and a growth-stage-dependent plant N concentration. Plants take up ammonium and nitrate in proportion to their concentrations in soil. Sensitivity analyses of the maximum daily N-uptake parameter in the Michaelis–Menten equation show that this parameter needs to be calculated directly from experimental data rather than calibrated. Since N leaching is small compared with plant N uptake, a 4% error in simulated N uptake will result in 30% error in N loss in tile drainage.

In chapter 13, use of the Michaelis–Menten equation for N uptake in watershed models is exemplified by Wang et al. (2008) in the SWAT (Soil and Water Assessment Tool) and HSPF (Hydrological Simulation Program—FORTRAN) models. In these models, the users specify monthly maximum N uptake, which is then scaled to daily or even finer time steps. This approach works better for forested areas than for cropland because the N uptake is more stable during the period of a month in forestland than in cropland. By taking into account the soil moisture effects on N uptake at both above and below field capacity, the modified Michaelis–Menten equation proposed by Wang et al. (2008) correctly simulates the relationship between N uptake and N export in watersheds in both dry and wet years, as compared with the target-based plant N uptake, first-order kinetic plant N uptake, and the concentration-based Michaelis–Menten approaches in the HSPF model and the plant N-demand approach in the SWAT model. This modified Michaelis–Menten equation should work even better if its maximum N uptake is linked to plant biomass production or plant N demand.

1.3 FUTURE RESEARCH NEEDS

There are many system models in the modeling community. Comparison among the models in terms of N uptake is difficult, if not impossible, because differences exist in simulating N-related processes besides N uptake. Many comparisons in the literature do not generate new insight into the N-uptake mechanism *per se*. To effectively develop a good N-uptake model, all the processes in Figure 1.1 need to be evaluated independently. Frequently, when a model fails to predict N uptake, it is difficult to identify which processes are contributing to the simulation errors. For example, when we attribute errors to soil N supply, it is only fair if we have continuous soil N concentration measurements (especially around the root surface) and rooting depth data. When we blame plant N demand simulation for the N-uptake error, we should have continuous plant N measurement and daily N demand. With advances in *in situ* microscopic plant and soil measurements, plant N-uptake models are expected to be evaluated more rigorously.

Computing power is currently not the main limitation to the number of processes that can be included in agricultural system models. The challenge is how to parameterize these processes. Biological processes function in multidimensional space due to multifactor effects on a single process, and due to feedback mechanisms in response to environmental conditions. However, in most N-uptake models, quantification of processes is empirical, assuming steady state, and lacks feedback mechanisms. In addition, many of these processes are only qualitatively understood and are approximated in model formulations. The research outlined in this book suggests that further research may be useful in the following areas:

1. Plant N demand is the upper limit for plant N uptake. Thus, it is important to make its estimation realistic. Usually, each plant tissue is considered to have three preset N concentrations (minimum, optimum, and maximum) that vary with phenology and physiological age. However, this concept needs to be reexamined by dividing crop biomass into different components with characteristic N concentrations, such as structural material, green tissue, and storage (Jamieson et al. 2008). Remobilization of N between tissues

also needs to be further studied so that plant N demand can be correctly estimated (Malagoli et al. 2008).

2. Although passive and active (diffusive) N uptakes are used in RZWQM and other models (Malone and Ma 2008; Timlin et al. 2008; Wu et al. 2008), this separation is challenged by BassiriRad et al. (2008) and Hansen and Abrahamsen (2008). They showed that plant N uptake can occur without water uptake and that diffusion of N to the root surface may be more important than mass flow. In theory, all N entering the plant roots is carried by a N transporter on the root surface. However, the role of water is important for N diffusion in soil to the root surface and N translocation inside the plant (Timlin et al. 2008).

3. Active N uptake can be in the mathematical form of the Michaelis–Menten (Malone and Ma 2008) or diffusive equations (Timlin et al. 2008; Hansen and Abrahamsen 2008). The practice of using constant coefficients for the Michaelis–Menten equation needs to be reconsidered (Malagoli et al. 2008). The maximum uptake rate for the Michaelis–Menten equation has a major role in simulating plant N uptake and soil residual N (Malone and Ma 2008; Wang et al. 2008; BassiriRad et al. 2008).

4. Plant root distributions are important in estimating N supply from the soils. A 3-D root distribution model improved plant N uptake compared with a 1-D root distribution (Wu et al. 2008). However, a 3-D root distribution model requires many more parameters than most users can supply. Most importantly, rooting depth should be correctly simulated so that deep-rooted crops can access soil N at a lower soil profile (Kristensen and Thorup-Kristensen 2008). Root functionality as affected by physiological age and longevity of root may be another issue for improvement in modeling plant N uptake (Kristensen and Thorup-Kristensen 2008; BassiriRad et al. 2008).

5. Uptake of dissolved soil organic nitrogen (DON) needs to be part of simulation models in systems where DON constitutes a large fraction of plant available N (Wu et al. 2008; BassiriRad et al. 2008; Jones et al. 2008). Including DON will not only improve N-uptake simulations, but also N balance in the environment (e.g., DON in leaching and runoff waters) (Jones et al. 2008).

6. Carbon assimilation determines N assimilation when using a predefined C/N ratio (Timlin et al. 2008). However, the amount of C allocated to the root as determined by the root/shoot ratio is also important in plant N uptake (BassiriRad et al. 2008). The feed-forward and feedback between C and N assimilations and N uptake under elevated CO_2 need further investigation (Timlin et al. 2008).

7. Further understanding of soil microbial contribution to root functionality and soil N supply is needed (Jones et al. 2008; BassiriRad et al. 2008; Boote et al. 2008). Soil microbes are the mediators for soil organic C/N dynamics and N fixation. Very few models explicitly simulate soil microbial populations.

8. N uptake may be different at field and watershed scales. There is a need to scale plant N uptake from the field scale (Malone and Ma 2008) to the watershed scale (Wang et al. 2008), especially when similar N-uptake

mechanisms are assumed. There is also a need to upscale the N-uptake mechanisms from the transporter level to single-root and to whole-plant-root systems.

ACKNOWLEDGMENT

The editors wish to thank all the contributors for their fine chapters and the editorial staff at Taylor & Francis Group for their patience and assistance. Each chapter was reviewed by two editors and one contributor to the book, and we are also deeply indebted for their work.

REFERENCES

BassiriRad, H., V. Gutschick, and H. L. Sehtiya. 2008. Control of plant nitrogen uptake in native ecosystems by rhizospheric processes. In *Quantifying and understanding plant nitrogen uptake for systems modeling*, ed. L. Ma, L. R. Ahuja, and T. W. Bruulsema. Boca Raton, FL: Taylor & Francis Group.

Boote, K. J., G. Hoogenboom, J. W. Jones, and K. T. Ingram. 2008. Modeling N-fixation and its relationship to N uptake in the CROPGRO model. In *Quantifying and understanding plant nitrogen uptake for systems modeling*, ed. L. Ma, L. R. Ahuja, and T. W. Bruulsema. Boca Raton, FL: Taylor & Francis Group.

Hansen, S., and P. Abrahamsen. 2008. Modeling water and N-uptake using a single root concept: Exemplified by the use in the Daisy model. In *Quantifying and understanding plant nitrogen uptake for systems modeling*, ed. L. Ma, L. R. Ahuja, and T. W. Bruulsema. Boca Raton, FL: Taylor & Francis Group.

Jamieson, P. D., R. F. Zyskowski, F. Y. Li, and M. A. Semenov. 2008. Water and N uptake and responses in models of wheat, potatoes and maize. In *Quantifying and understanding plant nitrogen uptake for systems modeling*, ed. L. Ma, L. R. Ahuja, and T. W. Bruulsema. Boca Raton, FL: Taylor & Francis Group.

Jones, D. L., J. F. Farrar, A. J. Macdonald, S. J. Kemmitt, and D. V. Murphy. 2008. Dissolved organic nitrogen and mechanisms of its uptake by plants in agricultural systems. In *Quantifying and understanding plant nitrogen uptake for systems modeling*, ed. L. Ma, L. R. Ahuja, and T. W. Bruulsema. Boca Raton, FL: Taylor & Francis Group.

Kristensen, H. L., and K. Thorup-Kristensen. 2008. Roots below one meter depth are important for uptake of nitrate by annual crops. In *Quantifying and understanding plant nitrogen uptake for systems modeling*, ed. L. Ma, L. R. Ahuja, and T. W. Bruulsema. Boca Raton, FL: Taylor & Francis Group.

Malagoli, P., F. Meuriot, P. Laine, E. Le Deunff, and A. Ourry. 2008. Modeling nitrate uptake and N dynamics in winter oilseed rape (*Brassica napus* L.). In *Quantifying and understanding plant nitrogen uptake for systems modeling*, ed. L. Ma, L. R. Ahuja, and T. W. Bruulsema. Boca Raton, FL: Taylor & Francis Group.

Malone, R. W. and L. Ma. 2008. N uptake effects on N loss in tile drainage as estimated by RZWQM. In *Quantifying and understanding plant nitrogen uptake for systems modeling*, ed. L. Ma, L. R. Ahuja, and T. W. Bruulsema. Boca Raton, FL: Taylor & Francis Group.

Smith, P., J. U. Smith, D. S. Powlson, W. B. McGill, J. R. M. Arah, O. G. Chertov, K. Coleman, U. Franko, S. Frolking, D. S. Jenkinsov, L. S. Jensen, R. H. Kelly, H. Klein-Gunnewiek, A. S. Komarov, C. Li, J. A. E. Molina, T. Mueller, W. J. Parton, J. H. M. Thornley, and A. P. Whitmore. 1997. A comparison of the performance of nine soil organic matter models using datasets from seven long-term experiments. *Geoderma. 81*: 153–225.

Timlin, D. J., M. Kouznetsov, D. H. Fleisher, S.-H. Kim, and V. R. Reddy. 2008. Simulation of nitrogen demand and uptake in potato using a carbon-assimilation approach. In *Quantifying and understanding plant nitrogen uptake for systems modeling*, ed. L. Ma, L. R. Ahuja, and T. W. Bruulsema. Boca Raton, FL: Taylor & Francis Group.

Wang, P., A. Sadeghi, L. Linker, J. Arnold, G. Shenk, and J. Wu. 2008. Simulated soil water content effect on plant nitrogen uptake and export for watershed management. In *Quantifying and understanding plant nitrogen uptake for systems modeling*, ed. L. Ma, L. R. Ahuja, and T. W. Bruulsema. Boca Raton, FL: Taylor & Francis Group.

Wu, L., I. J. Bingham, J. A. Baddeley, and C. A. Watson. 2008. Modeling plant N uptake using 3-dimensional and 1-dimensional root architecture. In *Quantifying and understanding plant nitrogen uptake for systems modeling*, ed. L. Ma, L. R. Ahuja, and T. W. Bruulsema. Boca Raton, FL: Taylor & Francis Group.

Zhu, Y., W. G. Li, H. B. Ye, G. S. McMaster, and W. X. Cao. 2008. Modeling grain protein formation in relation to nitrogen uptake and remobilization in rice. In *Quantifying and understanding plant nitrogen uptake for systems modeling*, ed. L. Ma, L. R. Ahuja, and T. W. Bruulsema. Boca Raton, FL: Taylor & Francis Group.

2 Modeling Nitrogen Fixation and Its Relationship to Nitrogen Uptake in the CROPGRO Model

Kenneth J. Boote, Gerrit Hoogenboom,
James W. Jones, and Keith T. Ingram

CONTENTS

2.1   Introduction 14
      2.1.1   Background to the CROPGRO-Legume Model 14
      2.1.2   Scheme and Order of Plant N Balance and N Fixation in
               CROPGRO 16
      2.1.3   N Fixation and Nodule Growth in CROPGRO 19
      2.1.4   Consequences of N Deficiency on Tissue N Levels, Vegetative
               Growth, Partitioning, Photosynthesis, and Seed Growth 19
      2.1.5   Initialization and Parameters for Nodule Growth and N
               Fixation in CROPGRO 20
2.2   Model Performance 21
      2.2.1   Dynamics—N Deficit Is Needed to Trigger Nodule Growth
               and N Fixation 21
      2.2.2   Dynamics of Nodule Growth and N Fixation: Priority for
               Assimilate 23
      2.2.3   Sensitivity of Nodule Growth and N Fixation to Water Deficit 24
      2.2.4   Seasonal Pattern of Nodule Growth, N Fixation, and N
               Accumulation in Rain-Fed vs. Irrigated Conditions 25
      2.2.5   Sensitivity Analysis to Discover Probable Temperature
               Sensitivities of Nodule Growth and N Fixation 30
      2.2.6   Comparative Analysis of Temperature Sensitivities of
               Different Grain Legumes 33
      2.2.7   Sensitivity Analysis: Effect of Source:Sink Relations on N
               Fixation 34
      2.2.8   Sensitivity Analysis: Growth and Yield Response to Applied
               N Fertilization 36

      2.2.9   Robustness for Different Grain Legumes, Soils, and Climate
            Conditions 38
2.3   Summary and Future Improvements 39
References...39
Appendix 2.1...42
A2.0 Nitrogen Fixation (NFIX)..42
A2.1 Run Initialization 42
A2.2 Seasonal Initialization..42
A2.3 Rate/Integration 42

2.1 INTRODUCTION

One of the main differences between the simulation of grain legumes and grain cereals is the prediction of nitrogen (N) fixation. The objective of this chapter is to describe how nodule growth and N fixation are simulated in the CROPGRO-legume model in relation to N-uptake processes. We introduce the theory of N deficit that causes the increased carbohydrate availability that is used to drive nodule growth and nitrogenase activity. Where feasible, literature values are used for setting parameters, and where information does not exist, sensitivity analyses are used to set relationships to obtain accurate and reasonable crop growth, N concentration, carbohydrate concentrations, and nodule growth under various temperature and water-deficit conditions. Simulations of nodule growth, N-fixation rate, specific nitrogenase activity, tissue N concentrations, crop N accumulation, and crop dry-matter accumulation are compared with available data for soybean under water-limited versus fully irrigated conditions. Sensitivity of processes, growth, and yield to N fertilization, temperature, water deficit, and source-sink treatments are also shown.

2.1.1 BACKGROUND TO THE CROPGRO-LEGUME MODEL

The CROPGRO-legume model is a process-oriented, mechanistic model with subroutines that simulate crop development, carbon (C) balance, crop and soil nitrogen (N) balance, and soil water balance, which has been described in detail by Boote et al. (1998a, 1998b). For full details on the CROPGRO growth and partitioning modules, see the 2004 DSSAT V4 crop model documentation (Boote et al. 2004, 1–102). Crop development includes processes such as vegetative and reproductive development, which determine life cycle duration, the duration of root and leaf growth, and the onset and duration of reproductive organ growth. Thus, crop development processes influence dry-matter partitioning among plant organs over time. Crop C balance includes daily inputs from photosynthesis, conversion of C into crop tissues, C losses through abscission, and growth and maintenance respiration. The C balance also includes leaf area expansion, growth of vegetative tissues, pod addition, seed addition, shell growth, seed growth, nodule growth, tissue senescence, and carbohydrate mobilization. The crop N balance includes daily soil N uptake, N_2 fixation, mobilization and translocation of N from old vegetative tissues to new growth of vegetative or reproductive tissues, rate of N use for new tissue growth, and rate of N loss in abscised parts. Soil N balance processes are those described by Godwin and Jones (1991) and Godwin and Singh (1998).

Soil water balance processes include infiltration of rainfall and irrigation, run-off, soil evaporation, distribution of root water uptake from soil layers, drainage of water below the root zone, and crop transpiration (Ritchie 1998). The time step in CROPGRO is mostly daily for carbon and nitrogen balance processes, but is hourly for some processes, such as calculation of thermal time and leaf-to-canopy assimilation. The time step for nodule growth and N fixation is daily, and matches the timing of daily C/N balance and dry-matter growth. Model state variables are predicted and output on a daily basis for crop, soil water, and soil N balance processes.

CROPGRO is a generic model that uses one common FORTRAN code to predict the growth of a number of different grain legumes, including soybean (*Glycine max* L. Merr.), peanut (*Arachis hypogaea* L.), dry bean (*Phaseolus vulgaris* L.) (Boote et al. 1998a, 1998b), cowpea (*Vigna unguiculata* L.) (Boote 1998, unpublished), faba bean (*Vicia faba* L.) (Boote et al. 2002), velvet bean (*Mucuna pruriens* L.) (Hartkamp et al. 2002a, 2002b), and chickpea (*Cicer arietinum* L.) (Singh and Virmani 1994). This versatility is achieved through input files that define species traits and cultivar attributes. The species file includes cardinal temperatures, tissue compositions, conversion costs, coefficients, relationships, and sensitivities to stresses for processes such as photosynthesis, nodule growth, N_2 fixation, maintenance respiration, leaf area growth, pod addition, and seed growth. The model code, along with information input from the species, ecotype, and cultivar files, is based on our understanding of crop–soil–weather relationships. More information on CROPGRO's generic nature and file input structure is available in Hoogenboom et al. (1994), Boote et al. (1998a, 1998b), and chapter 2 of the DSSAT V4 CROPGRO crop growth and partitioning module (Boote et al. 2004, 1–102).

The CROPGRO model originated from the SOYGRO 4.2 model developed by Wilkerson et al. (1983). The SOYGRO model did not explicitly grow nodules or fix N, nor did it explicitly take up mineral N, although it did account for the cost of N assimilation (cost of mineral N and N fixation was the same). Between 1989 and 1994, major model innovation occurred; the soil N balance and N-uptake features as well as N fixation were added (Hoogenboom et al. 1989), resulting in the release of the first version of CROPGRO V3.0 in 1994 (Hoogenboom et al. 1991, 1992, 1993). This version of CROPGRO was made a generic model capable of simulating several different legumes (soybean, peanut, and dry bean) with the same common FORTRAN source code, thus eliminating the need to make parallel changes in source code of multiple grain legume models. To the extent possible, species-related features were removed from the code and placed into a species parameter file along with a cultivar traits file, both of which are read as external input files. Major code improvements in nodule growth and N fixation were made by K. J. Boote for the V3.1 release of the CROPGRO model, followed by additional minor changes, culminating in the 1998 release of CROPGRO V3.5. Concurrent with the 1994 release of CROPGRO V3.0, all DSSAT models including CROPGRO began to use the same soil N balance (Godwin and Singh 1998) and soil water balance models (Ritchie 1998). Information on the N-fixation module is available in the CROPGRO documentation (Boote et al. 2004). Appendix 2.1 at the end of this chapter reprints pp. 44–47 on the N-fixation module from that documentation. The latest release of CROPGRO, V4.0, as a generic grain legume module in the Cropping System Model (CSM) has no further improvements to nodule growth and N fixation (Jones et al. 2003; Hoogenboom et al. 2004).

2.1.2 SCHEME AND ORDER OF PLANT N BALANCE AND N FIXATION IN CROPGRO

It is important to present the overall scheme and order of C and N balance processes in the CROPGRO model before we describe the particular processes of nodule growth, N fixation, and N uptake (Figure 2.1). Daily N supply, which is N available for new growth, is the sum of soil N uptake by roots, N fixation by nodules, and N mobilized from vegetative tissues. Daily C supply is the sum of gross canopy assimilation minus maintenance respiration, plus carbohydrate mobilized today, minus the costs of N assimilation and N fixation. These processes occur daily, and there is a daily update of the state variables of N mass and N concentration in different organs. The DSSAT V4 documentation (Boote et al. 2004) gives the various state variables associated with dry matter (mass), N mass, and N concentration, as well as the sub-routines and calling orders.

The soil N balance processes in CROPGRO are identical for all the DSSAT models, and include root N uptake, mineralization, immobilization, nitrification, denitrification, and N leaching processes, with options of using either the Godwin (Godwin and Singh 1998) or the CENTURY soil N balance, the latter of which was incorporated by Gijsman et al. (2002). Daily uptake of nitrate and ammonium by roots depends on soil and root factors and on crop N demand, as described below.

The order of C and N balance operations and order of called subroutines in CROPGRO (Figure 2.1) are as follows:

1. Simulate C mobilization, photosynthesis, and maintenance respiration.
2. Compute N demand from today's dry-matter growth increment and "target" N concentrations in each tissue (in DEMAND).
3. Compute potential N mobilization from vegetative tissues as a function of thermal time. The N demand is decreased by the amount of N mobilization that can go to seed growth (in DEMAND).
4. Simulate N uptake, which depends on soil and root factors, but with the constraint that actual N uptake cannot exceed N demand (NUPTAK).
5. Compute actual N mobilization for seed growth (MOBIL), which can be less than potential N mobilization if C gain and thus N demand are low on a given day.
6. Simulate C allocation to nodules, which is the amount of carbohydrate sufficient to fix N needed to meet the computed daily N deficit. In addition, a continuous, but small fraction of root assimilates goes to nodule growth and N fixation (in CROPGRO main).
7. Simulate N fixation (nitrogenase activity), possibly limited by nodule mass and temperature effect on specific activity (NFIX).
8. Simulate nodule growth using carbohydrate that is not used for N fixation, but this is limited by temperature and maximum relative growth rate. Nodule growth is limited if water deficit is concurrently limiting nitrogenase activity (NFIX).

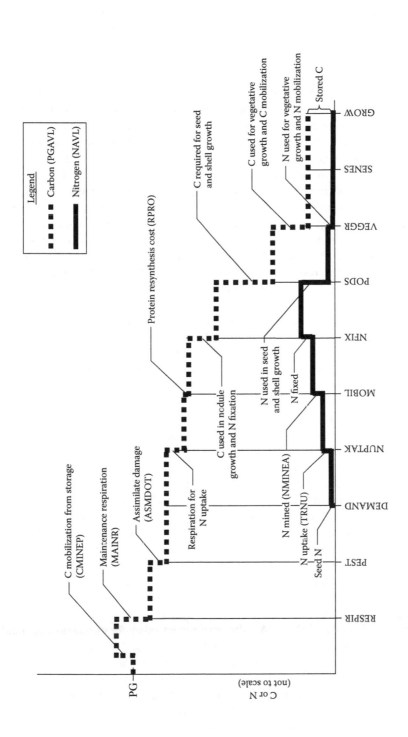

FIGURE 2.1 Schematic of the flow of C (PGAVL) and N (NAVL). The calculations for each process are shown in the order that they are computed through subroutine calls to appropriate modules. PG is gross photosynthesis per land area per day. [Figure from Boote et al. (2004). With permission.]

9. Simulate tissue growth, which depends on C and N supplies that are allocated in order of priority to seeds, pods, then to vegetative tissues (PODS and VEGGR).

The order of N balance operations, taken from the DSSAT V4 documentation (Boote et al. 2004, Figure 1), follows the logic that today's photosynthesis, maintenance respiration, and C mobilization are based on the values of state variables at the end of yesterday, but using today's environmental conditions to compute rates. On an hourly step, leaf photosynthesis is simulated and scaled to canopy assimilation as described by Boote and Pickering (1994), and depends on leaf area index, canopy height, canopy width, row spacing, leaf N concentration, specific leaf weight, genetic potential leaf rate, solar radiation, temperature, and crop water status. Crop water status is computed from the ratio of actual crop transpiration to potential transpiration. The C mobilization depends on the current carbohydrate pool and a mobilization rate that is crop-specific and temperature-dependent, and that accelerates after seeds begin to grow. The rate of N mobilization depends on the size of the available protein pool and a crop-specific, temperature-dependent, day-length-dependent mobilization rate. The N mobilization rate increases up to twofold as actual partitioning of assimilate to seed reaches 100%, but mobilization is restricted if there is insufficient C for seed growth to use that mobilized N. Using today's C supply, the model calculates the potential growth of each type of tissue. Then crop N demand is computed based on the growth increment of each tissue today, multiplied by a "target" high N concentration of each tissue, which assumes that N is nonlimiting. The target N concentration for each tissue type is given in the species file. Potential growth is summed over all tissue types. In addition, crop N demand includes N to replenish any N deficiencies in vegetative tissues, which occur if N concentration in vegetative tissue is less than the target N concentration. This deficiency demand or ability to refill vegetative tissue N deficit declines after setting seeds. In principle, N deficiency demand is an attractive feature that is similar in theory and approach to CERES models (Ritchie et al. 1998), but its construction differs from CERES models and its functionality in CROPGRO has not been well tested.

With the computed crop N demand as an upper limit, the model computes root uptake of N from the soil. The N uptake per unit root length is a Michaelis–Menten function of concentration of nitrate and ammonium in each soil layer. Potential crop N uptake per unit land area is thus computed from the nitrate and ammonium concentration of each soil layer, the root length density in each layer, and the soil water availability of that layer. Actual N uptake is the minimum of potential N uptake and N demand. Thus, actual N uptake from each layer is proportionately scaled back from potential N uptake per layer. An alternative mechanism for nitrate uptake could be one based on mass flow of solution nitrate arriving at the roots, as the model computes water uptake for each individual soil layer, and the nitrate concentration of the water in that layer is known. Uptake of ammonium, being an exchangeable cation, would still need to remain as a function of root length density. Potential model improvements in CROPGRO that should be evaluated include: (a) the use of a mass flow mechanism for root N uptake, (b) N deficiency demand for N uptake, and (c) alternative methods of computing N stress.

2.1.3 N Fixation and Nodule Growth in CROPGRO

If actual N uptake meets the N demand, there will be no N fixation today. If N uptake is less than N demand, the amount of today's N deficit is computed and then multiplied by the glucose equivalent cost of fixing N (7.06 g glucose g^{-1} N fixed) to compute the amount of carbohydrate partitioned to nodules for N fixation today. The mineral N uptake is likewise charged a cost of N reduction and assimilation into amino acids. The N fixation (NFIXN, see appendix 2.1) occurs up to the limit of allocated carbohydrate supply, nodule mass, a species-defined nodule-specific activity, and soil temperature. The N fixation is a two-stage computation, in which the first step is to calculate the potential rate of N fixation, considering the minimum of either limiting C supply or limiting the rate of nitrogenase (considering both nodule mass and soil temperature). In the second step, this potential rate was limited to actual, depending on minimum of either flooding or plant turgor effect. If nodule mass is too low to use all the carbohydrate for N fixation at the temperature-limited specific activity, then that "leftover" carbohydrate is used for nodule mass growth (NODGRO, see appendix 2.1) at a species-defined and temperature-dependent relative growth rate, a defined tissue cost, and the minimum of plant water stress or excess water stress. The relative growth rate for nodules is 0.17 g g^{-1} d^{-1} for soybean. The effects of environmental factors of temperature, water deficit, flooding, and nodule age act on the nodule growth or N fixation on a land-area basis, rather than affecting just nodule growth rate (RGR) or N fixation (SNA).

The code allows a thermal lag phase after plant emergence for infection and nodule growth, after which nodule mass is initialized at a species-dependent initial nodule mass. The current thermal lag of zero physiological days after emergence for nodule infection is consistent with data of Bergersen and Goodchild (1973) showing that small nodules were present on soybean at 8 days after sowing, leghemoglobin was observed at 10 days after sowing, and N fixation began at 12 days after sowing. Model initiation of nodule mass occurs at emergence, which is typically 5 to 8 days after sowing, so any time difference between simulated and actual onset of N fixation would be small. Sensitivity analyses conducted early during model development showed that we needed to create a continuous partial allocation of assimilate to nodules in order to prevent excessive delay in nodule growth and N fixation. Thus, we currently have a partial allocation (bypass) to nodules that amounts to 5% of the assimilate allocated to root growth. The specific nitrogenase activity (SNA) is 45 g N kg DW nod^{-1} d^{-1} for soybean. Nitrogenase activity, in summary, is dependent on soil temperature, plant water status (TURFAC), soil aeration (excess soil water), and nodule age.

2.1.4 Consequences of N Deficiency on Tissue N Levels, Vegetative Growth, Partitioning, Photosynthesis, and Seed Growth

What if today's N supply is deficient for today's growth, even after considering today's N fixation? Vegetative tissues are grown at a normal rate, but with a lower N concentration, so long as N concentration in new tissues remains above the minimum N level required for growth (Ng). To the extent that N concentration of new tissue would fall below Ng, dry-matter growth of tissues is limited proportionately to maintain tissue N equal to Ng. CROPGRO allows any excess carbohydrate from photosynthesis to accumulate in leaves, petioles, and stems, and this carbohydrate is slowly mobilized for later use. In addition, a ratio called NSTRES is computed from the ratio of today's N supply/(N demand × 0.70). The NSTRES factor causes increased partitioning of dry matter to root when the ratio of today's N supply/N demand falls below 0.7. NSTRES in CROPGRO is computed very differently from that in the cereal models, as it is based on today's supply-to-demand ratio, while that in CERES is computed on cumulative vegetative N concentration relative to upper and lower critical limits. For CERES models, the NSTRES is used to directly reduce dry-matter growth (PCARB). In contrast, in CROPGRO, the decline in leaf N concentration causes a decline in leaf and canopy photosynthesis. The photosynthesis response follows a quadratic function defined by the minimum N concentration at which the rate is zero and the optimum N concentration at which the rate is maximum, based on leaf photosynthesis versus leaf N concentration, with data for soybean coming from Boote et al. (1978), Boote et al. (1977, unpublished), and Boon-Long et al. (1983).

What is the priority for N, in the event of N deficiency, and does this change during seed filling? Nodule growth and N fixation have priority over seed growth for both assimilate (for fixation and growth) and N (to grow nodules), in order to make up any N deficit during seed filling. While nodules have high priority for N, their mass is so small that this is unimportant relative to reproductive use of N. CROPGRO gives seed growth priority over pod wall growth, which, in turn, has priority over vegetative tissue. If N deficiency is large, then seeds are grown at lower N concentration, an approach justified by data of Hayati et al. (1996). Under N deficiency, the reduction in seed N concentration is primary, but seed growth rate is also progressively reduced, following a Michaelis–Menten growth response to the N supply/N demand ratio. On the other hand, seed N concentration is allowed to increase if N is surplus, which can happen during seed fill if a stress, such as a severe drought, reduces C supply when N is available from N mobilization. The rate of N mobilization during grain filling in CROPGRO is a function of the photothermal age of the crop canopy (and is faster during grain filling), but it is not regulated (demanded) by seed filling rate, nor can luxury N fertilization substantially reduce this N mobilization from foliage. As a consequence of the decline in foliar N concentration, the leaf and canopy photosynthesis will decline during grain filling, regardless of N fertilization.

2.1.5 Initialization and Parameters for Nodule Growth and N Fixation in CROPGRO

Parameters for nodule growth and N fixation for soybean consist of an initialized nodule mass per plant (0.14 g) at emergence, a thermal lag time of zero physiological days after emergence, a relative growth rate of 0.17 g g^{-1} d^{-1}, and a specific nitrogenase activity (SNA) of 45 g N kg^{-1} d^{-1} for soybean, based on the mean of two literature values, 368 and 552 mmol e$^-$ (kg nodule DW h)$^{-1}$ for soybean nodule activity measured in flow-through systems (Layzell and Moloney 1994). The conversion is based on 1 mole N_2 (28 g) fixed per 6 mole of e^-. Converting to a daily rate gives a range from 41.2 to 61.8 (mean 51.5) g N fixed kg DW nod^{-1} d^{-1}. We initially used a value of 50 g N fixed kg DW nod^{-1} d^{-1}, which we reduced to 45 in 1995 to improve the match of simulated-to-observed nodule mass over time for a 1984 Bragg soybean data set. The cost of N fixation is 7.06 g glucose g^{-1} N fixed, which is halfway between the measured values of 6.75 and 7.43 g glucose g^{-1} N fixed as reported for soybean by LaRue and Patterson (1981) and Ryle et al. (1979). Based on theory, the cost should be 5.18 g glucose g^{-1} N fixed for Hup^+ and 6.43 g glucose g^{-1} N fixed for Hup^-. Going from Hup^- to Hup^+ increases simulated soybean yield 2.7%. So, the cost we use is higher than even the maximum efficiency of Hup^-. The species file actually defines the cost of N assimilation or N fixation in terms of glucose cost per gram of protein synthesized (2.556 or 2.830 g glucose per g protein, respectively). Thus, the model effectively sends glucose to nodules, and nodules return the full complement of amino acids (we do not simulate amides or production of C-skeletons explicitly). The growth synthesis cost of growing nodules is charged from the glucose allocated to nodules, but the maintenance costs are charged elsewhere "off the top."

2.2   MODEL PERFORMANCE

A major goal in testing model performance was to demonstrate reasonable behavior of the individual model processes (N_2 fixation, N-uptake rate) and model state variables (nodule mass, total biomass, N accumulation, N concentration, carbohydrate concentrations) under a range of climatic and soil conditions. We particularly wanted robust model behavior on soils with a full range of soil N supply from very limiting (sands with no N fertilization) to less N limiting, under different levels of soil water deficit and soil temperature. In addition, model performance was tested, where possible, against actual data on nodule mass, nitrogenase activity, N-fixation rate, N accumulation, and tissue N concentrations measured in field studies on Bragg soybean conducted on a Millhopper fine sand at Gainesville, Florida, by DeVries et al. (1989a, 1989b) and Albrecht et al. (1984). These studies satisfied our goal to evaluate potential N fixation on a fine sand with no N fertilization, and also under rain-fed versus irrigated conditions.

2.2.1   DYNAMICS—N DEFICIT IS NEEDED TO TRIGGER NODULE GROWTH AND N FIXATION

Our first goal in sensitivity analyses was to determine the type and robustness of simulated response for an N-limited situation. Therefore, we simulated the time-course dynamics of nodule growth, N fixation, N concentration, and nonstructural carbohydrate accumulation under a developing N deficiency for Bragg soybean sown on 5 May 1976 at 13 pl m^{-2} in 30-cm rows on an unfertilized sandy soil at Gainesville, Florida (Figures 2.2A and 2.2B). At emergence, initial nodule mass is small, N fixation is insufficient, and daily N uptake from the soil soon becomes insufficient. Thus the model attempts to send sufficient assimilate to nodules to fix the required N, but N fixation is still insufficient because nodule mass is small and SNA is saturated. This carryover assimilate is used for nodule growth, but this is

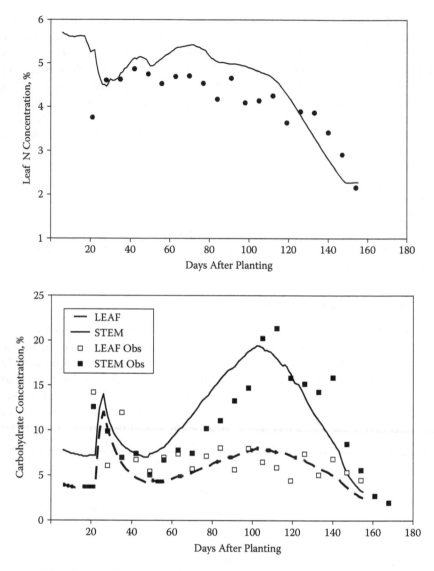

FIGURE 2.2 (A) Leaf N concentration and (B) leaf total nonstructural carbohydrate concentration of irrigated Bragg soybean grown with no fertilizer N on a fine sand soil at Gainesville, Florida, in 1976. [Adapted using data points from Boote et al. (1998b) with updated simulated values from CROPGRO V4.0.]

limited by nodule RGR, existing nodule mass, and soil temperature. As a result, there is an N deficiency, new tissue is grown at less than "target" N concentrations, and carbohydrate that cannot be used for growth or N fixation begins to accumulate in leaves and stems. While that carbohydrate pool is mobilized daily at a temperature-dependent rate, the net carbohydrate accumulation will continue to increase until N fixation and nodule growth catches up. Two characteristic responses occur, as shown in Figure 2.2:

1. There is a dip in leaf N concentration caused by growing tissue at less than "target" N concentration.
2. There is a peak in carbohydrate accumulation in the leaf and stem that concurrently matches the dip in leaf N concentration.

Once the crop has grown sufficient nodule mass to fix N at the defined SNA, then leaf N concentration begins to recover and the carbohydrate concentration begins to decline. This dip in leaf N concentration, peak in carbohydrate, and beginning recovery occurs about 24–28 days after sowing, but nodule growth and total N-fixation rate continue to increase over time because the plant is still in its exponential growth phase. The accumulation of carbohydrate starting at 50 days in Figure 2.2 is not related to N deficit, but rather is a "programmed" accumulation of carbohydrate that begins at anthesis and slows only when a full seed load has been set. Beginning at anthesis, there is a ramp function that allows up to 26% of the growth allocated to stems after anthesis to be stored as mobilizable carbohydrate in stems by the time the last pod is set, although there is concurrent mobilization as well.

2.2.2 DYNAMICS OF NODULE GROWTH AND N FIXATION: PRIORITY FOR ASSIMILATE

Early in model formulation, we had allowed carbohydrate to nodules only when there was a "daily" shortage of N supply, and this carbohydrate was used for SNA first, and then for growth. But sensitivity analysis by James Hansen during his Ph.D. program (Hansen 1996) showed that this approach caused too much delay in onset of nodule growth and N fixation, excessive dip in N concentration and accumulation of carbohydrate, and slow early growth. To solve this, we provided a continuous partial allocation of assimilate to nodules every day, amounting to a fraction of the daily assimilate allocated to roots. Based on sensitivity analyses, this allocation fraction was set at 0.05 of assimilate allocated to roots, and the initial nodule mass was increased to 0.014 g/plant, sufficient to minimize the dip in leaf N concentration and to keep the carbohydrate accumulation in leaf and stem to less than 30% by comparison with data from 1976 (see Figure 2.2B), 1979, and 1984 at Gainesville. The change was also needed to minimize dip in SLA, which was not seen during that time in our data, and to minimize decline in leaf photosynthesis as a result of low N concentration, which we also did not see (Boote 1987, unpublished data). The amount of "bypass" flow depends on root growth and is added to the amount of assimilate computed to be sent to nodules to fix sufficient N to make up today's deficit if nodule mass is adequate and if stresses do not limit SNA. The logic to justify this "bypass" allocation is that nodules are somewhat like parasites that directly access the phloem and can grow slowly, even with low but sufficient root N uptake for the crop. There is scientific literature to support limited nodule growth despite the presence of adequate N for crop growth.

We also learned via sensitivity analyses that we had to give nodule growth and N fixation first priority for assimilate over seed growth. The senior author discovered this more than 20 years ago, in interactions with Dr. Basil Acock during programming sessions to develop the GLYCIM soybean model. The GLYCIM model runs hourly, with hourly partitioning to organs including seeds. During one of these modeling sessions, Dr. Acock found that the model's seeds were aborting every night for lack of N. We decided that his model was giving seeds first priority for carbohydrate, thereby starving the nodules and, thus, running out of N for seed growth at night. Later work by Dr. Dennis Egli and his students also showed that seeds were not the "demanding" organs as suggested by the self-destruct hypothesis of Sinclair and deWit (1975), but were rather more passive receptacles for arriving assimilate (Hayati et al. 1995, 1996). Hayati et al. (1996) showed that under N-limiting conditions, seeds continued growing nearly as rapidly as under N-sufficient conditions, but with reduced N concentration. We also concluded that nodules were similar in some ways to root-knot nematodes (*Meloidogyne* spp.) that were shown by Stanton (1986) to be strongly plumbed into the root phloem and which were found to be first-priority users of assimilate, over seed or vegetative growth.

2.2.3 SENSITIVITY OF NODULE GROWTH AND N FIXATION TO WATER DEFICIT

For general use at multiple locations around the world, we wanted to have a crop model that responds appropriately over a range of temperatures and soil water levels, but primary data on relationships for sensitivity to those parameters were lacking. Initially for model V3.1 of CROPGRO, we had the nodule growth rate (RGR) and N fixation (SNA) sensitive to the fraction of available soil water in the nodule zone. A current hypothesis prevailing at that time was that soil water had a direct effect on nodule function. This approach did not work well because the upper nodule zone of topsoil (upper 30 cm) is very prone to drying, even when there is adequate water below 30 cm to meet plant needs for transpiration. For V3.5 of CROPGRO, we changed the nodule growth and nitrogenase activity to be sensitive only to plant water status as shown in Figure 2.3. This gave a more stable and better response and continues to be used in the most recent model, version V4.0. Plant TURFAC (affecting vegetative growth) is now the driver of the SWFACT variable and the 8-day running-memory variable, SWMEM8, as discussed in Appendix 2.1. The goal of the 8-day running memory is to allow a damage effect of past plant water deficit. Having nodules sensitive to plant water status is consistent with the data of Sprent (1972), who found that nodules in dry soil gained water from roots in moist soil. In addition, Albrecht et al. (1984) reported that SNA and nodule relative water content did not fully recover until the plant water status was fully recovered by irrigating deeply, despite good water status in the nodule zone. CROPGRO allows crop species to have differential sensitivity of nodule RGR and SNA to plant water status (Figure 2.3). For soybean, the sensitivity of nodules to water deficit is intermediate to the sensitivities of photosynthesis (SWFAC) and vegetative growth (TURFAC). On the other hand, peanut nodule sensitivity to water deficit is presumed to have the same sensitivity as photosynthesis response to SWFAC.

Nodule growth and N fixation are sensitive to excess soil water (flooding) through anaerobic stress computed from the portion of pore space that is filled with water (EPORS and FLDSUM, discussed in Appendix 2.1). At present, when the pore space becomes completely water-filled, both nodule growth and N fixation are restricted, and nodule senescence is enhanced. The flood sensitivity aspects of the model have

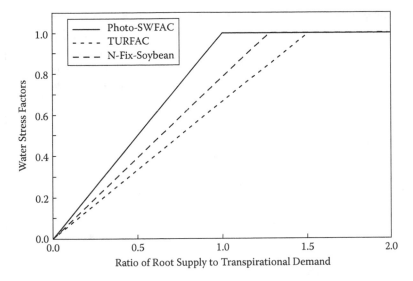

FIGURE 2.3 Sensitivity of photosynthesis (SWFAC), nodule growth, and nitrogenase activity (N-Fix-Soybean), and expansive growth (TURFAC) to ratio of root water supply to transpirational demand for water.

not been tested sufficiently to set the minimum pore space that must remain air-filled in the nodule zone to prevent anaerobic stress.

Nodule death rate (RNDTH, discussed in Appendix 2.1) was set to a given daily fraction under stress, and this fraction is modified by the maximum of flooding (anaerobic) stress, water stress, or C shortage stress effects. Nodule death rate under irrigated and water deficit conditions was calibrated to the nodule mass of the 1984 rain-fed Bragg soybean experiment (Figure 2.4).

2.2.4 Seasonal Pattern of Nodule Growth, N fixation, and N Accumulation in Rain-Fed vs. Irrigated Conditions

We tested the CROPGRO-Soybean model against data on nodule dry-matter growth, N fixation, crop dry-matter growth, and N accumulation from a 1984 experiment that had irrigated and rain-fed treatments (DeVries et al. 1989a, 1989b). Figure 2.4 shows the simulated nodule mass accumulation over time for the two treatments compared with observed nodule mass. With the value of maximum SNA from the literature, the predicted nodule mass was within 20% of observed nodule mass for the soybean crop, although we did reduce the potential SNA from 50 to 45 g N fixed kg DW nod^{-1} d^{-1} in order to increase simulated nodule mass about 10%. The code is designed to allow "carryover" carbohydrate left over from nitrogenase to be used for nodule growth as a way of catching up on nodule mass if N is limiting. This feature worked poorly under water deficit, allowing excessive nodule growth that affected future nitrogenase capacity. Therefore, when water deficit limited specific nitrogenase activity, we had to concurrently limit nodule growth to use only 10% of that "carryover" carbohydrate (that not used for SNA if the plant is under water deficit). This was a feature learned by sensitivity analyses and calibration against N fixation

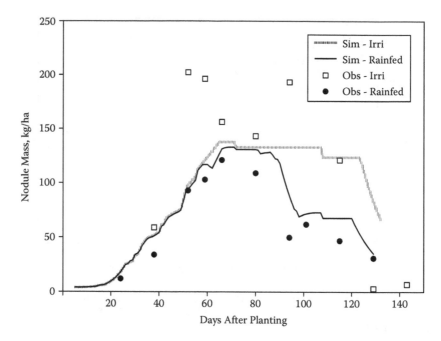

FIGURE 2.4 Simulated and observed nodule mass over time for irrigated and rain-fed Bragg soybean grown on a fine sand at Gainesville, Florida, in 1984. [Observed data are from DeVries et al. (1989a, 1989b).]

and nodule mass over time measured in the 1984 rain-fed Bragg soybean study. Nodule senescence under water deficit functioned well, but required minor adjustments to stress sensitivity.

The simulated N fixation closely matched the magnitude and life-cycle pattern of measured N fixation over time (Figure 2.5). The very rapid 1-day or 2-day dips in N fixation simulated for both irrigated and rain-fed crops is caused by cloudy, low-radiation days. Maximum N-fixation rate was 4 to 5 kg N [ha·day]$^{-1}$. This agreement was good to see in two respects: First, from a modeling standpoint, we were able to use a specific activity from the literature that predicted nodule mass within 20% of observed values. Second, the simulated rates of N_2 fixation were close to field measurements of N fixation despite using excavated whole root systems of individual plants and simply correcting (multiplying) from acetylene reduction to N_2 fixation (DeVries et al. 1989a). The model correctly simulated the decrease in N fixation during the 21-day water deficit period (78–99 days after sowing) during reproductive growth. However, it did not predict the low rate measured on rain-fed plants during days 61–65, when a transient water deficit period occurred, although the model did not indicate much stress. Perhaps the good prediction of N fixation under irrigation could have been anticipated, since the sandy soil provided a low N supply, and the model was already known to accurately predict dry matter and N accumulation over time for this site (Figure 2.6 shows N accumulation over time). Thus, the model should have predicted that N fixation would contribute the remaining (major) amount of N required.

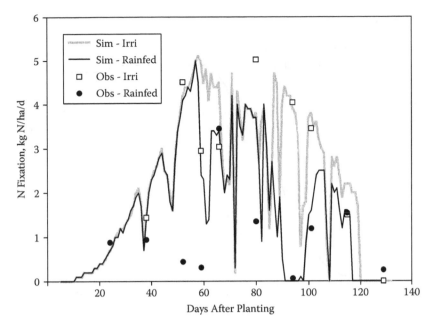

FIGURE 2.5   Simulated and observed N-fixation rate over time for irrigated and rain-fed Bragg soybean grown on a fine sand at Gainesville, Florida, in 1984. Sharp but short dips are usually associated with cloudy days. [Observed data are computed from DeVries et al. (1989a).]

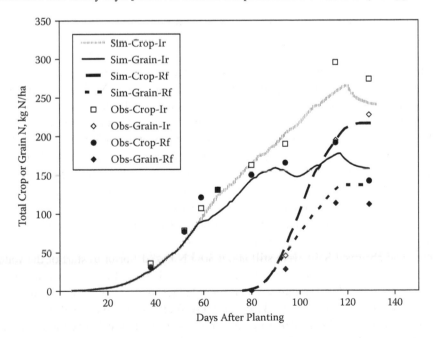

FIGURE 2.6   Simulated and observed crop and grain N accumulation over time for irrigated and rain-fed Bragg soybean grown on a fine sand at Gainesville, Florida, in 1984. [Observed data are from DeVries et al. (1989b).]

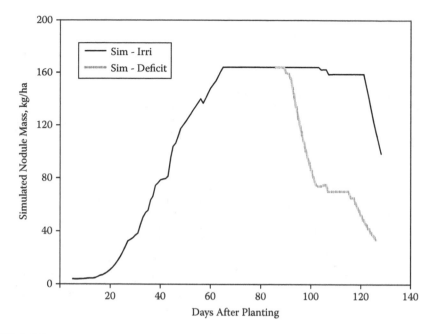

FIGURE 2.7   Simulated nodule mass over time for irrigated treatment versus a 19-day water deficit during seed filling for Bragg soybean grown on a fine sand at Gainesville, Florida, in 1980.

Model simulations were also compared with data from a 1980 experiment in which nitrogenase activity, canopy CER, stomatal conductance, and plant water status of Bragg soybean were measured periodically during a 19-day water deficit period initiated after the R5 beginning seed stage (Albrecht et al. 1984). This crop was sown 13 June 1980 at 30 plants per m^2 in 76-cm rows in Millhopper fine sand soil with no N fertilizer. In this experiment, simulated nodule mass reached a peak of 160 kg ha^{-1} and then senesced rapidly during the 19-day water deficit period to about 60 kg ha^{-1} (Figure 2.7). Simulated N fixation, initially at about 5 kg N [ha·d]$^{-1}$, first began to decline 8 days after water-withholding (in concert with simulated onset of water deficit), and then fell to zero by 11 days after water-withholding, and recovered to about 1.3 kg N [ha·d]$^{-1}$ after water deficit was relieved (Figure 2.8). Albrecht et al. (1984) measured specific nitrogenase activity on recovered nodules, but did not measure total nodule mass nor total N fixation. The simulated SNA approached 40 g N kg^{-1} d^{-1} during vegetative growth while nodule mass was still increasing. The predicted and observed SNA were still about 30 g N kg^{-1} d^{-1} prior to start of the water deficit experiment. Simulated SNA fell to zero at 11 days after water-withholding, but the observed SNA maintained somewhat longer, only reaching zero at 15 days after withholding (Figure 2.9). Based on these findings, we may need to decrease the sensitivity of soybean SNA to plant turgor, making its sensitivity more like the sensitivity of photosynthesis to plant turgor. After re-irrigating, the observed and simulated SNA recovered to about 20 g N kg^{-1} d^{-1} within 3 days of re-irrigating, and reached a value as high as the irrigated control. However, both SNA and N fixation

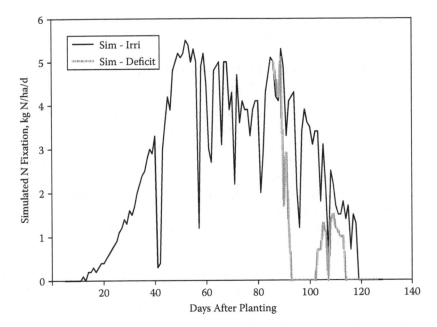

FIGURE 2.8   Simulated N fixation over time for irrigated treatment versus 19-day water deficit during seed filling for Bragg soybean grown on a fine sand at Gainesville, Florida, in 1980.

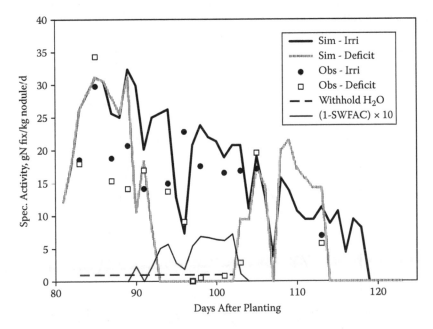

FIGURE 2.9   Simulated and observed specific nitrogenase activity over time for irrigated treatment versus 19-day water deficit during seed filling for Bragg soybean grown on a fine sand at Gainesville, Florida, in 1980. The value "1-SWFAC" is the degree of water deficit, where 0.0 is no stress and 1.0 is total stress. [Observed data are computed from Albrecht et al. (1984).]

were also declining in the irrigated control over time, as the crop was more than halfway through seed fill and had only another 15–20 days to maturity.

2.2.5 SENSITIVITY ANALYSIS TO DISCOVER PROBABLE TEMPERATURE SENSITIVITIES OF NODULE GROWTH AND N FIXATION

As shown in Appendix 2.1, potential N fixation is a function of C supply, nodule mass times specific activity, and soil temperature. Likewise, nodule growth rate is a function of C supply (left after N fixation), nodule mass times relative growth rate, and soil temperature. Both are also sensitive to TURFAC and excess water. ACSTF and ACSTG, discussed in Appendix 2.1, are functions of soil temperature in the nodule zone and are used to affect N fixation and nodule growth rate, as shown in Figure 2.10. There was inadequate published data to parameterize the sensitivity of nodule growth rate (RGR) and specific nitrogenase activity (SNA) to soil temperature. Therefore, we used sensitivity analyses of model performance to parameterize the sensitivity of nodule growth and specific nitrogenase activity to soil temperature; particularly since the first release of CROPGRO-Soybean with N fixation (V3.1) was initially developed with data from warm locations, such as Florida. Subsequently, version 3.1 of the model was tested with four seasons of data on growth, tissue N concentration, and N accumulation from relatively cool seasons at Ames, Iowa (Sexton et al. 1998), and in a high-elevation location in northwest Spain where the nighttime minimum temperatures frequently reached 13°C (Sau et al. 1999). We discovered, for moderately cool seasons in Iowa, that model version 3.1 simulated large dips in leaf N concentration (as low as 3.0%), large nodule mass (exceeding 600 kg ha⁻¹), leaf carbohydrate pools of 34% to 40%, and very inhibited early growth associated with low leaf N, which caused low photosynthesis, resulting in low biomass accumulation and low yield in 3 of 4 years (Sexton et al. 1998). Simulated early crop growth

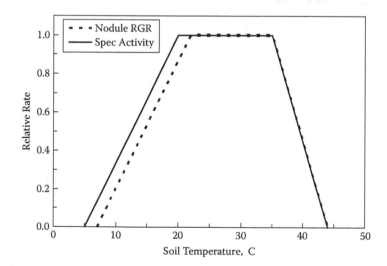

FIGURE 2.10 Sensitivity of nodule growth rate (Nodule RGR) and specific nitrogenase activity (Spec. Activity) to soil temperature for soybean.

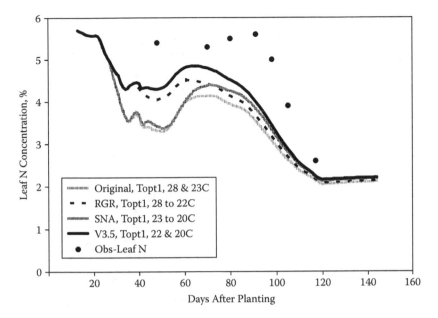

FIGURE 2.11. Simulated leaf N concentration for Kenwood soybean grown in Ames, Iowa, in 1995 with different temperature functions affecting nodule RGR or nodule SNA. [Observed data are from Sexton et al. (1998).]

in those fields was much less than observed growth, and simulated leaf N dipped much more than in the Sexton et al. (1998) data.

We know that soybean grows well in that location, as shown in the observed growth measurements. Therefore, as part of the recalibration for the 1998 V3.5 model release, we reduced the cardinal temperatures for the lowest limit of optimum temperature (Topt1) for RGR from 28°C to 22°C, and for SNA, the Topt1 from 23°C to 20°C (Sexton et al. 1998). Figure 2.10 shows the values for temperature sensitivity of nodule growth and SNA currently used in V3.5 and V4.0 model releases, which resulted in the improved simulation (see solid line for V3.5–4.0) of leaf N concentration (Figure 2.11), leaf nonstructural carbohydrate (Figure 2.12), nodule mass (Figure 2.13), and crop N accumulation (Figure 2.14).

Our goals in setting these cardinal temperatures were to hold nodule mass at a maximum of 200 to 300 kg ha^{-1}, to minimize the drop in leaf N concentration, to limit the carbohydrate accumulation to less than about 25%, and to mimic the actual dry-matter growth rate observed in those fields where soybean grew well (Sexton et al. 1998; Sau et al. 1999). While Sau et al. (1999) also tested other temperature-dependent functions such as photosynthesis, they found the temperature effects on nodule RGR and SNA were the dominant temperature-response functions with respect to effects on dry-matter growth and yield.

We also learned that the modification of those temperature functions on RGR and SNA had to be balanced. Poor model behavior, such as lower leaf N concentration (Figure 2.11) or high leaf carbohydrate accumulation (Figure 2.12) remained even if we only reduced the Topt1 for SNA. Excessive simulated nodule growth

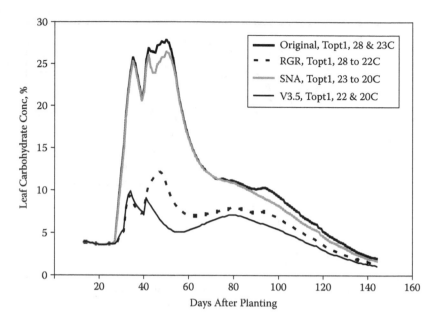

FIGURE 2.12   Simulated leaf nonstructural carbohydrate concentration for Kenwood soybean grown in Ames, Iowa, in 1995 with different temperature functions affecting nodule RGR or nodule SNA.

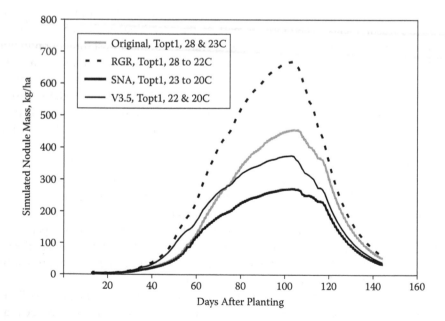

FIGURE 2.13   Simulated nodule mass for Kenwood soybean grown in Ames, Iowa, in 1995 with different temperature functions affecting nodule RGR or nodule SNA.

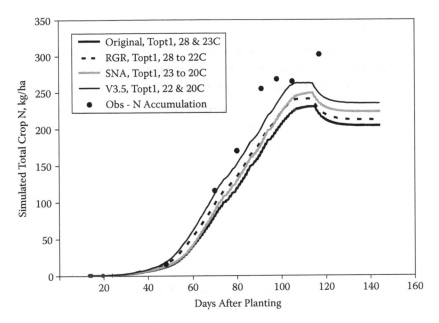

FIGURE 2.14 Simulated and observed total crop N accumulation for Kenwood soybean grown in Ames, Iowa, in 1995 with different temperature functions affecting nodule RGR or nodule SNA. [Observed data are from Sexton et al. (1998).]

as large as 1000 kg ha^{-1} occurred if we only reduced the Topt1 for RGR (Figure 2.13). But combining the changes to both temperature functions gave more balanced behavior and increased yield by 362 kg ha^{-1}, biomass by 612 kg ha^{-1}, leaf area index by 0.59 units, and harvest index by 0.009, and improved the prediction of N accumulation for the Iowa soybean data sets (Figure 2.14).

2.2.6 COMPARATIVE ANALYSIS OF TEMPERATURE SENSITIVITIES OF DIFFERENT GRAIN LEGUMES

Since the CROPGRO model simulates a number of different grain legumes, there is a similar problem of how to estimate the temperature sensitivities and other parameters for the various legumes in the absence of data. Data limitations forced the same type of sensitivity analyses for the other legumes to ensure reasonable early season growth, nodule mass, minimal decline in leaf N concentration, and carbohydrate accumulation under various scenarios of soil N supply and soil temperature (climatic regions). Boote et al. (2002) reviewed how the process was done during the modification of the CROPGRO model for the ability to predict growth and yield of faba bean. They showed data on nodule mass and leaf N concentration predictions of faba bean in that paper. Table 2.1 shows comparative cardinal temperatures for nodule growth rate and specific nitrogenase activity for the grain legume species simulated by the CROPGRO model. It is important to note that faba bean is a cool-season, somewhat frost-hardy grain legume grown in the winter season, while dry bean and soybean are relatively cool-tolerant species, and peanut and velvet bean are warm-season legumes. The cardinal temperatures reflect these differences.

TABLE 2.1

Cardinal Temperatures for Nodule Growth Rate and Specific Nitrogenase Activity for Several Crops Simulated by the CROPGRO V4 Grain Legume Model

	Temperature (°C)							
	Nodule Growth Rate				Specific Nitrogenase Activity			
Crop	Tbase	Topt1	Topt2	Tceiling	Tbase	Topt1	Topt2	Tceiling
Faba bean	1	16	25	40	1	16	25	40
Dry bean	6	21	35	44	4	19	35	44
Soybean	7	22	35	44	5	20	35	44
Cowpea	7	22	35	44	5	20	35	44
Peanut	9	25	34	44	7	23	34	44
Velvet bean	7	26	35	44	5	23	35	44

2.2.7 SENSITIVITY ANALYSIS: EFFECT OF SOURCE:SINK RELATIONS ON N FIXATION

It is also important to document that simulated nodule growth and N fixation respond realistically to variation in source supply (light, CO_2, insect defoliation) as well as variation in sink strength, such as insect depodding, that may occur under field conditions. Therefore, we evaluated how the CROPGRO-Soybean model performed relative to the published data of Lawn and Brun (1974), who measured nodule mass and N fixation under these treatments. They sowed Clay (MG 0) and Chippewa (MG I) soybean cultivars at 25 plants m^{-2} in 0.5-m row spacing on 28 May 1974 at Saint Paul, Minnesota. The Waukegan silt loam soil, having 1.5% organic carbon, was amended with 11,200 kg ha^{-1} of corn cobs one month prior to sowing to immobilize mineral N to ensure complete reliance of the soybean crop on N fixation. The DSSAT soil organic matter module (Godwin and Singh 1998) linked to CROPGRO simulated that there was no net release of mineral N until the end of the season, in general agreement with aims of the researchers. On 11 August, four source:sink treatments were applied:

1. supplemental light (12,500 lux all day long, estimated equal to 4.5 MJ m^{-2} d^{-1})
2. partial shade (50%)
3. partial depodding (50%, although the net amounted to 25%)
4. partial defoliation (60%)

Figures 2.15A and 2.15B illustrate model simulations of N fixation over time for these five treatments on both cultivars. Data were converted from acetylene reduction per plant to N_2-fixation rate per unit land area, using the plant population and the same conversion assumptions as described for the data of deVries et al. (1989a, 1989b). The simulated nodule mass was more than 500 kg ha^{-1} (data not shown), which was nearly twofold higher than observed (assuming 20% dry matter for fresh nodules). Possibly the temperature adjustments on nodule mass and N fixation need further modification, as this is a cool location. The simulated N fixation was 5 to 6 kg N ha^{-1} d^{-1} and was comparable to that observed (Figure 2.15). The model per-

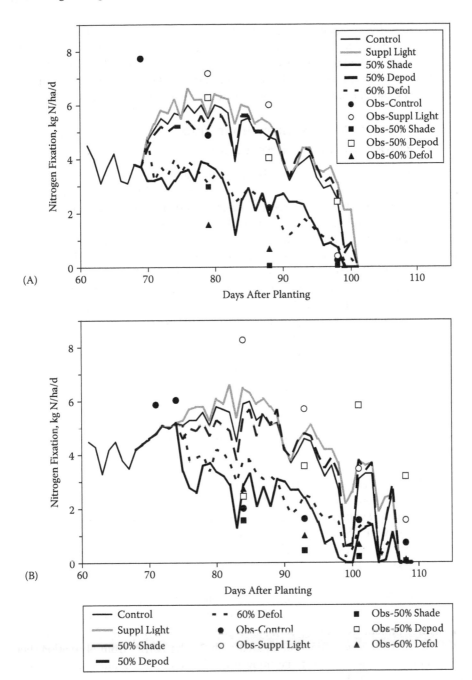

FIGURE 2.15 Simulated rate of N fixation of: (A) Clay and (B) Chippewa Upper soybean sown on 28 May 1974 at Saint Paul, Minnesota, and exposed to five source:sink treatments on 11 August: (1) control, (2) supplemental light (12,500 lux all day long, equal to 4.5 MJ m^{-2} d^{-1}), (3) partial shade (50%), (4) partial depodding (50%), and (5) partial defoliation (60%). [Points are observed data from Lawn and Brun (1974).]

formed generally as expected in response to variation in source (supplemental light, shading, and defoliation), with better maintenance of N fixation with supplemental light, and severe rapid reductions under shading and defoliation. While the observed depodding treatment enhanced N fixation, the simulated depodding initially did not enhance N fixation because of lack of sink tissues, but later N fixation was higher.

2.2.8 SENSITIVITY ANALYSIS: GROWTH AND YIELD RESPONSE TO APPLIED N FERTILIZATION

Another important sensitivity analysis was to evaluate seed yield response to applied N fertilizer, to document the consistency of response, and to verify that simulated soybean seed yield is not increased too much by fertilizer N application. This was done to mimic the reality that exists in field situations where nodulation is effective. It was important to set the N-fixation responsiveness of the legume model such that it did not show any "stumbles," i.e., reduced growth rates or low leaf N situations for variable initial soil mineral N, for variable N fertilization, or for variable soil N mineralization situations. Having robustness of this response is important to modeling of soybean growth because model users typically do not know the residual mineral N in a soil, and they may have a poor description of the amount of N mineralized over time for their particular soil. This robust model response of N fixation in a very-low-N sandy soil in Florida was illustrated in Figures 2.2 through 2.9. Robust response to soil water deficits (Figures 2.4 to 2.9) and regional variation in temperature (Figures 2.10 to 2.14) were shown. Figure 2.16 illustrates the CROPGRO-Soybean model response to application of N fertilization on Millhopper fine sand soil in Florida. Soybean grain yield was increased from 3650 to about 3860 kg ha^{-1} as N fertilizer increased from zero to 640 kg N ha^{-1}. There are two reasons the soybean model predicted this small 5%–6% yield increase with high N fertilization:

1. The cost of nitrate reduction is less than the cost of N fixation (5.36 versus 7.06 g carbohydrate [g N fixed] $^{-1}$)
2. The extra N uptake allows leaf N to increase slightly as in luxury consumption above the N concentrations of a strictly N-fixing soybean plant (Figure 2.17), and this additional N both increases photosynthesis and maintains photosynthesis longer when plants mobilize N during seed-filling

The simulations show delayed onset of nodule growth and N fixation with higher N fertilization application (Figure 2.18). For this sandy soil, the simulated uptake of "native" mineral N coming from soil organic matter mineralization was only 16 kg N ha^{-1}, which is very small compared with 331 kg N ha^{-1} from N fixation. For a higher organic matter soil in the midwestern United States, simulated mineral N uptake would be much higher.

With these modifications, the CROPGRO model potentially can simulate the amount of N fixed in most fields in the midwestern United States, with the proviso that the prediction is only as good as the user's description of the initial mineral N (nitrate and ammonium), the soil organic matter (SOM) concentration, and the appropriate fractionation of SOM into the three pools of soil organic matter to

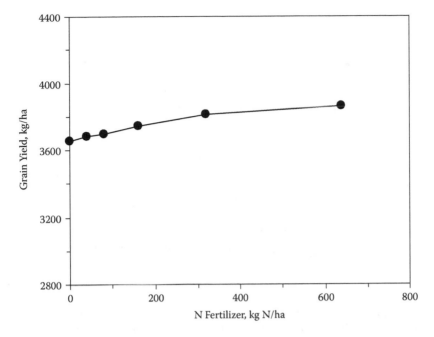

FIGURE 2.16 Simulated grain yield of Bragg soybean in response to N fertilization on Millhopper fine sand soil in Gainesville, Florida, with 1979 weather.

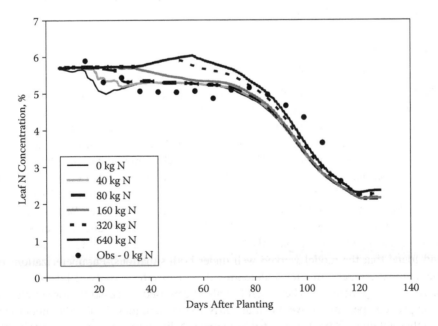

FIGURE 2.17 Simulated leaf N concentration over time for Bragg soybean in response to N fertilization on Millhopper fine sand soil in Gainesville, Florida, with 1979 weather. This shows luxury uptake where fertilizer N is applied. Observed leaf N concentration values are for unfertilized control.

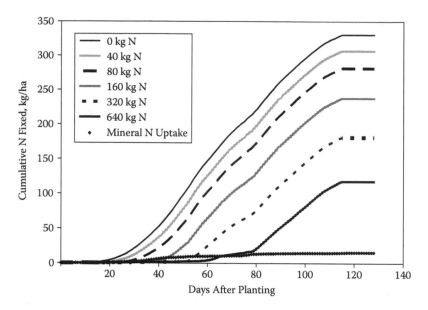

FIGURE 2.18 Simulated cumulative N fixation over time for Bragg soybean as affected by N fertilization on Millhopper fine sand soil in Gainesville, Florida, with 1979 weather. The dashed line is the simulated mineral N uptake without N fertilizer application.

adequately simulate the CENTURY SOM module that is linked to the DSSAT crop models (Gijsman et al. 2002).

2.2.9 ROBUSTNESS FOR DIFFERENT GRAIN LEGUMES, SOILS, AND CLIMATE CONDITIONS

The CROPGRO-Legume model has been tested extensively and been found to work well with N fixation "turned on" under low N fertilization and low N soils for 40 to 50 data sets on soybean, 30 or more on peanut, and 12 or more on faba bean. Nearly all of these data are for legumes grown without fertilizer N, but only a few have data on actual nodule growth or N fixation. The 1984 Bragg data, including nodule mass, N fixation, N concentrations, and N accumulation, are part of the DSSAT V4.0 database (Jones et al. 2003). The N-fixation and nodule mass data for the 1980 Bragg study (Albrecht et al. 1984) and the Minnesota study of Lawn and Brun (1974) are not yet in that database. In the case of faba bean, Boote et al. (2002) tested simulations of treatments relying on only N fixation compared with N fertilization only and found that the model worked well under both situations. Parameterizations of the cardinal temperatures for nodule growth and specific activity of faba bean are, of course, very different. The model has also been used successfully for simulating chickpea, cowpea, and velvet bean under N-fixing situations. While the model will simulate adequate N fixation for dry bean under N limitation, we are less certain that this approach is correct for dry bean, which is generally considered to be a poor N-fixing plant. In most available studies, N fertilizer was applied, and for this reason, the simulations were usually made with N considered nonlimiting.

2.3 SUMMARY AND FUTURE IMPROVEMENTS

In summary, we attempted to model N_2 fixation as mechanistically as practical in the CROPGRO model by considering explicit nodule growth and specific nitrogenase activity. The strategy is built around carbohydrate being sent to nodules for N fixation and/or nodule growth if today's N supply is deficient. Some of the lessons we learned along the way, with sensitivity analyses and by comparison with data, show that to properly simulate N fixation:

1. The processes need to be on a land-area basis.
2. There is a zero thermal lag time from emergence to initializing nodule mass.
3. There needs to be a direct partial allocation of assimilate to nodules to minimize lags and ensure minimal nodule growth.
4. Temperature functions can be solved by comparison with dynamics of processes under cool soil regions when there is insufficient data to minimize the extent of leaf N decline and leaf carbohydrate accumulation.
5. Nodules need first priority for assimilate over seeds.
6. Water stress must be based on plant water status rather than nodule-zone soil water.
7. If water stress limits nitrogenase activity, only a small fraction of the "carryover" carbohydrate sent to nodules can be used for nodule growth.
8. Responsiveness to low or high soil mineral N and fertilizer N must be programmed such that lags in nodule growth, N fixation, and crop growth are minimized.
9. The benefit of N fertilization on yield must mimic the rather minor benefit observed in field situations.

Future improvements are needed to include the effects of soil pH on nodulation, nodule growth, and N fixation. The crop models do not yet consider soil pH or salinity effects on growth and yield. Flooding effects also need to be better parameterized and considered. There is a strategy (and input) in the model to set degree of "ineffective" versus "effective" nodulation, but it is limited to a decimal value between 0 and 1, and it has not been tested well for such intermediate cases. This is needed for dry bean cultivars and some peanut cultivars.

REFERENCES

Albrecht, S. L., J. M. Bennett, and K. J. Boote. 1984. Relationship of nitrogenase activity to plant water stress in field-grown soybeans. *Field Crops Res.* 8: 61–71.

Bergersen, F. J., and D. J. Goodchild. 1973. Cellular location and concentration of leghaemoglobin in soybean root nodules. *Aust. J. Biol. Sci.* 26: 741–756.

Boon-Long, P., D. B. Egli, and J. E. Leggett. 1983. Leaf N and photosynthesis during reproductive growth in soybeans. *Crop Sci.* 23: 617–620.

Boote, K. J., R. N. Gallaher, W. K. Robertson, K. Hinson, and L. C. Hammond. 1978. Effect of foliar fertilization on photosynthesis, leaf nutrition, and yield of soybeans. *Agron. J.* 70: 787–791.

Boote, K. J., J. W. Jones, and G. Hoogenboom. 1998a. Simulation of crop growth: CROPGRO model. In *Agricultural systems modeling and simulation*, ed. R. M. Peart and R. B. Curry, chap. 18. New York: Marcel Dekker.

Boote, K. J., J. W. Jones, G. Hoogenboom, W. D. Batchelor, and C. H. Porter. 2004. DSSAT v4 CROPGRO crop growth and partitioning module. In *Decision support system for agrotechnology transfer version 4.0 (DSSAT v4)*. Vol. 4. *Crop model documentation*, ed. J. W. Jones, G. Hoogenboom, P. W. Wilkens, C. H. Porter, and G. Y. Tsuji, chap. 2. Honolulu: Univ. of Hawaii.

Boote, K. J., J. W. Jones, G. Hoogenboom, and N. B. Pickering. 1998b. The CROPGRO model for grain legumes. In *Understanding options for agricultural production*, ed. G. Y Tsuji, G. Hoogenboom, and P. K. Thornton, 99–128. Dordrecht: Kluwer Academic.

Boote, K. J., M. I. Mínguez, and F. Sau. 2002. Adapting the CROPGRO legume model to simulate growth of faba bean. *Agron. J.* 94: 743–756.

Boote, K. J., and N. B. Pickering. 1994. Modeling photosynthesis of row crop canopies. *Hort-Science* 29: 1423–1434.

DeVries, J. D., J. M. Bennett, S. L. Albrecht, and K. J. Boote. 1989a. Water relations, nitrogenase activity, and root development of three grain legumes in response to soil water deficits. *Field Crops Res.* 21: 215–226.

DeVries, J. D., J. M. Bennett, K. J. Boote, S. L. Albrecht, and C. E. Maliro. 1989b. Nitrogen accumulation and partitioning by three grain legumes in response to soil water deficits. *Field Crops Res.* 22: 33–44.

Gijsman, A. J., G. Hoogenboom, W. J. Parton, and P. C. Kerridge. 2002. Modifying DSSAT crop models for low-input agricultural systems using a soil organic matter–residue module from CENTURY. *Agron. J.* 94: 462–474.

Godwin, D. C., and D. C. Jones. 1991. Nitrogen dynamics in soil-plant systems. In *Modeling plant and soil systems*, ed. J. Hanks and J. T. Ritchie. Agronomy Monograph No. 31, 287–321. Madison, WI: ASA, CSSA, and SSSA.

Godwin, D. C., and U. Singh. 1998. Nitrogen balance and crop response to nitrogen in upland and lowland cropping systems. In *Understanding options for agricultural production: System approaches for sustainable agricultural development*, ed. G. Y. Tsuji , G. Hoogenboom, and P. K. Thornton, 55–77. Dordrecht: Kluwer Academic.

Hansen, J. W. 1996. A systems approach to characterizing farm sustainability. Ph.D. diss., Univ. of Florida, Gainesville.

Hartkamp, A. D., G. Hoogenboom, R. Gilbert, T. Benson, S. A. Tarawali, A. Gijsman, W. Bowen, and J. W. White. 2002a. Adaptation of the CROPGRO growth model to velvet bean as a green manure cover crop, I: Model evaluation and testing. *Field Crops Res.* 78: 27–40.

Hartkamp, A. D., G. Hoogenboom, and J. W. White. 2002b. Adaptation of the CROPGRO growth model to velvet bean as a green manure cover crop, I: Model development. *Field Crops Res.* 78: 9–25.

Hayati, R., D. B. Egli, and S. J. Crafts-Brandner. 1995. Carbon and nitrogen supply during seed filling and leaf senescence in soybean. *Crop Sci.* 35: 1063–1069.

Hayati, R., D. B. Egli, and S. J. Crafts-Brandner. 1996. Independence of nitrogen supply and seed growth in soybean: Studies using an *in vitro* culture system. *J. Exp. Bot.* 47: 33–40.

Hoogenboom, G., J. W. Jones, and K. J. Boote. 1989. Nitrogen fixation, uptake, and remobilization in legumes: A modeling approach. *Agronomy Abstracts* 16.

Hoogenboom, G., J. W. Jones, and K. J. Boote. 1991. Predicting growth and development with a generic grain legume model. Paper No. 91-4501. St. Joseph, MI: ASAE.

Hoogenboom, G., J. W. Jones, and K. J. Boote. 1992. Modeling growth, development, and yield of grain legumes using SOYGRO, PNUTGRO, and BEANGRO: A review. *Trans. ASAE* 35: 2043–2056.

Hoogenboom, G., J. W. Jones, K. J. Boote, W. T. Bowen, N. B. Pickering, and W. D. Batchelor. 1993. Advancement in modeling grain legume crops. Paper No. 93-4511. St. Joseph, MI: ASAE.

Hoogenboom, G., J. W. Jones, P. W. Wilkens, W. D. Batchelor, W. T. Bowen, L. A. Hunt, N. B. Pickering, U. Singh, D. C. Godwin, B. Baer, K. J. Boote, J. T. Ritchie, and J. W. White. 1994. Crop models. In *DSSAT Version 3*, ed. G. Y. Tsuji, G. Uehara, and S. Balas, vol. 2, 95–244. Honolulu: Univ. of Hawaii.

Hoogenboom, G., J. W. Jones, P. W. Wilkens, C. H. Porter, W. D. Batchelor, L. A. Hunt, K. J. Boote, U. Singh, O. Uryasev, W. T. Bowen, A. J. Gijsman, A. du Toit, J. W. White, and G. Y. Tsuji. 2004. *Decision Support System for Agrotechnology Transfer Version 4.0* [CD-ROM]. University of Hawaii, Honolulu, HI.

Jones, J. W., G. Hoogenboom, C. H. Porter, K. J. Boote, W. D. Batchelor, L. A. Hunt, P. W. Wilkens, U. Singh, A. J. Gijsman, and J. T. Ritchie. 2003. DSSAT cropping system model. *Eur. J. Agron.* 18: 235–265.

Larue, T. A., and T. G. Patterson. 1981. How much nitrogen do legumes fix? *Adv. Agron.* 4: 15–38.

Lawn, R. J., and W. A. Brun. 1974. Symbiotic nitrogen fixation in soybeans, I: Effect of photosynthetic source–sink manipulations. *Crop Sci.* 14: 11–16.

Layzell, D. B., and A. H. M. Moloney. 1994. Dinitrogen fixation. In *Physiology and determination of crop yield*, ed. K. J. Boote, J. M. Bennett, T. R. Sinclair, and G. M. Paulsen, 311–335. Madison, WI: ASA, CSSA, SSSA.

Ritchie, J. T. 1998. Soil water balance and plant water stress. In *Understanding options for agricultural production*, ed. G. Y. Tsuji, G. Hoogenboom, and P. K. Thornton, 41–54. Dordrecht: Kluwer Academic.

Ritchie, J. T., U. Singh, D. C. Godwin, and W. T. Bowen. 1998. Cereal growth, development and yield. In *Understanding options for agricultural production*, ed. G. Y. Tsuji, G. Hoogenboom, and P. K. Thornton, 79–98. Dordrecht: Kluwer Academic.

Ryle, G. J. A., C. E. Powell, and A. J. Gordon. 1979. The respiratory costs of nitrogen fixation in soybean, cowpea and white clover, II: Comparisons of the cost of nitrogen fixation and the utilisation of combined nitrogen. *J. Exp. Bot.* 30: 145–153.

Sau, F., K. J. Boote, and B. Ruiz-Nogueira. 1999. Evaluation and improvement of CROP-GRO-soybean model for a cool environment in Galicia, northwest Spain. *Field Crops Res.* 61: 273–291.

Sexton, P. J., W. D. Batchelor, K. J. Boote, and R. Shibles. 1998. Evaluation of CROPGRO for prediction of soybean nitrogen balance in a Midwestern environment. *Trans. ASAE* 41: 1543–1548.

Sinclair, T. R., and C. T. deWit. 1975. Comparative analysis of photosynthate and nitrogen requirements in the production of seeds by various crops. *Science* 18: 565–567.

Singh, P., and S. M. Virmani. 1994. Modeling growth and yield of chickpea (*Cicer arietinum* L.). *Field Crops Res.* 46: 1–29.

Sprent, J. I. 1972. The effects of water stress on nitrogen-fixing root nodules, 4: Effects on whole plants of *Vicia faba* and *Glycine max*. *New Phytol.* 71: 603–611.

Stanton, M. A. 1986. Effects of root-knot nematodes (*Meloidogyne* spp.) on growth and yield of "Cobb" soybean (*Glycine max* (L.) Merrill). M.S. thesis, Univ. of Florida, Gainesville.

Wilkerson, G. G., J. W. Jones, K. J. Boote, K. T. Ingram, and J. W. Mishoe. 1983. Modeling soybean growth for crop management. *Trans. ASAE* 26: 63–73.

APPENDIX 2.1*

A2.0 NITROGEN FIXATION (NFIX)

Subroutine NFIX calculates nitrogen fixation by nodules and is called if both the ISWNIT (nitrogen simulation) and ISWSYM (nitrogen fixation) switches are set to "Y." The routine contains sections for data "Run Initialization," "Seasonal Initialization," and "Rate/Integration," and is thus called three times from the CROPGRO routine. Table A2.1 lists input parameters used by NFIX, and Table A2.2 lists input and output arguments used by NFIX.

A2.1 RUN INITIALIZATION

Table A2.1 lists input parameters that are read from FILEIO and FILEC by the NFIX routine. Variable definitions and units can be found in Appendices Q and I of the DSSAT V4 documentation. The input data includes curve information for determining the effect of temperature, canopy age, dry soil, and wet soil on N fixation and nodule growth rates.

A2.2 SEASONAL INITIALIZATION

Several variables are set to zero at the beginning of each season of simulation in the "Seasonal Initialization" section.

A2.3 RATE/INTEGRATION

The N_2 fixation component in CROPGRO has been changed considerably since the earlier versions in 1992–1993. There is a thermal-time requirement for initiating first-nodule mass. When N uptake is sufficient, nodule growth is slow, receiving a minimum fraction (bypass flow) of the total assimilate that is allocated to roots.

TABLE A2.1
Input Parameters Read by NFIX Routine

FILEIO Parameters (e.g., DSSAT30.INP):

EFINOC	EFNFIX

FILEC Parameters (e.g., SBGRO030.SPE):

CNODCR	FNFXW	PRONOD	TYPFXD
DWNODI	FNNGT	RFIXN	TYPFXT
FNFXA	NDTHMX	SNACTM	TYPFXW
FNFXD	NODRGM	TYPFXA	TYPNGT
FNFXT			

---

* Reprinted from pp. 44–47, DSSAT V4 Crop Growth and Partitioning Module (Boote et al. 2004).

TABLE A2.2

Interface for NFIX Routine

Variable Name	Definition	Units
	Input Arguments	
AGRNOD	CH_2O requirement for nodule growth	g[CH_2O]/g[nodule]
CNODMN	Minimum CH_2O reserved for nodule growth	g[CH_2O]/m^2/day
CTONOD	CH_2O to allocate to nodules to fix N needed for reproductive and vegetative growth	g[CH_2O]/m^2/day
DLAYR(L)	Soil depth in layer L	cm
DXR57	Relative time between first seed (NR5) and physiological maturity (NR7)	—
FILECC	Pathname plus filename for SPE file	—
FILEIO	Filename for INP file	—
NLAYR	Number of soil layers	—
NR7	Day when 50% of plants first have yellowing or maturing pods	days
PLTPOP	Plant population	# plants/m^2
SAT(L)	Saturated soil water content in layer L	cm^3/cm^3
ST(L)	Soil temperature by soil layer	°C
SW(L)	Soil water content in layer L	cm^3/cm^3
TURFAC	Water stress factor for expansion	—
YRDOY	Current day of simulation	YYDDD
YRSIM	Beginning date of simulation	YYDDD
	Output Arguments	
CNOD	CH_2O used in N fixation and nodule growth (including respiration costs) today	g[CH_2O]/m^2/day
DWNOD	Nodule mass	g[nodule]/m^2
DWNODA	Nodule mass	g[nodule]/m^2
NDTH	Nodule death rate	g[nodule]/m^2/day
NFIXN	Amount of N fixed during the day	g[N]/m^2/day
NODGR	New nodule growth	g[nodule]/m^2/day
SENNOD	Daily senesced matter from nodules in soil layer L	kg[dry matter]/ha
WTNFX	Cumulative weight of N fixed	g[N]/m^2

When N uptake is deficient (less than N demand) for growth of new tissues, carbohydrates are used for N_2 fixation to the extent of the nodule mass and the species-defined nodule-specific activity. If nodule mass is insufficient, then assimilates are used for nodule growth at a rate that is dependent on soil temperature and species-defined nodule relative growth rate. The N_2-fixation rate is further influenced by soil temperature, soil water deficit, soil aeration (flooding), and plant reproductive age. The N_2-fixation process is decreased as the plant water stress factor (TURFAC) is decreased below a species-dependent value.

The first call to the "Rate/Integration" section of NFIX occurs when plant development has reached a specified physiological age (VSTAGE greater than TTFIX). On this first call, the nodule mass is initialized based on user-supplied initial nodule mass (DWNODI) and the current plant population (PLTPOP).

Next, three factors (ACSTF, ACSTG, and FLDSUM) are summed over the layers of the soil profile. These factors affect the N-fixation rates and/or nodule growth rates. All three are computed using the CURV function, which returns a factor between 0.0 and 1.0 based on four critical curve definition points. ACSTF is a soil temperature factor that affects potential and actual N-fixation rates. ACSTG is a soil temperature factor that affects nodule growth rate. FLDSUM is related to excess soil moisture and affects N-fixation rates.

```
        ACSTF = ACSTF + DLAYR(I) * FLAYR *
CURV(TYPFXT,FNFXT(1),FNFXT(2),FNFXT(3),FNFXT(4),ST(I))

        ACSTG = ACSTG + DLAYR(I) * FLAYR *
CURV(TYPNGT,FNNGT(1),FNNGT(2),FNNGT(3),FNNGT(4),ST(I))

        EPORS = MAX(SAT(I) -SW(I),0.0)
        FLDSUM = FLDSUM + DLAYR(I) * FLAYR *
CURV(TYPFXW,FNFXW(1),FNFXW(2),FNFXW(3),FNFXW(4),EPORS)
```

These factors are then divided by DNOD, the depth of the nodule zone, to yield the nodule growth and fixation factors, TNFIX, TNGRO, and FLDACT:

```
        TNFIX = ACSTF / DNOD
        TNGRO = ACSTG / DNOD
        FLDACT = FLDSUM / DNOD
```

Other factors that are computed are SWFACT, a drought factor based on TURFAC (water stress factor), and NFXAGE, which determines the effect of canopy age on N growth rates:

```
        SWFACT = CURV(TYPFXD,FNFXD(1),FNFXD(2),FNFXD(3),
            FNFXD(4),TURFAC)

        NFXAGE = CURV(TYPFXA,FNFXA(1),FNFXA(2),FNFXA(3),
            FNFXA(4),DXR57)
```

A running average of the last 8 days of water deficit factor (SWFACT) is kept. Both the 8-day average and the current day's value of SWFACT are used to reduce the N-fixation rate.

CLEFT is a running value of CH_2O left over as the various processes occur in the following order:

1. Set aside a minimum (CNODMN) for nodule growth,
2. Fix N_2,
3. Grow nodule mass with any CH_2O not used.

The CH_2O left for nodule growth after N fixation is:

```
CLEFT = CTONOD - CNODMN
```

The specific nodule activity (SNACT) is computed taking into account the maximum activity of the nodules (SNACTM) and strain effects (EFNFIX) for N fixation plus the amount CNODMN (except in the case where N_2 fixation was limited by water deficit).

The nodule death rate, RNDTH is computed as a function of the soil water deficit factor (SWFACT), soil water excess factor (FLDACT), and carbon deficit factor (CNFACT):

```
RNDTH = NDTHMX * MAX((1.-FLDACT),(1.-SWFACT),
        (1.-CNFACT))
NDTH = MIN(1.0,RNDTH) * DWNOD
```

NDTHMX is the maximum nodule death rate for flooded or dry soil conditions. NDTH is the nodule death rate per unit area.

If physiological maturity has not been reached (DAS < NR7), potential N fixation is computed

```
PNFIXN = MIN((CLEFT * 0.16 / RFIXN), (DWNOD *
         SNACT)) * TNFIX
```

RFIXN is defined as the respiration required for biological N fixation in $g[CH_2O]/g[protein]$. The 0.16 factor converts from g[protein] to g[N]. TNFIX is the soil temperature factor affecting N fixation. So, the potential rate of N fixation is the nodule activity rate, limited by available carbon and soil temperature.

Actual N fixation, NFIXN, is computed with the soil water drought factors (SWFACT and SWMEM8) and soil water excess factor (FLDACT).

```
NFIXN = PNFIXN * MIN(SWFACT, SWMEM8, FLDACT)
```

Normally, CH_2O not used for N fixation (because of limited nodule mass or activity) can be used to grow more nodule mass, thereby growing the mass needed to overcome this limitation. In the case of plant water deficit that decreases N-fixation-specific activity, we do not want the extra CH_2O to be used for nodule growth. For this purpose, we compute actual N fixation (NFIXN) limited by water deficit and potential N fixation with no water deficit (PNFIXN) and their respective CH_2O costs, CUSFIX and PCSFIX.

```
PCSFIX = (PNFIXN / 0.16) * RFIXN
CUSFIX = (NFIXN / 0.16) * RFIXN
```

The amount of CH_2O not used for potential N fixation (CNOFIX) is set aside and is not used for nodule growth in the case of water stress.

```
CNOFIX = PCSFIX - CUSFIX
```

Actually, 10% of it can be used for nodule growth plus the amount CNODMN.

```
CLEFT = MAX(0.0,CLEFT - CUSFIX- 0.9*CNOFIX) +
        CNODMN
```

The maximum nodule relative growth rate is reduced by EFINOC, the inoculation effectiveness (or rhizobium density factor), and EFNFIX, the strain efficiency to obtain the effective nodule relative growth rate (NODRGR).

```
IF (DAS .LT. NR7) THEN
        NODRGR = NODRGM * EFNFIX * EFINOC
ELSE
        NODRGR = 0.0
ENDIF
```

The new nodule growth rate (NODGR) is limited by either supply or demand for C.

```
NODGR = MIN(CLEFT/AGRNOD,DWNOD*NODRGR)
        * TNGRO * MIN(SWFACT,FLDACT) * NFXAGE
```

The C required for new nodule growth today and the C used in N fixation and nodule growth (including respiration costs) today are calculated as:

```
CNODGR = NODGR * AGRNOD
CNOD = CUSFIX + CNODGR
```

3 Modeling Nitrate Uptake and Nitrogen Dynamics in Winter Oilseed Rape (*Brassica napus* L.)

Philippe Malagoli, Frédéric Meuriot, Philippe Laine, Erwan Le Deunff, and Alain Ourry

CONTENTS

3.1   Introduction 48
3.2   Experimental and Modeling Methods 50
    3.2.1   Growth Conditions 50
        3.2.1.1   Controlled Conditions 50
        3.2.1.2   Field Conditions 51
    3.2.2   Experimental Design 51
        3.2.2.1   Experimental Treatments for NO_3^- Influx
                Measurements 51
        3.2.2.2   Experimental Treatment, Labeling, and Harvest for
                Field 52
    3.2.3   Analysis and Computing Methods 52
        3.2.3.1   Total Nitrogen and Isotopic Analyses 52
        3.2.3.2   N Flow Calculations for the Field Experiment 52
    3.2.4   Modeling Method 53
        3.2.4.1   Kinetic Equations of Nitrate Transport Systems 53
        3.2.4.2   Endogenous and Environmental Effects on HATS
                and HATS+LATS 54
        3.2.4.3   Introduction of Auxiliary Variables in the Model 54
        3.2.4.4   Calculation of Unregulated and Regulated Uptake 54
        3.2.4.5   Sources of Input Variables 55
        3.2.4.6   Basic Assumptions for Model Construction 55
3.3   Results and Discussion 56
    3.3.1   Kinetics of NO_3^- Influx 56
        3.3.1.1   In Induced and Noninduced Plants 56
        3.3.1.2   Effects of N Deprivation on Influx Rates and Gene
                Expression 58
    3.3.2   Modeling NO_3^- Uptake during the Growth Cycle 58

              3.3.2.1   Effect of Light/Darkness Cycle and PAR on NO_3^-
                        Influx 58
              3.3.2.2   Effect of Root Temperature on NO_3^- Influx 60
              3.3.2.3   Effect of Ontogeny on NO_3^- Influx 60
              3.3.2.4   Cumulative Effect of Regulative Variables and
                        Impact of N Fertilization Levels on N Uptake 61
      3.3.3   Partitioning of N Uptake and N Remobilization to Vegetative
              and Reproductive Tissues 62
              3.3.3.1   Partitioning of Uptaken N 62
              3.3.3.2   Partitioning of Mobilized N 65
              3.3.3.3   Mobilization of N between Senescing Leaves and
                        the Need for a Compartmental Model 65
3.4   Conclusion 66
References 67

3.1   INTRODUCTION

Nitrogen (N) represents a central element involved in most important physiological plant processes (growth and development, photosynthesis, enzymatic functions). Limited availability of nitrogen within intensive agrosystems frequently leads farmers to apply additional N fertilizer amounts to avoid any N deficiency and depressed crop yield. However, both the N fertilization rate and schedule have to accurately match crop N demand along the growth cycle to minimize leaching losses and subsequent pollution of the wider environment (Boelcke et al. 1991; Sieling and Christen 2001).

Because of its widespread use in human (oil) and animal-centered (high-protein-content seed residues) food industry and more recently as a biofuel, the area of winter oilseed rape culture has greatly increased in Northern Europe, especially in France, and in China over the last decade. From an agronomical viewpoint, *Brassica napus* L. is commonly cycled in cereal crop rotations as a valuable break crop. Winter oilseed rape culture is also well known for its high capacity to take up mineral N from the soil. Accordingly, it is widely used as a catch/cover crop to reduce nitrate leaching from arable cropping systems during autumn and early winter (Boelcke et al. 1991; Sieling and Christen 2001). Nonetheless, despite oilseed rape's high nitrate-uptake capacity, yields are often far below those expected with such high N application rates (up to 150 kg N·ha⁻¹ applied for a seed yield averaging 3 t·ha⁻¹) when compared with cereal yields with similar applied N input levels. Typically, less than 50% of the applied N fertilizer is recovered in the harvested seeds (Boelcke et al. 1991). This low N recovery rate suggests that a significant N amount still remains in vegetative tissues at harvest and in the soil and is ultimately released back to the environment. This plant N release from oilseed rape occurs primarily during spring and fall, when leaves contain high amounts of N (usually more than 2% of their dry weight [DW], compared with wheat's minimal leaf N content, which is less than 1% (Schjoerring et al. 1995). In fact, it is well established that a discrepancy occurs for N remobilization in vegetative tissues until the pod-filling stage in *B. napus* L. (Malagoli et al. 2005a). Moreover, whether this high amount of N remaining in the leaves is due to an early leaf fall (i.e., source limitation) or to an incomplete N mobilization (sink-strength limitation) is still unclear.

This issue needs to be addressed to improve the N use efficiency (NUE) of winter oilseed rape crop (Schjoerring et al. 1995). For example, it can be questioned whether increased N storage capacity into buffer compartments (such as stem and taproot) would improve leaf N remobilization efficiency, which would allow matching of the seed N demand. Currently, data on the internal allocation patterns of N taken up within winter rape remains relatively scarce. Further, it is likely that the interrelationships of N flows between many N source sinks will be complex, varying both ontogenetically and along the main plant axis. It is still not known how a given leaf behaves in terms of dependence on N newly taken up (and subsequent allocation) or to remobilized N from other tissues, and how these processes are affected by leaf nodal position within the stem, i.e., its trophic nodal position (insertion in source/ sink network) as well as by environmental conditions (i.e., light and temperature).

Crop-level models have already been built to predict N uptake and dynamics in winter oilseed rape (Gabrielle et al. 1998). They usually rely on a demand/supply scheme, where plant N supply is predicted from both soil N concentration and active/ diffusive mechanisms. Plant N demand is usually based on a growth-rate-driven equation. While a plant N dilution curve may provide the most generic description of plant N demand, this concept may not be relevant when high N rates are applied to the soil (Gabrielle et al. 1998). Indeed, this demand/supply scheme does not take into account root physiological processes involved in both N uptake and associated regulations. Thus, it has been shown that maximal rate of nitrate root uptake in *B. napus* far exceeds that of many other cultivated species (Lainé et al. 2002). Based on physiological studies and kinetic equations, Forde (2002) reported that four main classes of nitrate transport systems are involved in nitrate root uptake in oilseed rape, namely, low- or high-affinity transport systems (LATS and HATS) with a constitutive (CHATS, CLATS) or an inducible (IHATS, ILATS) component.

Many studies have established that some plant metabolites (amino acids, nitrate, sugars, etc.) up- and/or down-regulate LATS or HATS activities and/or corresponding nitrate transporter gene expression under controlled conditions (Touraine et al. 2001). However, no study has yet predicted to what extent changes of environmental factors would quantitatively and qualitatively alter the relative contribution of each N transport system to total N uptake during plant development at the whole plant or field level. For instance, how N fertilization rates would affect the contribution of individual N transport systems, if any, under natural conditions remains unknown (Huang et al. 1999; Ono et al. 2000; Fraisier et al. 2001). This lack of knowledge led us to propose a novel N-uptake model to provide a mechanistic and powerful tool linking and extending recent molecular and physiological advances in N transport system regulation (Schjoerring et al. 1995).

To achieve this goal, the effect of several environmental (low root temperature and photosynthetically active radiation [PAR]) and endogenous factors (day/night cycle and ontogenetic stages) on HATS and LATS activities were determined by measuring ^{15}N influx under controlled conditions. Associated response curves were described and combined with basic kinetic equations in the model. The availability of independent field experimental data (from the INRA Oilseed Rape databank; http://www.bioclim.inra.grignon.fr) provided model inputs to enable simulation of N uptake by a winter oilseed rape crop during the whole growth cycle. The model reliability (i.e., model structure and assumptions) was then evaluated by comparing model outputs with harvested exported N by this crop under field conditions. Finally, simulations through this field-level mechanistic model gave a unique chance to assess (a) the contribution of each nitrate (NO_3^-) transport system to total N uptake with increasing N fertilizer application rates, (b) the impact of each environmental or endogenous factor on N uptake during the growth cycle, and (c) the sensitivity of key model parameters regulating plant N acquisition.

3.2   EXPERIMENTAL AND MODELING METHODS

All treatments that were experimentally used—^{15}N labeling procedures, mass spectrometry analysis, and associated methods—have been previously described (Malagoli et al. 2004). Analysis of gene expression is detailed in Faure-Rabasse et al. (2002) and Beuve et al. (2004). Data from field experiments were extracted from results published by Noquet et al. (2003) and Malagoli et al. (2004, 2005a, 2005b), the latter explaining precisely how modeling of nitrate uptake was performed. They are summarized in the following subsections.

3.2.1   GROWTH CONDITIONS

3.2.1.1   Controlled Conditions

3.2.1.1.1   Kinetics of NO_3^- Uptake, Light/Darkness,
Temperature, and PAR Effect Experiments

Seeds of *Brassica napus* L. cv. Capitol were germinated and grown in hydroponic solution (25–50 seedlings per plastic tank) in a greenhouse in 1999. The aerated nutrient solution contained 1 mM KNO_3 (except for experimental plants used for establishing kinetics), 0.40 mM KH_2PO_4, 1.0 mM K_2SO_4, 3.0 mM $CaCl_2$, 0.50 mM $MgSO_4$, 0.15 mM K_2HPO_4, 0.20 mM Fe-Na EDTA, 14 μM H_3BO_3, 5.0 μM $MnSO_4$, 3.0 μM $ZnSO_4$, 0.7 μM $CuSO_4$, 0.7 μM $(NH_4)_6Mo_7O_{24}$, and 0.1 μM $CoCl_2$ and was renewed every two days. pH was maintained at 6.5 ± 0.5 by adding $CaCO_3$ (200 mg L^{-1}). The natural light was supplemented with phytor lamps (150 μmol m^{-2} s^{-1} of photosynthetically active radiation at the height of the canopy) for 16 h per day. The thermoperiod was 24°C ± 1°C (day) and 18°C (night).

3.2.1.1.2   N-Deprivation Experiment

Seeds were sown into six culture units (three plants/unit) of a flowing solution culture (FSC) system incorporating automatic control of concentrations of NO_3^-, K^+, and H^+ in solution (Clement et al. 1974; Hatch et al. 1986). Nutrient concentrations in each culture unit were initially (μM): NO_3^-, 250; K^+, 250; $H_2PO_4^-$, 50; Mg^{2+}, 100; SO_4^{2-}, 325; Fe^{2+}, 5.4, with micronutrients as previously described by Clement et al. (1974). Nutrient solutions were drained and refilled until day 18. Then the nutrient solution was allowed to deplete by plant uptake until day 24. From this date, K^+ concentrations were maintained at 20 ± 2 μM, and other nutrients (except NO_3^-) were supplied automatically in fixed ratios to the net uptake of K^+, for 1 mol of K, 0.645 and for 0.057, 0.045, 0.00075, and 0.522 mol of S, Mg, P, Fe, and Ca, respectively. Micronutrients were supplied as described by Clement et al. (1974). An external

NO_3^- concentration of 20 ± 2 μM was maintained automatically in each culture unit from day 24 until the start of the N-deprivation period (day 26).

3.2.1.1.3 Developmental Stage Effect Experiment

Seeds of *Brassica napus* L. cv. Capitol plants were taken from a field plot located in Saint-Aubin d'Arquenay (Normandy, France) when they were vernalized at the three-to-four-leaf stage. Plants with a well-developed taproot were harvested cautiously at the bolting stage (three to four leaves), taking care not to damage the root system. The roots were gently rinsed with deionized water before their transfer to a hydroponic system, and plants were then grown in the greenhouse as previously described.

3.2.1.2 Field Conditions

A winter oilseed rape crop (*Brassica napus* L. cv. Capitol) was sown on a clay loam soil on 10 September 2000 at Hérouvillette, 10 km north of Caen (49"10' N, 00"27' W), France. Nitrogen fertilizer was applied as NH_4NO_3 at the start of stem extension (GS 2.5; 75 kg N ha^{-1}) and at the bud visible stage (GS 3.3; 150 kg N ha^{-1}). Soil N was in excess, and water stress was not observed during the experimental period due to regular rainfall during spring. Plant density was 37–40 plants m^{-2} during the experiment (from stem extension to harvest).

3.2.2 EXPERIMENTAL DESIGN

3.2.2.1 Experimental Treatments for NO_3^- Influx Measurements

Different experimental treatments (i.e., N deprivation, light/darkness cycle, low root temperatures, ontogenetic effect during the growth cycle, and photosynthetically active radiation) were realized for NO_3^- influx measurements.

When a factor was tested on NO_3^- uptake, the others remained constant during the experiment, except for the day/night cycle experiment. The thermoperiod was 20°C and 15°C during the light and the dark periods, respectively. Given that temperature had no effect on NO_3^- uptake between 15°C and 20°C, it can be assumed that its incidence on NO_3^- uptake was limited during the light/darkness cycle experiment.

NO_3^- influx rate was measured from (a) three batches of 50 seedlings (kinetics of NO_3^- uptake) or 25 seedlings (day/night cycle, low temperature, and PAR effect), (b) from six batches of three plants (N deprivation), and (c) from six plants (ontogenetic effect). The root system was rinsed twice with 1 mM $CaSO_4$ solution for 1 min and then placed in a complete nutrient solution for 5 min containing either 100 μM or 5 mM $K^{15}NO_3^-$ (^{15}N excess of 99%). For the kinetics of the NO_3^- uptake-related experiment, nitrate concentrations ranged from 0 to 7.5 mM (10, 25, 50, 75, 100, 135, 250, 1000, 2500, 5000, 7500 μM). The extent of NO_3^- depletion from these solutions during the influx assays was less than 4% in each case. At the end of feeding, roots were given two 1-min washes in 1 mM $CaSO_4$ at 4°C before being harvested. At harvest, shoots and roots were sampled separately, weighed, dried, ground into a fine powder, and kept in a vacuum with $CaCl_2$ until total nitrogen and isotopic analyses. For the experiment to measure the ontogenetic effect, the root system of harvested plants was separated into taproot and lateral roots.

3.2.2.2 Experimental Treatment, Labeling, and Harvest for Field

The ^{15}N labeling experiment was performed from stem extension (at the beginning of March) to seed maturity (at the beginning of July). Seven days before each weekly harvest date, 12 plants at the same developmental stage were randomly selected within the canopy. The petiole of each senescing leaf was attached to the stem by a nylon thread to facilitate collection of the fallen leaves. Then, 750 mL of labeled nitrogen (1 mM K^{15}NO$_3$, ^{15}N excess = 10%) was applied to the soil surface (about 400 cm^2) around each plant. Seven days after ^{15}N labeling, the plants were harvested and the root system in the top 30-cm layer of the soil was recovered carefully. The 12 plants were pooled in three sets of four plants. At each harvest, the plants were separated into lateral roots, taproot, green and dead leaves, stem, flowers, and pods. However, because of the difficulty in recovering lateral roots quantitatively under field conditions, this component was omitted from further analysis. The green leaves were numbered and then sampled individually as a function of their insertion along the stem, measured by counting leaf scars.

3.2.3 ANALYSIS AND COMPUTING METHODS

3.2.3.1 Total Nitrogen and Isotopic Analyses

Total nitrogen and ^{15}N in the plant samples were determined with a continuous flow isotope mass spectrometer (Twenty-twenty, PDZ Europa Scientific Ltd., Crewe, U.K.) linked to a C/N analyzer (Roboprep CN, PDZ Europa Scientific Ltd., Crewe, U.K.). Influx of NO$_3^-$ was calculated from ^{15}N contents of roots and shoots.

3.2.3.2 N Flow Calculations for the Field Experiment

Dry weights and N contents of each tissue were subjected to polynomial regression ($r^2 \geq 0.90$) to minimize variation between each harvest date, and N uptake was estimated from the difference in plant N content between two harvest times. The partitioning of absorbed N was calculated from the excess ^{15}N in each tissue combined with the previously calculated total plant N uptake. Based on the assumption that unlabeled N from the soil was taken up and allocated in different plant tissues in a similar way to labeled N, the real N uptake by each tissue could be calculated as

$$\frac{(N_{d+7days} - N_d) \times {}^{15}N \text{ excess in each tissue}}{\text{total } {}^{15}N \text{ excess in plant}}$$

where

N_d = total nitrogen content of the plant (mg per plant) at day d when ^{15}N fertilizer was applied

$N_{d+7days}$ = total nitrogen content in the plant (mg per plant) at time d+7 days after ^{15}N fertilizer was applied

After calculating the N in each organ derived from uptake, the pattern of net translocation of endogenous unlabeled N (N absorbed before the beginning of labeling) among plant parts could be used to estimate N mobilization within the plant. At each harvest, the amount of N mobilized from or to each tissue (N_{mob}) was calculated by subtracting the total amount of N ($N_{d+7days}$) from (a) the N derived from uptake (N uptake) and (b) the previous amount of N in this tissue seven days before (N_d):

$$N_{mob} = N_{d+7days} - N_d - N \text{ uptake}$$

Thus positive values of N_{mob} represent nitrogen that was mobilized to the tissue, whereas negative values correspond to a net mobilization of N from the tissue. For each leaf insertion, the following values were calculated:

1. The cumulative amounts of both N derived from N taken up, and endogenous N mobilized from source tissues
2. Percentage of mobilization of N = $(N_{max} - N_{min}) \times 100/N_{max}$, where N_{max} and N_{min} correspond to the highest and lowest values of total nitrogen (mg per plant), respectively
3. Dates of appearance, loss or abscission, and start of mobilization of endogenous N (calculated when N mobilization reached negative values) expressed in thermal time (°C days)

3.2.4 Modeling Method

A mechanistic single-root model that explains nitrate uptake of plants mainly based on NO_3^- concentration around the root system was developed. The proposed thermal time-step model simulates total NO_3^- uptake by rape crops from the root transport processes formalized by kinetic equations of the different NO_3^- transport systems involved in N uptake.

3.2.4.1 Kinetic Equations of Nitrate Transport Systems

Faure-Rabasse et al. (2002) have determined the kinetics of the constitutive and inducible components of HATS and HATS+LATS in 15-day-old seedlings of rape using ^{15}N labeling experiments. These authors demonstrated that influx rates approximated Michaelis–Menten kinetics below 200 µM (CHATS, IHATS), while at higher NO_3^- concentrations (>1 mM), influx rates of CHATS+CLATS and IHATS+ILATS exhibited nonsaturable kinetics. A Michaelis–Menten-type equation ($I = I_m \times [NO_3^-]/([NO_3^-] + K_m)$ where I = nitrate influx rate; I_m = maximum nitrate influx rate; K_m = affinity constant) for HATS components (constitutive and inducible) and linear equations ($I = a \times [NO_3^-] + b$; a, b are constants) for HATS+LATS constitutes the basis of the model. Influx values of CLATS and ILATS were estimated in our model by subtracting the I_m value of CHATS and CHATS+CLATS influx and IHATS from IHATS+ILATS influx, respectively. It is noteworthy that LATS is considered to operate when soil NO_3^- concentration is above 900 µM and 1 mM for the CLATS and ILATS, respectively. At lower NO_3^- concentrations, LATS activities were integrated into the functioning of HATS because it is not physiologically possible to distinguish the HATS and LATS activities. Indeed, only the use of LATS mutants would allow quantification of the real contribution of LATS under 1 mM of NO_3^-, as previously reported (Wang et al. 1998; Liu et al. 1999).

3.2.4.2   Endogenous and Environmental Effects on HATS and HATS+LATS

The response curves of the effects of different factors such as the light/darkness cycle (16/8 h), ontogeny, application of low temperatures (from 24°C to 4°C) to the root system, or variations of photosynthetically active radiation (from 0 to 500 μmol m^{-2} s^{-1}) on HATS and HATS+LATS activities were obtained by measuring NO$_3^-$ influx at 100 μM and 5 mM K^{15}NO$_3$, respectively. This made it possible to calculate HATS (CHATS+IHATS) activities at the initial concentration (100 μM) and LATS (CLATS+ILATS) activities from the difference in uptake rates measured for the two substrate concentrations used. Variations of influx for each transport system as a function of the studied factors were subsequently fitted with polynomial equations.

3.2.4.3   Introduction of Auxiliary Variables in the Model

Environmental and endogenous factors, introduced into the model as auxiliary variables, allowed integration of regulations by N demand on NO$_3^-$ uptake. Indeed, light/darkness cycle and ontogeny were chosen and incorporated into the model in order to take short- (light/darkness cycle) and long-term (ontogeny) regulations acting on N transport systems into account. Environmental variables such as temperature and radiation were introduced because of their well-known impact on growth and N uptake. Consequently, endogenous factors can be considered as "metaregulation mechanisms" that are influenced by climatic factors (PAR and temperature) permitting access to a higher level of nitrate-uptake regulation by N demand.

To integrate the effects of these different factors on HATS and HATS+LATS activities, a standard influx (SI) value was determined for each studied factor: SI$_{light-darkness}$, SI$_{ontogeny}$, SI$_{temperature}$, and SI$_{PAR}$. Each value was defined for the following conditions: 12 p.m. for a light/darkness cycle of 16/8 h, 20°C for root temperature, 300 μmol s^{-1} m^{-2} for PAR, and B4 stage for ontogeny (i.e., four-leaf stage). These conditions were similar to those used by Faure-Rabasse et al. (2002) to determine nitrate influx kinetics. Because effects of these factors were measured in different experiments, SI values allowed us to adjust the fluctuations of measured influx between these experiments. Thus, a corrected influx (CI) was determined for each factor studied by application of a correction factor defined as the ratio between the value of nitrate influx, obtained by the above-cited adjusted polynomial equations (section 3.2.4.2), and the SI value for each studied factor. For example, the average influx values per day (1884 and 4889 μmol NO$_3^-$ d^{-1} g^{-1} root DW for HATS and HATS+LATS, respectively) obtained by integrating equations for HATS and for LATS from 0 to 24 hours were divided by respective SI$_{light/darkness}$ values. (For detailed equations, see Malagoli et al. 2004.)

3.2.4.4   Calculation of Unregulated and Regulated Uptake

An unregulated uptake (expressed in kg N–NO$_3^-$ ha^{-1}) was calculated from kinetic equations by taking into account soil nitrate concentration and root biomass at different depths and plant densities throughout the growth cycle, and by operating a change of time scale (from hours to days) by multiplying by 24.

The regulated uptake in the model was determined by multiplying the kinetic equations by correction factors. Integration of light/darkness and ontogenetic cycle factors allowed a change of time scale from hour to day and from days to growth cycle, respectively. The last auxiliary variables, temperature and PAR, were then integrated to simulate NO_3^- uptake in environmental field conditions. Changes in day length (minutes) during the year were also taken into account in the model with a day-length reference value equal to 960 min.

3.2.4.5   Sources of Input Variables

Input variables (soil nitrate concentrations and root biomass at different soil depths, temperature, and PAR) needed to run the model were obtained from the INRA oilseed rape database of experiments carried out at Grignon/Châlons/Laon/Reims (http://www-bioclim.inra.grignon.fr). Details about these experiments can be found in Gosse et al. (1999).

The main difficulty encountered when running the model was to estimate the lateral root biomass, which was assumed to be the only part of the root system involved in N uptake, this assumption being based on the fact that NO_3^- uptake by the taproot was found to be insignificant (about 1%, unpublished results). Independently of the studied developmental stage, previous experiments carried out under controlled conditions have shown that lateral root biomass represents an approximately constant proportion of 43% of the total root biomass (taproot + lateral root). This value was introduced as a parameter (p) in the model. To estimate lateral root biomass, a pattern of lateral root biomass distribution among soil layers was estimated from the frequencies of lateral root impact as a function of soil depth available in the database. Using this distribution and the total calculated lateral root biomass, lateral root biomass in each soil layer was assessed. Nitrogen uptake (kg ha^{-1}) in each soil layer was obtained by multiplying the regulated uptake by the root biomass calculated in each soil layer.

Nitrate concentrations in the different soil layers were determined every 15 days using the Skalar method (Gosse et al. 1999). Soil NO_3^- limitation was defined as the maximum soil nitrate stock available for N uptake. Thus, soil nitrate limitation was calculated every 15 days over the whole growth cycle from the database corresponding to soil NO_3^- concentration and soil water content in the different soil layers and N uptake by plants between two harvest dates. An interpolation was made between these two dates.

Finally, model output (i.e., predicted N uptake by the crop) is the sum of N uptake along the root profile. This model was tested to compare observed and predicted N uptake by oilseed rape plants with three levels of N fertilization (N0: 0 kg N ha^{-1}; N1: 135 kg N ha^{-1}; N2: 273 kg N ha^{-1}). The highest soil NO_3^- concentrations were found in the first soil layer. The variation scale in this soil layer ranged from 0.23 to 4.1, from 0.15 to 4.1, and from 0.24 to 7.0 mM for N0, N1, and N2 fertilization, respectively. The model was built using Model Maker software (Cherwell Scientific, Oxford, U.K.).

3.2.4.6   Basic Assumptions for Model Construction

Both nitrate (NO_3^-) and ammonium (NH_4^+) can be used for N nutrition by many crop species. However, it has been reported that Brassicaceae are characterized as NH_4^+-

sensitive plants (see review by Britto and Kronzucker 2002). Even if NH_4NO_3 is used to fertilize plots, ammonium in the soil is readily oxidized to NO_3^- by nitrifying bacteria present in the soil. NO_3^- is the prominent form of N available to most cultivated plants grown under normal field conditions. Moreover, availability of NO_3^- in the soil is often considered as rate limiting for plant growth (Redinbaugh and Campbell 1991). For these reasons, NO_3^- was assumed to be the sole N source used for N nutrition in our work. No NO_3^- efflux was considered under field conditions. Kinetic parameters—K_m and I_m for HATS, a and b for HATS+LATS—were assumed to remain constant during a growth cycle, while NO_3^- transporters were assumed to have a homogenous spatial distribution along lateral roots. A minimum temperature of 4°C was assumed to be the lowest temperature at which growth may occur. Consequently, nitrate uptake by transport systems when temperature was below 4°C was considered as negligible.

It was hypothesized that the taproot/lateral root ratio assessed under controlled conditions was similar under field conditions and remained constant throughout the growth cycle. No competition for water, light, or mineral nutrient acquisition was considered between oilseed rape plants.

Concerning auxiliary variables, plants at the vegetative stage (B4) were used to study the effects of root temperature, light/darkness cycle, and PAR on NO_3^- uptake. Effects of these factors were formalized by polynomial equations and assumed to be the same for all developmental stages. Up- and down-regulations of nitrate transport systems that may occur at the plant level through the effects of different phloem or root compounds issued from nitrate assimilation or photosynthetic activity (amino acids, organic acids, sugars) were implicitly included through the light/darkness cycle or ontogeny. The two constitutive and inducible components of each transport system (high or low affinity) were assumed to be similarly regulated by these compounds. The effect of PAR was taken into account in the model from bolting to harvest according to Chapman et al. (1984) and Mendham et al. (1981), who have demonstrated a decrease of about 60%–80% of PAR transmitted inside the rape canopy at the beginning of flowering. No interaction between climatic factors (temperature and PAR) was taken into account.

3.3 RESULTS AND DISCUSSION

3.3.1 Kinetics of NO_3^- Influx

3.3.1.1 In Induced and Noninduced Plants

On day 15 after sowing, half of the total number of plants previously grown without N were supplied with 1 mM KNO_3 for 24 h to induce the NO_3^- uptake system. Nitrate influx rate was measured on day 16 for both induced and noninduced plants.

At least three different transport systems for NO_3^- uptake were distinguished in *Brassica napus* L. on the basis of a kinetic characterization (Figure 3.1). Two of them were constitutive systems in noninduced seedlings. For NO_3^- concentrations below 200 µM, influx rates approximated Michaelis–Menten kinetics (Figure 3.1A, CHATS), with an estimated I_{max} of 26.3 µmol·h⁻¹·g⁻¹ DW and a K_m of 15.9 µM. A second low-affinity system (Figure 3.1B, CLATS) exhibited nonsaturable

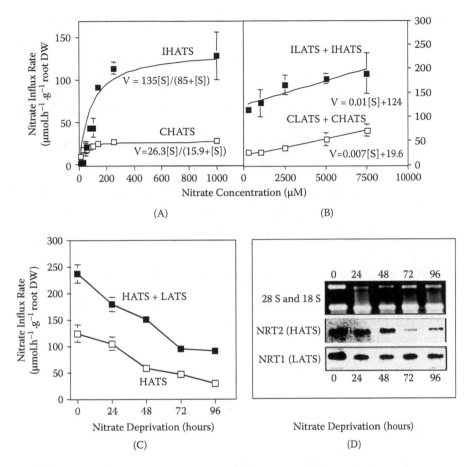

FIGURE 3.1   Kinetic analysis of ^{15}N-nitrate influx by *Brassica napus* L. roots with low (A) or high (B) nitrate concentrations in the nutrient solution. CHATS and IHATS: constitutive and inducible high-affinity transport system, respectively. CLATS and ILATS: constitutive and inducible low-affinity transport system, respectively. (C): Effect of nitrate deprivation on low (CLATS+ILATS) or high (CHATS+IHATS) affinity transport system for nitrate, estimated by ^{15}N labeling. (D): Changes in NRT1 and NRT2 gene expression estimated by Northern blotting during nitrate deprivation. [Redrawn from Faure-Rabasse et al. (2002).]

kinetics between 1 to 7.5 mM NO_3^-. When seedlings were induced by exposure to 1 mM NO_3^- for 24 h prior to assaying influx, NO_3^- uptake increased across the entire range of concentrations. The inducible high-affinity system approximated Michaelis–Menten kinetics at substrate concentrations lower than 1 mM (Figure 3.1A, IHATS), with an I_{max} of 135 µmol·h^{-1}·g^{-1} DW and a K_m of 85 µM. Nitrate influx attributable to the IHATS was fivefold higher than the one associated with the CHATS. The kinetics of NO_3^- uptake determined at high concentrations (1–7.5 mM NO_3^-) suggests that the LATS is devoid of an inducible component. These results show that a simple mathematical description (two Michaelis–Menten plus two linear equations) can be used to describe nitrate transporter activities as a function of external nitrate concentration.

3.3.1.2 Effects of N Deprivation on Influx Rates and Gene Expression

On day 26 after sowing, the automatic supply of NO_3^- was switched off to culture units, and the concentrations of NO_3^- in these units were allowed to deplete by plant uptake to <1 µM over 2 h. This point was taken as time zero for the N-deprivation treatment. NO_3^- influx and mRNA abundance were measured on N-deprived plants at intervals during the period of N deprivation (0, 24, 48, 72, and 96 h). NO_3^- influx was measured 2 h prior to the end of the photoperiod for six culture vessels per treatment.

Plant growth (i.e., dry-matter production) was not significantly affected during the first four days of NO_3^- deprivation, implying that the N deprivation effect on NO_3^- influx and gene expression during this period were unrelated to changes in rates of growth or senescence. Nitrate uptake through high-affinity systems (Figure 3.1C, 100 µM) decreased progressively over the four days of N deprivation from 125 to 30 µmol·g^{-1} root DW h^{-1}. Meanwhile, regarding the importance of HATS in total nitrate influx rate, this decrease also reduced the HATS+LATS uptake from 240 to 100 µmol·g^{-1} root DW h^{-1}. Seventy percent of this decline was observed during the first 48 h of N starvation (see Faure-Rabasse et al. 2002 for a full description). The abundance of mRNA encoding for the *BnNRT2* NO_3^- transporters (i.e., HATS) decreased as a function of the duration of NO_3^- deprivation (Figure 3.1D), demonstrating that this gene is nitrate inducible. The expression of the *BnNRT1* gene was less strongly affected by nitrate deprivation.

3.3.2 Modeling NO_3^- Uptake during the Growth Cycle

For each studied developmental stage (C2, D2, E, F2, G2, G4, G5), plants were acclimated for 1.5 h to KNO_3 concentrations (100 µM or 5 mM), as previously described, before influx measurements were made at 12 p.m.

A mechanistic model was built combining the mathematical description of NO_3^- influx kinetics (from data in Figure 3.1) and data from field experiments (root biomass, soil nitrate concentrations, temperature, PAR) obtained from the INRA Châlons Oilseed Rape database (France) to simulate NO_3^- uptake. When total N uptake was controlled only by soil NO_3^- concentrations (i.e., unregulated uptake), model outputs were largely overestimated (17 times higher) compared with the measured data. Further, nitrate systems regulations (a) occurring during the day/night cycle or during development and (b) resulting from the differential effect of temperatures or photosynthetically active radiation availability were successively introduced into the model.

3.3.2.1 Effect of Light/Darkness Cycle and PAR on NO_3^- Influx

Fifteen-day-old seedlings were transferred from the greenhouse to a culture room for one week. Light was provided by high-pressure sodium lamps (300 µmol m^{-2} s^{-1} of photosynthetically active radiation at the height of the canopy), and the thermoperiod was 20°C (day) and 15°C (night). Before each measurement, plants were acclimated for 1.5 h in a nutrient solution containing either 100 µM or 5 mM KNO_3. NO_3^- influx was then determined at $t = 3, 6, 9,$ and 12 h after the beginning of the diurnal period (i.e., 9 a.m., 12 p.m., 3 p.m., and 6 p.m.) and at $t = 0, 2, 4, 6,$ and 8 h after the beginning of the dark period (i.e., 10 p.m., 12 a.m., 2 a.m., 4 a.m., and 6 a.m.). The same procedure was undertaken for the PAR experiment, and different PAR values (ranging from 0 to 500 $\mu mol\ m^{-2}\ s^{-1}$) were obtained by varying the height between the top of the canopy and the lamps. After acclimation for 1.5 h in a nutrient solution containing either 100 μM or 5 mM KNO_3, NO_3^- influx was measured at 12 p.m.

NO_3^- influx displayed a marked diurnal cycle, as shown by minimal and maximal values reported in Figure 3.2A. From the start of the light period (6 a.m.), HATS activity increased about 1.5-fold to reach a maximum value (about 100 $\mu mol\ NO_3^-$ g^{-1} root DW h^{-1}) at 12 p.m. By the end of the light period HATS influx then decreased progressively back to a similar value measured at the start of the light period (70 $\mu mol\ NO_3^-\ g^{-1}$ root DW h^{-1}). During the dark period, HATS influx decreased to a nearly constant value of about 50 $\mu mol\ NO_3^-\ g^{-1}$ root DW h^{-1}. The pattern of HATS+LATS influx exhibited two peaks (about 266 $\mu mol\ NO_3^-\ g^{-1}$ root DW h^{-1}) at 9 a.m. and 6 p.m. Integration of this short-term effect was of major importance when

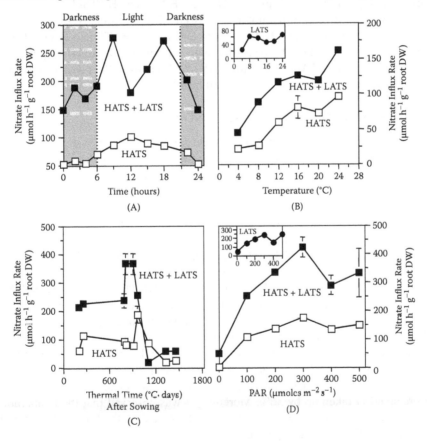

FIGURE 3.2 Variations of HATS (■) and both HATS+LATS (□) activities during the light/darkness cycle (A), as a function of root temperature (B), during a growth cycle (C), and as a function of photosynthetically active radiation ($\mu mol\ m^{-2}\ s^{-1}$) (D) in *Brassica napus* L. var. Capitol. Insets in Figures 3.2B and 3.2D represent the LATS activity, obtained by subtracting HATS activity from HATS+LATS activity. Vertical bars indicate ±SD for $n = 3$ when greater than the symbol. [Redrawn from Malagoli et al. (2004).]

scaling from hourly to daily influx, as it decreased simulated N taken up by 50%. When the day length was further taken into account, it led to a more pronounced decrease during winter, and a final 32% decrease compared with the unregulated N uptake was observed at harvest.

Light availability limitation occurs during winter (i.e., PAR intensity decreases at the canopy level) and spring (i.e., leaf, flower, pod shields [up to 80%] leading to a decreased light penetration through leaf layers within the canopy and subsequent low light radiation values for leaves). This effect becomes quite important, as a lower availability of photosynthetically active radiation will consequently down-regulate the photosynthesis and evapotranspiration, as well as decrease both the energy supply and the water fluxes to the roots, leading to reduced N-uptake capacities. It was then important to take these daily PAR variations into account and to consider changes of HATS and HATS+LATS activities as a function of PAR values (ranging from 0 to 500 μmol m^{-2} s^{-1}) (Figure 3.2D). Both HATS and HATS+LATS activities increased with PAR up to 300 μmol m^{-2} s^{-1} (from 3.7 to 177 μmol NO$_3^-$ g^{-1} root DW h^{-1}, and from 50 to 450 μmol NO$_3^-$ g^{-1} root DW h^{-1} for HATS and HATS+LATS, respectively). LATS activity followed the same pattern, with a saturation point (250 μmol NO$_3^-$ h^{-1} g^{-1} root DW) at 300 μmol m^{-2} s^{-1}. When this effect was included into the model, simulated N uptake was reduced by 19%.

3.3.2.2   Effect of Root Temperature on NO$_3^-$ Influx

Two days before the experiments, 15-day-old seedlings were transferred from the greenhouse to a control room under the previously described conditions. NO$_3^-$ influx was measured after acclimation for 1.5 h in a nutrient solution containing either 100 μM or 5 mM KNO$_3$ at different root temperatures (4, 8, 12, 16, 20, and 24°C) maintained with a cryostat. All influx measurements lasted about 45 min from 12 p.m. Temperatures of the solution used for NO$_3^-$ influx measurements were similar to those applied during the pretreatment.

The time course of the NO$_3^-$ influx of HATS and HATS+LATS showed a similar pattern for the range of tested temperatures (from 4°C to 24°C; Figure 3.2B). However, LATS influx (obtained by subtracting HATS activity from HATS+LATS activity) was barely altered by low root temperature, except at 4°C (inset in Figure 3.2B). In addition, decreasing the root temperature from 24°C to 4°C resulted in a sharp reduction of HATS activity (from 100 to 25 μmol NO$_3^-$ g^{-1} root DW h^{-1}). Overall results show that the N transport activities will be affected differently by temperature, the LATS being less sensitive than the HATS, i.e., when low temperatures occur in winter or in early spring. The effect of varying temperature decreased the predicted N taken up by 36%. Moreover, it was observed that the temperature factor decreased the simulated N uptake (−80%) more than the PAR and light/darkness cycle (−50%) during this period.

3.3.2.3   Effect of Ontogeny on NO$_3^-$ Influx

The time course of the NO$_3^-$ influx of HATS and HATS+LATS was different with developmental stages (Figure 3.2C). NO$_3^-$ influx of both transport systems was more or less unchanged from the two-leaf stage (B2) to the bolting stage (C2) (about 130 and 240 μmol NO_3^- g^{-1} root DW h^{-1} for HATS and HATS+LATS, respectively). From the bolting stage (C2) to the initiation of bud development (E), HATS+LATS activity increased 1.4-fold, while HATS activity remained constant (about 80 μmol NO_3^- g^{-1} root DW h^{-1}; Figure 3.2C). A drastic increase of HATS influx was observed from the E stage (81 μmol NO_3^- g^{-1} root DW h^{-1}) to the F2 (flowering) stage (187 μmol NO_3^- g^{-1} root DW h^{-1}), whereas HATS+LATS influx dropped abruptly from 366 to 250 μmol NO_3^- g^{-1} root DW h^{-1}. Both HATS and HATS+LATS activities decreased thereafter to a minimal value (30 μmol NO_3^- g^{-1} root DW h^{-1}). As a consequence, it appears that both transport systems are differentially regulated during plant growth (i.e., with an up- and a down-regulation during bolting and after flowering, respectively). The latter result suggests that N content in the pods at maturity will be mostly derived from the mobilization of N previously taken up early in the growth cycle for the growth of vegetative tissues. Taking into account this long-term effect allows us to scale up from daily N uptake to the complete growth cycle. When regulations occurring simultaneously or successively on transport system activities along the whole life cycle and at the whole plant level (i.e., long-term effect) were integrated into the model, their impact was only observed at the end of the growth cycle (−24%; Figure 3.3A) during the seed-filling stage (G2 to G5).

3.3.2.4 Cumulative Effect of Regulative Variables and Impact of N Fertilization Levels on N Uptake

Both light/darkness and temperature effects were responsible for 66% of the overall decrease in predicted nitrate uptake, emphasizing the major role played by these variables on NO_3^- uptake by plants when no fertilizer (N0 treatment) was applied (Figure 3.3B). Integrating all these variables in the model resulted in a 5.8-fold reduction of the simulated total N uptake at harvest compared with the unregulated uptake (Figure 3.3B). Comparison of the amounts of the measured and the predicted N uptake at harvest shows that the simulated N uptake, defined as the regulated uptake without soil N limitation, was still three times higher than the observed N uptake, taking soil nitrate availability into account (Figure 3.3B). When the model was run with increasing N inputs (from N1 to N2 treatment), outputs showed that the model was responsive to N fertilizer application compared with the N0 treatment (Figure 3.4).

Integration of the four variables decreased the amount of simulated total N uptake by about a factor of 5.5 for N0 (Figure 3.4A) and N1 (Figure 3.4B) treatments and 3.5 for the N2 treatment (Figure 3.4C). However, an overestimation of the modeled N uptake after the flowering stage was observed for all N treatments. When plant N supply was limited by the soil N availability, the model more accurately matched measured data for N0, N1, and N2 (comparison of regulated uptake with and without soil N limitation; Figures 3.4A, 3.4B, and 3.4C). Indeed, in spite of the high potential activity of the N transport system described in the model (i.e., regulated uptake without soil N limitation), soil N supply from the start of the flowering (F2) to the pod filling (G2) stage was not sufficient to match plant N requirements. Based upon satisfactory model outputs, sensitivity analysis was then performed to identify key parameters that need to be improved through genetic engineering or better management practices (Malagoli et al. 2005b). The analysis showed that (a)

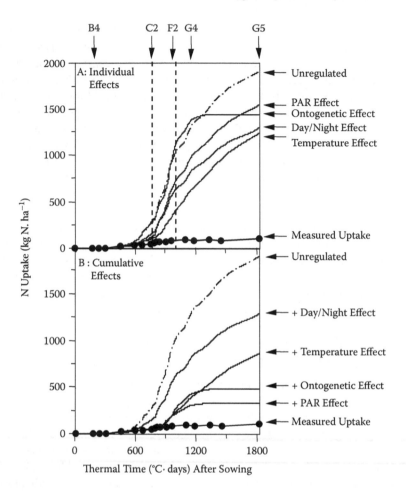

FIGURE 3.3 Simulation of N uptake by an oilseed rape crop (*Brassica napus* L. var. Capitol) with no fertilizer when introducing the impact of the light/darkness cycle, root temperature, ontogeny, and PAR effects on NO_3^- uptake during the growth cycle, either individually (A) or cumulatively (B). [Redrawn from Malagoli et al. (2004).]

the affinity constant of saturable transporters is nonlimiting within the common range of concentrations found in the field, and (b) the contribution of low-affinity nitrate transport to total N remained low and is important solely during periods of high-organic N mineralization and immediately after N fertilization.

3.3.3 PARTITIONING OF N UPTAKE AND N REMOBILIZATION TO VEGETATIVE AND REPRODUCTIVE TISSUES

3.3.3.1 Partitioning of Uptaken N

A weekly $^{15}NO_3^-$ labeling under field conditions allowed us to monitor cumulative N uptake and track uptake- and remobilization-derived N into plant compartments from bolting to harvest (Figure 3.5A). The amounts of ^{15}N increased in green leaves,

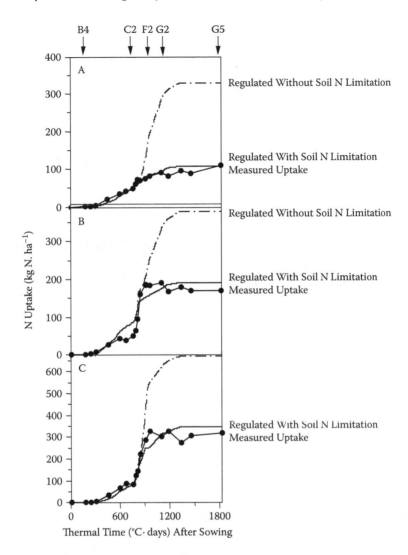

FIGURE 3.4 Simulation of N uptake by an oilseed rape crop (*Brassica napus* L. var. Capitol) with and without the soil N limitation during a growth cycle as a function of three N fertilizer levels: (A) 0 kg N ha⁻¹ (treatment N0); (B) 135 kg N ha⁻¹ (treatment N1); (C) 273 kg N ha⁻¹ (treatment N2). [Redrawn from Malagoli et al. (2004).]

stem, taproots and flowers until the end of the flowering (F2). Nitrogen uptake and further allocation (Figure 3.5A) to taproot and flowers reflects growth, relative mass, and N content. In contrast, the green leaf compartment showed a constant increase of N allocation until the end of flowering, while its biomass remained close to a steady-state value from 1700 to 2350 °C·days. This probably reflects a high N turnover in green leaves and subsequently a high re-export of N. A similar conclusion can be made for the stem after flowering: a continuous increase of allocation of N derived from uptake (Figure 3.5), while its biomass remained the same. From the G1 stage, pods became the main organs where uptaken ¹⁵N was allocated, while no significant

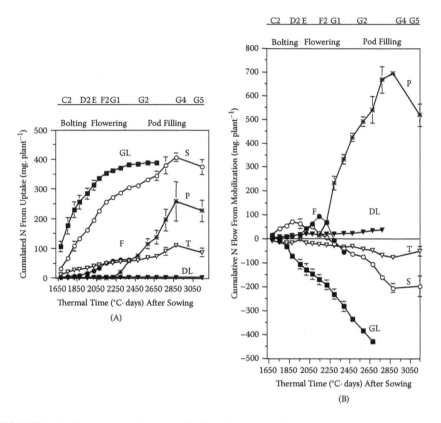

FIGURE 3.5 Cumulative N uptake, further allocation (A), and cumulative endogenous N flows (B) estimated by weekly ¹⁵N labeling in *Brassica napus* L. var. Capitol plants grown under field conditions to taproot (T), green leaves (GL), dead leaves (DL), stem (S), flowers (F), and pods (P). Vertical bars indicate ±SD for $n = 3$ when larger than the symbol. [Redrawn from Malagoli et al. (2005a).]

increase in ¹⁵N was observed in green leaves, although ¹⁵N continued to accumulate in the stem and taproot until 2850 °C·day.

The ¹⁵N flux in each organ (expressed in µg N·plant⁻¹·[°C·day]⁻¹), i.e., the slope of the different straight lines defined between the inflection points in each curve, was calculated to determine the rate of ¹⁵N allocation associated to each compartment. From the start of bolting (C2) to the start of pod filling (G1), organs can be classed as a function of their decreasing order in priority: leaves, stem, taproot, and flowers. The amount of ¹⁵N allocated to the stem was relatively constant (490 ± 13 µg N·plant⁻¹·[°C·day]⁻¹). In contrast, sink strength of leaves and flowers varied during bolting and flowering. From D2/E to G1, the amount of ¹⁵N allocated to green leaves declined (from 1174 ± 132 during C2/D2 to 380 ± 64 µg N·plant⁻¹·[°C·day]⁻¹ during D2/G1) and increased in flowers (from 39 ± 7 during C2/D2 to 157 ± 9 µg N·plant⁻¹·[°C·day]⁻¹ during D2/G1), reflecting a switch in ¹⁵N allocation priority to the benefit of the new growing sinks. During pod filling (from G1 to G5), N uptake was maintained at a significant level—30% of the total N taken up by the crop, with the pods becoming the main sink for uptaken N (Figure 3.5A).

3.3.3.2 Partitioning of Mobilized N

Taking all data together allowed us to determine endogenous N flows before and during pod filling (Figure 3.5B). An overview of source–sink relationships for N was obtained by summing endogenous N influx (positive values of N_{mob}) and efflux (negative value of N_{mob}) for each tissue, so that a sink organ for N had an increase in cumulative endogenous N flow, whereas a source organ had a decrease in cumulative endogenous N flow. Figure 3.5B illustrates the transition from sink to source behavior of a tissue when the maximum cumulative flow from remobilization is reached and then declines. This demonstrates that the leaves and the taproot, to a lesser extent, were permanent sources of endogenous N during the studied period. From the start of bolting (C2) to the visible buds stage (E), endogenous N coming from leaves and taproot were mainly allocated to the stem (86%) and later to flowers (14%), although a portion remained in dead leaves. The status of the stem changed from sink to source during floral transition at about 1850 °C·days. During the flowering period (from E to F2), flowers became the only sinks for endogenous N, supplied by the leaves (57%), the stem (38%), and the taproot (5%), before behaving as a source at 2150 °C·days. During pod filling, all vegetative tissues behave as sources for endogenous N. Indeed, about 690 mg of endogenous N were mobilized to the pods: 36%, 34%, 22%, and 8% being mobilized from leaves, stems, inflorescences, and taproot, respectively.

This N partitioning establishes for the first time that the contribution of labeled and unlabeled N flows to each tissue could be accurately determined under field conditions in *B. napus* L. Exogenous (i.e., labeled) N derived from concurrent uptake is the only source of N for a source tissue, while a sink organ will get N from concurrent uptake as well as from mobilization of N from source tissues. Endogenous (i.e., unlabeled) N represented 35%, 64%, and 73% of the total N allocated to the stem, the flowers, and the pods, respectively. Leaves were the most important source organ for endogenous N mobilization throughout the experiment, although contributions from stem and taproot increased between G3 and G4 during pod filling. Endogenous N mobilization rate was strongly increased during pod filling: in the stem (by 2.7-fold), taproot (by 2.5-fold), and to a lesser extent in the leaves (by 1.4-fold). Comparison of data for pods in Figures 3.5A and 3.5B shows that 73% of N in these reproductive tissues was derived from internal mobilization. However, it should be kept in mind that, during this field experiment, rainfall and mineral N in the soil remained relatively high, and this would probably minimize the role of internal recycling, as a significant N uptake occurred during pod development (Figure 3.5A).

3.3.3.3 Mobilization of N between Senescing Leaves and the Need for a Compartmental Model

With a large number of leaves, the oilseed rape plant represents a complex compartmental system. On a theoretical basis, it can be assumed that, within a leaf rank, changes of total N content during the leaves' life span (Figure 3.6) is a result of two incoming N fluxes of different origin—(a) the allocation of N directly derived from uptaken N, and (b) N allocated to this leaf rank that is derived from N remobilization from lower leaf insertions—and one outgoing flux corresponding to the

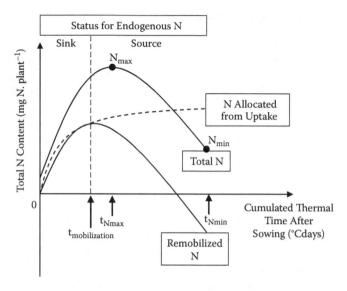

FIGURE 3.6 Conceptual description of N dynamics in each organ during source/sink transition by parameters characterizing N allocation and N mobilization. t_{Nmax} and t_{Nmin}: time when total N content ($mg \cdot plant^{-1}$) is maximal and minimal, respectively; N_{max} and N_{min}: maximal and minimal total N content value ($mg \cdot plant^{-1}$); $t_{mobilization}$: time when endogenous N mobilization starts. [Redrawn from Malagoli et al. (2005b).

N mobilized when each organ switches to a source status for N. Balance of these fluxes describes the sink–source transition that may occur, as described in Figure 3.6, for all compartments of the plant, and particularly for leaves. Malagoli et al. (2005b) have successfully modeled this system to have access to a highly mobile N pool (circulating amino acids, peptides, and other N translocation forms) that may behave as a signal for regulation of root N uptake (Forde 2002). In a second step, the effect of environmental changes (mostly light and temperature) has been studied to explain the capacity of the leaves to efficiently remobilize N. Such an approach has been used to understand how the canopy/morphogenesis of a given genotype will affect the local environment and subsequent leaf senescence and, consequently, the level of N remobilization to the seeds (Gombert et al. 2006).

3.4 CONCLUSION

The use of equations describing the kinetics of NO_3^- high- and low-affinity transport systems, HATS and LATS, respectively, and of the oilseed rape plant databank from INRA (providing inputs and outputs needed to run the model) facilitated the development of a basic mechanistic N-uptake model during the growth cycle. Characterization of endogenous (day/night cycle, ontogeny) and environmental (root temperature and photosynthetically active radiation) factors on transport system activities by ^{15}N-labeling experiments allowed the development of response curves. Integration of these factors led to the development of a model to quantify the impact of each studied factor, as well as the relative contribution of each transport system during the growth cycle. Simulations showed that HATS represents 89% of N taken up at harvest (71% and 18% for inducible and constitutive components of HATS, respectively) when no fertilizer was applied. LATS activity occurred early in the growth cycle. A fall N-fertilizer application increased both its duration of function and its contribution to total N uptake.

Weekly application of ^{15}N on individual plants under field conditions made it possible to discriminate N flux coming from current N uptake (labeled N) and N remobilization (unlabeled N) between each tissue from bolting to harvest. In our conditions, endogenous N coming from N mobilized from vegetative tissues represents the main N source for N pod filling (73% of total N allocated toward pods at harvest, with 36%, 34%, 22%, and 8% coming from leaves, stem, flowers, and taproot, respectively). Despite a decrease of N uptake occurring during the flowering period and associated with an increase of the endogenous N pool circulating at the whole-plant level, no correlation was clearly found between N uptake and N remobilization. In conclusion, our results showed that this macroelement is submitted to a complex dynamic within the plant, thereby opening new research areas in an effort to optimize the utilization of nitrogen by this crop.

It appears that optimization of both the N fertilization rate and schedule, combined with a genetic improvement of the N harvest index in winter oilseed rape, are probably the most readily available tools to improve the yield of this crop while minimizing N release to the environment. Accordingly, a better understanding of N uptake and N partitioning, along with the contribution of these key physiological processes to biomass and yield formation, is required as a basis for genetic improvement. However, genotypic characteristics are difficult to estimate due to the large interaction between genotype and the environment. The ultimate aim of our work is to discriminate genotypic responses in N efficiency and to what extent they are modulated by environmental changes and the availability of nitrogen resources. This approach will be helpful in identifying relevant indicators and, further, breeding plants with higher efficiency to mobilize N toward harvested tissues in N-limiting soil conditions. A simulation with the compartmental model based on N allocation and remobilization within the plant indicated that a drop of dead leaves with a N content as low as 1% DW would increase either seed yield or N content by 15%. Our work, now in progress, will strive to understand how efficient N uptake and/or remobilization levels to the seeds will be affected in contrasting genotypes differing by their shoot or even root architecture.

REFERENCES

Beuve, N., N. Rispail, P. Lainé, J. B. Cliquet, A. Ourry, and E. Le Deunff. 2004. Putative role of aminobutyric acid (GABA) as a long distance signal in up-regulation of nitrate uptake in *Brassica napus* L. *Plant Cell Environ.* 27: 1035–1045.

Boelcke, B., J. Leon, R. R. Shulz, G. Shroder, and W. Diepenbrock. 1991. Yield stability of winter oil-seed rape (*Brassica napus* L.) as affected by stand establishment and nitrogen fertilization. *J. Agron. Crop Sci.* 167:241–248.

Britto, D. T., and H. J. Kronzucker. 2002. NH_4^+ toxicity in higher plants: A critical review. *J. Plant Physiol.* 159: 567–584.

Chapman, J. F., D. H. Scarisbrick, and R. W. Daniels. 1984. Field studies on ^{14}C assimilate fixation and movement in oil-seed rape (*Brassica napus* L.). *J. Agric. Sci.* 102: 23–31.

Clement, C. R., M. J. Hopper, R. J. Canaway, and P. P. H. Jones. 1974. A system for measuring the uptake of ions by plants from solutions of controlled composition. *J. Exp. Bot.* 25: 81–99.

Faure-Rabasse, S., E. Le Deunff, P. Lainé, J. H. Macduff, and A. Ourry. 2002. Effects of nitrate pulses on *BnNRT1* and *BnNRT2* genes mRNA levels and nitrate influx rates in relation to the duration of N deprivation in *Brassica napus* L. *J. Exp. Bot.* 53: 1711–1721.

Forde, B. G. 2002. Local and long-range signalling pathways regulating plant responses to nitrate. *Annu. Rev. Plant Biol.* 53: 203–224.

Fraisier, V., M. F. Dorbe, and F. Daniel-Vedele. 2001. Identification and expression analysis of two genes encoding putative low-affinity nitrate transporters from *Nicotiana plumbaginifolia*. *Plant Mol. Biol.* 45: 181–190.

Gabrielle, B., P. Denoroy, G. Gosse, E. Justes, and M. N. Andersen. 1998. Development and evaluation of a CERES-type model for winter oilseed rape. *Field Crops Res.* 57: 95–111.

Gombert, J., P. Etienne, A. Ourry, and F. Le Dily. 2006. The expressions patterns of *SAG12/Cab* genes reveal the spatial and temporal progression of leaf senescence in *Brassica napus* L. with sensitivity to the environment. *J. Exp. Bot.* 57: 1949–1956.

Gosse, G., P. Cellier, P. Denoroy, B. Gabrielle, P. Laville, B. Leviel, E. Justes, B. Nicolardot, B. Mary, S. Recous, J. C. Germon, C. Hénault, and P. K. Leech. 1999. Water, carbon and nitrogen cycling in a rendzina soil cropped with winter rape: The Châlons oilseed rape database. *Agronomie* 19: 119–124.

Hatch, D. J., M. J. H. Hopper, and M. S. Dhanoa. 1986. Measurement of ammonium ions in flowing solution culture and diurnal variation in uptake in *Lolium perenne*. *J. Exp. Bot.* 37: 589–596.

Huang, N. C., K. H. Liu, H. J. Lo, and Y. F. Tsay. 1999. Cloning and functional characterization of an *Arabidopsis* nitrate transporter gene that encodes a constitutive component of low-affinity uptake. *Plant Cell* 11: 1381–1392.

Lainé, P., A. Ourry, J. Macduff, J. Boucaud, and J. Salette. 2002. Kinetic parameters of nitrate uptake by different catch crop species: Effect of low-temperatures or previous nitrate starvation. *Physiol. Plant.* 88: 85–92.

Liu, K. H., C. Y. Huaug, and Y. F. Tsay. 1999. CHL1 is a dual-affinity nitrate transporter of *Arabidopsis* involved in multiple phases of nitrate uptake. *Plant Cell* 11: 865–874.

Malagoli, P., P. Lainé, E. Le Deunff, L. Rossato, B. Ney, and A. Ourry. 2004. Modeling N uptake in *Brassica napus* L. cv. Capitol during the growth cycle using influx kinetics of nitrate transport systems and field experimental data. *Plant Physiol.* 134: 388–400.

Malagoli, P., P. Lainé, L. Rossato, and A. Ourry. 2005a. Dynamics of nitrogen uptake and mobilization in field-grown winter oilseed rape from bolting to harvest, 1: Global N flows between vegetative and reproductive tissues in relation to leaf fall and their residual N. *Ann. Bot.* 95: 853–861.

Malagoli, P., P. Lainé, L. Rossato, and A. Ourry. 2005b. Dynamics of nitrogen uptake and mobilization in field-grown winter oilseed rape from bolting to harvest, 2: A ^{15}N-labelling-based simulation model of N partitioning between vegetative and reproductive tissues. *Ann. Bot.* 95: 1187–1198.

Mendham, N. J., P. J. Shipway, and R. K. Scott. 1981. The effects of delayed sowing and weather on growth, development and yield of winter oil-seed rape (*Brassica napus*). *J. Agric. Sci.* 96: 417–428.

Noquet, C., J.-C. Avice, L. Rossato, P. Beauclair, M.-P. Henry, and A. Ourry. 2003. Effects of altered source-sink relationships on N allocation and vegetative storage protein in accumulation in *Brassica napus* L. *Plant Sci.* 166: 1007–1018.

Ono, F., W. B. Frommer, and N. von Wiren. 2000. Coordinated diurnal regulation of low- and high-affinity nitrate transporters in tomato. *Plant Biol.* 2: 17–23.

Redinbaugh, M. G., and W. H. Campbell. 1991. Higher plant responses to environmental nitrate. *Physiol. Plant.* 82: 640–650.

Schjoerring, J. K., J. G. H. Bock, L. Gammelvind, C. R. Jensen, and V. O. Mogensen. 1995. Nitrogen incorporation and remobilization in different shoot components of field-grown winter rape (*Brassica napus* L.) as affected by rate of nitrogen application and irrigation. *Plant Soil* 177: 255–264.

Sieling, K., and O. Christen. 2001. Effect of preceding crop combination and N fertilization on yield of six oil-seed rape cultivars (*Brassica napus* L.). *Eur. J. Agron.* 7: 301–306.

Touraine, B., F. Daniele-Vedele, and B. Forde. 2001. Nitrate uptake and its regulation. In *Plant nitrogen*, ed. P. J. Lea and J. F. Morot-Gaudry, 1–37. Berlin: Springer-Verlag.

Wang. R., D. Liu, and N. M. Crawford. 1998. The *Arabidopsis* CHL1 protein plays a major role in high-affinity nitrate uptake. *Proc. Natl. Acad. Sci.* 95: 15134–15139.

4 Control of Plant Nitrogen Uptake in Native Ecosystems by Rhizospheric Processes

Hormoz BassiriRad, Vincent Gutschick, and Harbans L. Sehtiya

CONTENTS

4.1   Nitrogen Availability..71
       4.1.1   Mineralization..71
       4.1.2   Atmospheric N Deposition 73
       4.1.3   Dissolved or Soluble Organic N..75
4.2   Modeling Root System Characteristics Important to N Uptake 76
       4.2.1   Leverage of Plant Attributes and Soil Environmental Factors
                in Nutrient Acquisition and Growth 76
                4.2.1.1   Control Exerted by Root Kinetics (V_{max}) and by Soil
                           Properties 77
                4.2.1.2   Bulk Concentration of Nutrient and Attendant
                           Concentration in Soil Solution 78
                4.2.1.3   Control Exerted by Allocation to Roots...........................79
                4.2.1.4   Lack of Substantial Effect of Mass Flow 80
4.3   Integrating Other Root System Characteristics 83
4.4   Concluding Remarks..85
Acknowledgment 86
References...86
Appendix 4.1   Optimization of Root:Shoot Ratio for Water Relations 91

4.1 NITROGEN AVAILABILITY

4.1.1 MINERALIZATION

Nitrogen (N) availability in crop systems is largely determined by fertilizer application rates. In relatively undisturbed native ecosystems, however, N becomes available largely as a result of internal cycling of this element. Therefore, conceptual or quantitative models of plant N uptake must effectively incorporate processes of N cycling such as decomposition and mineralization. Mineralization of organic N is particularly important because it is often positively associated with increased net primary productivity (Vitousek and Howarth 1991; Retch et al. 1997; Joshi et al. 2006). This correlation is not surprising, because even in plant communities where a significant proportion of the plant N budget may be provided via amino acids and amino sugars (plant uptake of which will be discussed later), mineral N is still the preferred form for plants.

The amount of N mineralization varies from one system to the next. In temperate forest ecosystems, microbial activities and subsequent N mineralization can release 20 to 120 kg ha^{-1} of inorganic N annually (Zak et al. 1993; Fan et al. 1998; Rastetter et al. 2005). Estimates of N mineralization in grassland ecosystems range from 30 to 70 kg ha^{-1} y^{-1} in xeric systems, to as much as 270 kg ha^{-1} y^{-1} in mesic systems. Even in crop production systems, N mineralization can account for as much as 50 kg ha^{-1} of inorganic N mineralization into the soil during a single growing season (Schomberg and Cabrera 2001).

Despite the clear role of mineralization in determining soil N availability, models of plant N uptake in native systems seldom integrate it with mechanistic parameters such as root uptake kinetics and morphology. This is partly because microbial release of inorganic N is under complex control by soil substrate/litter quality, moisture, temperature, and pH (Groffman and Tiedje 1989; Holmes and Zak 1994), which can vary substantially from one site to another. Additionally, accurate estimation of N mineralization is further complicated by the fact that the dominant microbial types, bacteria and fungi, differ markedly in how rapidly they mineralize organic N (Vinten et al. 2002). Nonetheless, we suggest that robust models of plant N uptake require incorporation of soil mineralization characteristics.

Incorporation of N mineralization into physiological uptake models offers serious challenges. Current models of soil N mineralization can be contrasted as following either a functional approach or a mechanistic approach (Benbi and Richter 2002). Current analysis of the literature does not provide evidence that one approach is consistently more reliable than the other (Wang et al. 2004). The simple functional approach is based on lab incubation results, and in cases where only one soil organic fraction is considered to contribute to mineralization, the results are described by a first-order kinetics. The functional model approach has also been extended to describe cases where two or more soil organic fractions are assumed to control the overall mineralization. To date, the first-order kinetic models, including the single and double exponential models, remain widely used (Wang et al. 2004).

While it is more realistic to assume that there are multiple soil organic fractions that mineralize at distinctly different rates, double- or multiple-compartment functional models remain nonmechanistic. Consequently, and, perhaps more importantly, they are highly site specific. Furthermore, these models do not assess the relative proportion of mineral N that is further nitrified. Therefore, the relative availability of the two inorganic N forms will not be known. Benbi and Richter (2002) suggested, however, that for a given site, a functional approach is a reliable tool to predict soil inorganic N release when it is used with at least two compartments of soil organic N: a rapidly mineralizing fraction and a more recalcitrant (residual) one.

In contrast to simple functional models, the mechanistic mineralization models are process-based, derived from current understanding of how soil moisture, temperature, pH, and texture affect the key processes of the N cycle that affect net mineralization (for example, gross mineralization, immobilization, and nitrification). The mechanistic models range from noncompartmental approaches, where soil organic substrate is considered as a continuum of qualities as opposed to distinct compartments (Agren and Bosatta 1987), to multicompartmental approaches. We know of at least one series of mechanistic models that is based on analytically distinct pools of C and N (including labile and stable plant material, microbial biomass, and labile and stable organic matter) that have explicit compartments for exchangeable ammonium (NH_4^+)–N and nitrate (NO_3^-)–N (Gaunt et al. 2001; Sallih and Pansu 1993).

In general, the mechanistic models are often criticized for lack of robust validation under realistic field conditions and, when compared with the simple functional model approaches, they do not always show a clear advantage. Because these models can incorporate N compartments including NH_4^+ and NO_3^-, we believe that the mechanistic models may be better suited for integration into a physiologically driven uptake model such as the one we present in this chapter. We also suggest that those mineralization models, be they simple functional or mechanistic models, could improve the reliability of the N-uptake models if they capture the high temporal and spatial variability.

4.1.2 ATMOSPHERIC N DEPOSITION

It is estimated that anthropogenic N fixation exceeds biological fixation of this growth-limiting nutrient. Galloway and Cowling (2002) reported that, in 1990, the annual global N production from anthropogenic sources was roughly 140 Tg, the majority of which, about 85 Tg N per year, came from Haber–Bosch processing of N_2 into fertilizers. Fossil fuel combustion and other industrial sources are the remaining major mechanisms of anthropogenic N creation (Galloway and Cowling 2002). On average, Western Europe receives five times more wet N deposition than the United States (Holand et al. 2005). Nevertheless, in the United States, many native communities experience 25 to 100 kg ha^{-1} y^{-1} ammonium and nitrate deposition due to farming, industrial activities, and combustion of fossil fuel (Galloway et al. 1995). Because native ecosystems are generally N-limited (Schlesinger 1997; Vitousek et al. 1997), chronic atmospheric deposition represents a novel source of fertilizer that can alter plant N uptake.

Aber et al. (1989) and Schulze (1989) proposed conceptual models that attributed temperate forest decline in North America and Western Europe to increased deposition of anthropogenic nitrogen. These seminal papers brought much needed attention to the alarming increase in atmospheric N deposition in native systems and how it may impact them. While the exact amount of atmospheric N deposition varies from region to region, intensive farming and industrial activities are major culprits in raising ecosystem N load above the background levels by many orders of magnitude. Since the 1950s, atmospheric N deposition has doubled much faster than any other component of the global change, e.g., CO_2. In addition to N load, the knowledge of the composition of deposition is critically important.

The sources of N pollution are important, as they determine which inorganic N form is deposited. While N deposition originating from industrial sources is dominated by NO_3^-, N deposition associated with farming activities is made of NH_4^+ (Lovett 1994; Schulze 1989; Skeffington 1990). For example, atmospheric deposition in Chicago consists of a 4:1 ratio of nitrate to ammonium (NADP 1997), while atmospheric deposition in the Netherlands consists of at least a 3:1 ratio of ammonium to nitrate (Wilson et al. 1995). Many studies addressing N deposition with long-term fertilization plots have used a 1:1 ratio of nitrate to ammonium (e.g., Tilman 1993; Magill et al. 1997).

The uptake and assimilation of nitrate and ammonium require distinct physiological mechanisms and have different consequences on plant growth (Fernandes and Rossiello 1995). Compared with ammonium nutrition, nitrate nutrition tends to result in greater uptake of cations, higher tissue concentrations of carbohydrates, and smaller root-to-shoot ratios (Fernandes and Rossiello 1995). Assimilation of nitrate may also have a higher energetic cost than assimilation of ammonium. For example, the yield of six perennial grass species supplied with nitrate as a sole nitrogen source was only 22%–48% of the yield when supplied with ammonium (Wiltshire 1973). Although most of the work comparing plant responses to nitrate and ammonium has been done with agricultural species, nonagricultural species also vary in their capacity to take up and assimilate nitrate (Gebauer et al. 1988; Falkengren-Greu 1995). Therefore, we recommend that models of plant uptake pay close attention to the relative availability of the inorganic N form in the deposited N.

It is also important to know how fast the deposited N becomes available for plant uptake. A number of studies have shown that atmospheric N deposition may quickly become tied up, so that the availability at the root surface does not match the deposition rate. McNulty et al. (1991) sampled the forest floor at 11 sites along an N-deposition gradient in New England. They showed that N deposition was positively correlated with percent N in the forest floor and with potential net nitrification and mineralization, but negatively correlated with C:N, lignin:N, and Mg:N ratios in spruce needles in the litter layer. The speed with which forest ecosystems reach the limits of the N they can accumulate depends on forest history. Forests that were clear-cut or farmed accumulated N for a longer time before nitrification rates increased (Aber and Driscoll 1997). A series of 15 burned, logged, or little-disturbed forest sites in northern New Hampshire suggested that the N mineralization to forest floor C:N ratio may be a useful diagnostic of system N accumulation (Goodale and Aber 2001).

Stable isotope experiments suggest that most N added to forests accumulates in the forest floor. The forest floor retained 42%–58% of ^{15}N added to *Quercus velutina/ rubra* or *Pinus resinosa* stands in the Harvard Forest, which were receiving 58 kg N ha^{-1} y^{-1}. The greatest amounts of ^{15}N were retained in fine roots (13%–25%) and in litter plus humus (18%–19%); much smaller amounts of ^{15}N were retained in wood, leaves, and mineral soil (Nadelhoffer et al. 1999). The forest floor also retained the largest amounts (20%–45%) of added ^{15}N in four European coniferous forests that were part of the NITREX experiments (Tietema et al. 1998). At deposition rates of 30–80 kg ha^{-1} y^{-1}, Tietema et al. (1998) found that less N was retained in the forest floor, and interpreted this to mean that inputs exceeded the capacity of the microbial and plant population to immobilize N.

4.1.3 DISSOLVED OR SOLUBLE ORGANIC N

The fact that soils contain a large pool of soluble organic N (SON) that can be absorbed by plants has been known for some time (Keeney and Bremner 1964). It is only during the last two decades, however, that there has been growing recognition that in many native ecosystems, at least, a portion of plant N demand may be met through the uptake of soluble or dissolved organic N (Lipson and Näsholm 2001; Näsholm and Persson 2001; Schimel and Bennett 2004; Xu et al. 2006). This recognition is partly motivated by the fact that in many ecosystems the best estimates of annual inorganic N pools and fluxes could not fully account for the yearly increase in standing N biomass (Kielland 1990; Chapin et al. 1988; Fisk and Schmidt 1995; Leadly et al. 1997; Chen and Xu 2006). Additionally, much evidence has emerged showing that plant roots, either in association with or without mycorrhizal fungi (Abuzinadah and Read 1988; Raab et al. 1996, 1999; Wallenda and Read 1999; Chapin et al. 1993; Henry and Jefferies 2003; Näsholm et al. 1998; Näsholm et al. 2000; Lipson and Monson 1998), can effectively take up small-molecular-weight organic N such as amino acids through larger molecules such as polypeptides and proteins (Abuzinadah and Read 1986a, 1986b), which have also been shown to be taken up by the root system. Therefore, it is critical that models of plant nitrogen acquisition adequately address the role of soil organic N.

Given that root uptake of amino acids has been shown to be mediated by transporters (Chapin et al. 1993; Wallenda and Read 1999), it should be possible to integrate kinetic parameters as well as concentration of amino acids into uptake models such as that described in the following section. However, incorporation of such parameters is outside the scope of this chapter. Nonetheless, we suggest that some important consideration must guide any such effort. For example, it is important to consider the potential interaction between amino acid and inorganic N uptake. The availability of amino acids and their uptake by plants have been shown to significantly affect uptake of inorganic N. For example, the high availability of amino acids in the root medium leads to reduced uptake as well as reduced subsequent assimilation of nitrate (Padgett et al. 1993; Aslam et al. 2001). Therefore, accurate modeling of N uptake by plants requires a robust understanding of the interactive effects of inorganic N and amino acids uptake.

The majority of the studies that address SON as a source of plant N focus on amino acids. Interestingly, most of these studies deal with plant communities from colder climates such as boreal forest, alpine, and arctic ecosystems. It is currently not known if SON is proportionally more important to the N economy of plants in the colder as opposed to moderate and warmer climates. While amino acids constitute a significant pool of available N in many native systems (1% to 25%) (Chen and Xu 2006; Yu et al. 2002; Senwo and Tabatabai 1998) where plant species exhibit a well-developed absorption capacity, the role of amino acids in the annual N budget of native plants remains quite uncertain. Our current knowledge is limited to those studies that simply illustrate the presence of amino acids in the soil and a plant's ability to take them up. Little quantitative data is available to evaluate the relative contribution of amino acids to the annual plant N budget. In fact, when these N budget studies are conducted in detail, serious doubts emerge as to whether amino acids are a major source of the plant N economy (Owen and Jones 2001; Jones et al. 2004, 2005; Bennett and Prescott 2004).

In a grazed coastal marsh system, Henry and Jefferies (2002) showed that the uptake of amino acids might only be important when soil inorganic N availability is low. Furthermore, the focus on soil amino acids has taken attention away from a large number of other organic compounds that are often much more prevalent in the soil. For example, in many ecosystems, amino sugars and peptides are much more prevalent in the soil than amino acids (Chen and Xu 2006; Amelung et al. 1999). At the present, we do not know the relative importance of these forms of SON to plant N economy. Even when availability and uptake of soil amino acids are characterized, too often, these studies focus on one or two amino acids (e.g., glycine and glutamine). We suggest that in targeting one or two amino acids, adequate justification be given to such a focus. That is, are the target amino acids available in disproportionately larger concentrations than other soil amino acids?

4.2 MODELING ROOT SYSTEM CHARACTERISTICS IMPORTANT TO N UPTAKE

4.2.1 LEVERAGE OF PLANT ATTRIBUTES AND SOIL ENVIRONMENTAL FACTORS IN NUTRIENT ACQUISITION AND GROWTH

A number of attributes of the plant affect the rate of nutrient acquisition per unit mass and, thus, the plant's growth rate, both relative and absolute. These include the root:shoot ratio (r) and the fraction of photosynthate (PSate) allocated to fine roots (f_{FR}); the kinetic parameters of the uptake carrier proteins, V_{max} and K_m, in the Michaelis–Menten formulation of uptake rate, v (variously per unit area or per unit dry mass), in terms of the concentration, c_a, of nutrient at the root surface

$$v = V_{max} \frac{c_a}{c_a + K_m} \qquad (4.1)$$

and the root geometry as mean root radius (a) and mean root spacing ($2b$). One may further consider the roles of mycorrhizae in "expanding" the effective root geometry. In the soil, the bulk concentration of the chosen nutrient (c_b) is important, as is the diffusibility (D), amended by the effect of cation adsorption expressed as the buffering factor (c). Soil water potential directly affects these factors and also the degree of root contact with the soil solution. The rate of renewal of soluble nutrients, particularly mineralization of N, is important for setting the width of depletion zones around roots and, thus, the length of the diffusive pathway. Root growth rates also affect the relative importance of depletion zones, though not as strongly as one might intuit (e.g., Yanai 1994, Figure 3). Finally, one may consider the plant transpiration rate, which generates mass flow of nutrients and is set by a combination of plant and soil factors.

The roles of all these factors have been discussed ably and at length by a number of authors, including Tinker and Nye (2000) and Yanai (1994). We wish to emphasize selected factors that are not as widely appreciated as the others, or that have often been misinterpreted, or that have effects that defy some intuitive understanding:

1. The generally high leverage of V_{max} for uptake rates, in common ranges of other factors, especially bulk nutrient concentration, c_b
2. c_b itself, in soils usually regarded as low in nutrients
3. The low leverage of root:shoot ratio in many conditions, and the apparent primacy of water rather than nutrients in setting r
4. The lack of any substantial effect of mass flow on rates of nutrient uptake

We consider each of these in the following subsections.

4.2.1.1 Control Exerted by Root Kinetics (V_{max}) and by Soil Properties

Uptake of nutrients ultimately occurs at the root surface, but how much do the carrier proteins control uptake rates when low nutrient concentration in soil (c_b) or low diffusivity of a nutrient in soil can enforce a low concentration at the root surface? The question is of long standing, and models parameterized with experimental data offer the best route toward a quantitative understanding. In contrast, one cannot adjust V_{max} and K_m in replicate plants at will in an experimental apparatus, nor the root geometry. Consequently, simulations have been pursued for decades, with an early synthesis presented by Nye and Tinker (1977). Another important effort is modeling the effect of varied root geometries (root diameter, spacing, and vertical distribution) on uptake and growth (Gardner 1960; Lynch and Brown 2001).

The most basic effort toward this goal is then modeling uptake and resultant plant growth rate as functions of important root and soil properties. We will not attempt to reproduce the wide range of simulations done to date, but we do provide a visualization of the sensitivity of growth rate that may be compelling. The focus will be the relative growth rate, RGR, of a young plant, for which plant RGR is a critical trait and one that is most sensitive to nutritional status, uncomplicated by reproductive allocation, stand density, and other effects.

We use as a growth model a functional balance model presented by BassiriRad et al. (2001). This model assumes a tight coupling of uptake to the use of the nutrient in photosynthesis. Thus, one must specify the root uptake capacity and the photosynthetic nutrient-use efficacy (p^*, as grams of photosynthate (PSate) per gram N in leaves per day), as well as an efficiency of converting raw photosynthate into dry matter (β, gDM gPSate^{-1}). Root-uptake capacity can be specified directly as the rate, v, or, for later discussion, derived from the combination of the root's Michaelis–Menten kinetics (V_{max}, K_m) with a soil nutrient transport model that requires specification of bulk concentration (c_b), diffusivity (D), root radius (a), and mean root spacing ($2b$). The allocations to roots and to leaves are specified as root:shoot ratio (r) and the fraction of shoot mass as leaf mass (α_L). Our model also uses two factors representing product suppression of photosynthesis (significant in some simulations at elevated CO_2) and enforcement of a maximum RGR (from limitation on meristem activity and number).

We choose a base case representing a fast-growing ruderal or crop: $p^* = 40$ g PSate g N d^{-1}, $\beta = 0.5$ gDM gPSate^{-1}, $v = 0.017$ g N g DM$_{root}^{-1}$ d^{-1}, $r = 0.4$, and $\alpha_L = 0.5$. The molar concentrations of photosynthate that half-repress photosynthesis and RGR are, respectively, 1.0 and 0.5 (Gutschick and Pushnik 2005, and references therein), which explain more of the starch and sucrose repression of, for example, Rubisco (ribulose-1,5-bisphosphate carboxylase/oxygenase) gene expression.

Subsequently, we explore the effect of varying soil and root properties. In order to generate a single useful plot, we consider variations in the two soil parameters, c_b and D, over ranges covering high stress to high nutrient availability (0.1–2.0 mol m^{-3} and 10^{-11} to 2×10^{-10} m^2 s^{-1}, respectively). We set a single value of V_{max}, 8×10^{-8} mol m^{-2} s^{-1}. With $a = 20$ μm and dry matter constituting 25% of fresh mass, this is equivalent to $V_{max} = 115$ μmol g DM$_{root}$$^{-1}$ h^{-1} or 0.032 g N g DM$_{root}$$^{-1}$ d^{-1} at an effective 20 h per day uptake, similar to that found experimentally on the ruderal, *Helianthus annus* (Gutschick 1993; Gutschick and Kay 1995). To test the effect of changes in V_{max}, we rerun the simulations with 50% larger V_{max}. We then compute the sensitivity,

$$S = \frac{d(\text{RGR}) / \text{RGR}}{d(V_{max}) / V_{max}} = \frac{d \ln(\text{RGR})}{d \ln(V_{max})} \tag{4.2}$$

In the functional balance model, the maximal value of S is 0.5, resulting from RGR varying as the square root of uptake rate (e.g., an increase in RGR by a factor $\sqrt{2} = 2^{0.5}$ requires another factor of $\sqrt{2}$ in tissue nutrient content, or a factor of $\sqrt{2}\sqrt{2} = 2$ in nutrient-uptake rate).

The results are shown in Figure 4.1. It is clear that *RGR* is sensitive to c_b and D only at rather low magnitudes of each. At the same time that RGR stabilizes in response to variations in c_b and D, RGR becomes very sensitive to V_{max} (S approaches 0.4, or 80% of its theoretical maximum). That is, V_{max} is the controlling factor over the major range of soil conditions. The simulations can be repeated for other choices of plant parameters. For a slower-growing plant, with or without lower V_{max}, the range of c_b and D where these exert strong control shrinks roughly in proportion to either factor (RGR or V_{max}; results not shown).

The conclusion is that V_{max} is a strong contributor to plant performance. This is supported by many experimental studies showing tight regulation of V_{max} under varying growth conditions (reviewed in Glass 2005). One might expect that the optimal value of V_{max} might be predictable, based on balancing the cumulative cost of acquiring and metabolizing nutrients against their declining utility when they are accumulated in excess. However, a number of simulations fail to generate optimal uptake rates of N and optimal tissue N contents in realistic ranges (Gutschick 1993; Gutschick and Kay 1995). Constraints on development and, hence, on nutrient uptake may well be important. So, too, might trade-offs of increased risk of herbivory become important as N content rises, but the quantitative formulation of herbivory as a stochastic risk involves yet another level of modeling.

4.2.1.2   Bulk Concentration of Nutrient and Attendant Concentration in Soil Solution

Studies of root-uptake kinetics consistently show that the high-affinity uptake system (HATS) for a given nutrient has K_m values in the tens of micromoles (<0.1 mol m^{-3}). One might infer that nutrient concentrations at the root surface are similarly

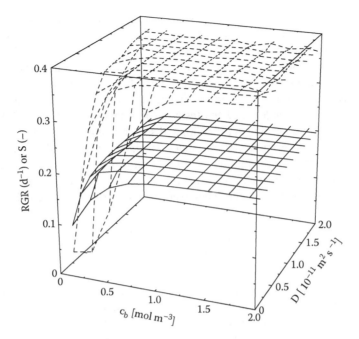

FIGURE 4.1 Modeled response of relative growth rate (RGR) of a ruderal or crop plant, in early growth, to various combinations of nutrient concentration in bulk soil (c_b) and diffusivity in soil (D). Solid lines are RGR (at fixed V_{max}); gray lines are logarithmic sensitivity to changes in root maximal uptake rate, V_{max}, as $d \ln(RGR)/d \ln(V_{max})$.

low. Klipp and Heinrich (1994) argue that natural selection tends to drive K_m to magnitudes at or below substrate concentrations for enzymes in general. We note briefly that c_b is commonly much higher, even in soils considered nutrient poor. The example of nitrogen in desert soils is offered. The data of Schaeffer et al. (2003) and Titus et al. (2002) indicate that, in North America's hottest and most N-poor desert, bulk concentrations average more (often much more) than 2 ppm as mass of N per mass of soil. At a bulk soil density of 1.3 tonnes m^{-3}, this is equivalent to 2.6 g N m^{-3}, or approximately 0.18 mol m^{-3}. The nutrient is actually carried in soil water that represents, on average, about 10% of the soil volume; consequently, concentration in this soil solution is on the order of 1.8 mol m^{-3} or greater. There are undoubtedly episodes when this concentration is depleted, including by rapid growth of plant (or microbial) communities. Otherwise, we should not observe responses of plant growth to added N that would only raise c_h even higher above saturation of the HATS. It is a challenge to resolve the large temporal excursions of c_b. Current experimental techniques, such as the use of bags of ion-exchange resin, are not up to the task. The relative importance of the low-affinity uptake system in natural conditions is also a challenge.

4.2.1.3 Control Exerted by Allocation to Roots

Generally, root:shoot ratios (r) are larger in ecosystems low in nutrients, which includes most "natural" ecosystems, in contrast to agricultural systems. It is also common that r increases with nutrient stress within a single genotype. Therefore, it is reasonable to assume that increased r has a significant net benefit and, moreover, that r is relatively close to its optimal value in both unstressed and stressed conditions. However, this assumption has not been rigorously tested on the basis of the fundamental physiology of growth. Such assessments can be addressed by using functional growth models. One simple model, the functional balance model, gives a negative answer, under the presumption that nutrient status is the dominant signal for root allocation. This model has been tested experimentally (Gutschick 1993; Gutschick and Kay 1995; Zerihun et al. 2000) and shown to have acceptable accuracy in explaining RGR and nutrient content under varying nutrient availability. However, the model indicates that the optimal root:shoot ratio is unity, for any growth conditions. We may inspect the final expression for RGR attained when uptake and use of a nutrient are in functional balance. This expression is derived in the cited references:

$$\text{RGR} = \frac{\sqrt{r}}{(1+r)} \sqrt{v \beta \alpha_L p^*} \qquad (4.3)$$

The variables have the same meaning as presented earlier. One can use simple calculus to show that the factor involving r is maximal at $r = 1$ (i.e., $(d/dr)[\sqrt{r}/(1+r)] = 0$ at $r = 1$). Note that changes of r by factors of 50% (to 0.5 and 1.5) change the whole factor by a much smaller amount (−6% and −2%, respectively). The cited references note this quandary in some detail.

This conclusion has some vulnerabilities. One assumption is that root properties (root fineness, spacing, V_{max}, K_m) do not vary as r varies. However, any variations in the other properties do not change the conclusion that RGR would improve under any choice of these properties, if r attains the value of 1. One might also consider that the fraction of root mass as fine roots (f_{FR}) can vary, both with age and with nutrient stress. However, this only introduces the factor f_{FR} under the radical in equation (4.3), separately from the factor r, and RGR should still increase until $r = 1$.

One possible resolution is that r may be set primarily by the need to balance water uptake to plant transpiration, E. The smaller the root system (smaller r), the smaller is the magnitude of E that can be supported. For a given shoot size, lower E requires lower stomatal conductance, g_s, which decreases the photosynthetic rate per leaf area, A. An increased root fraction can allow larger g_s and A. However, these reach plateau values at large r, while the diversion of mass from photosynthetic tissue (the factor $1/(1 + r)$ above) eventually curtails RGR. A more detailed argument is given in the appendix at the end of this chapter. Values of r well below unity are supported at typical magnitudes of physiological traits. The problem remains that r varies in response to nutrient stress at high water status, such as in hydroponics. One might invoke the common correlation between low nutrient status and low water status in the field to argue that natural selection has linked the two responses. There is no basis yet to make this assertion.

4.2.1.4 Lack of Substantial Effect of Mass Flow

The argument that we develop here may be succinctly stated: Mass flow sweeps nutrients toward the root surface, but it flattens (or even reverses) the diffusional gradient. The cancellation of the interference term (mass flow × diffusion) is almost exact. Consequently, nutrient uptake is expected to be identical with and without mass flow under a very wide range of conditions.

The simplest case to treat, and the case with the most likely contribution of mass flow, is that of a nonadsorbed nutrient such as nitrate. We treat the quasi-steady state, when diffusive gradients have become established; corrections for dynamics are discussed at the end.

Consider a soil with water moving with (vector) velocity $\vec{u}$ at a given location, at or away from the root surface. The soil offers a diffusivity D to the nutrient (taken as isotropic, with no real loss of generality). The nutrient has a concentration c at the given location.

Roots are typically nearly cylindrical and long relative to their diameter. In cylindrical coordinates, nutrient flow in the external solution is typically almost fully in the radial direction only. So, too, is the velocity of water, $\vec{u}$. We may then write a scalar equation for this radial component of the nutrient flux density. Using J as the magnitude of the total nutrient flux, we have

$$J = -uc - D\frac{\partial c}{\partial r} \rightarrow j = v_w c + D\frac{\partial c}{\partial r} \qquad (4.4)$$

On the right side, we have changed to a more intuitive convention, that positive flux ($j = -J$) and positive water flow ($v_w = -u$) are into the root. Let us assume that flow is essentially in steady state, i.e., that new root growth is not fast, nor are nutrient reserves readily depleted. In this case, there is no time dependence, and c depends only upon r; we may then replace $\partial c/\partial r$ with dc/dr, a total derivative. This formulation is an approximation of more complete equations (e.g., Darrah et al. 2006; Tinker and Nye 2000; Yanai 1994) that account for spatial variation in water content and solid–liquid equilibria of solutes.

In steady state, the flux across any radius (a shell at distance r) is equal to the flux at any other radius. The area that the flux crosses is proportional to r, so that the flux density multiplied by r is a constant. In particular, we can refer all flux densities (of water and of nutrients) to their values (j_a for nutrient, v_a for water) at the root surface, which we take as $r = a$, the root radius:

$$j(r)r = j(a)a \rightarrow j = \frac{j_a a}{r}; \text{ similarly}, v = \frac{v_a a}{r} \qquad (4.5)$$

With these substitutions, we may reorder equation (4.4) into a differential equation for $c(r)$:

$$D\frac{dc}{dr} = -\frac{v_a a}{r}c + \frac{j_a a}{r} \qquad (4.6)$$

This explicit (analytical) solution of this equation is a bit complicated, but it is readily derived, such as with the adjoint differential equation:

$$c(r) = c_a \left(\frac{r}{a}\right)^{-k} + \frac{j_a a}{kD}\left[1 - \left(\frac{r}{a}\right)^{-k}\right], \text{ with } k = \frac{v_a a}{D} \qquad (4.7)$$

An equivalent form was derived by Nye and Spiers (1964). There are interesting limiting cases of behavior. If there is no mass flow ($v_a = 0$; the case of zero transpiration), then $k \to 0$ (k goes to zero), and the limit of $(1/k)$ times the quantity in the square brackets reaches a limiting logarithmic form. Consider $(r/a)^{-k}$ as $\exp(-k^* \ln(r/a))$; for small k, this becomes $1 - k^* \ln(r/a)$, and we get simply

$$c(r) = c_a + \frac{j_a a}{D}\ln\left(\frac{r}{a}\right) \qquad (4.8)$$

That is, there is a logarithmic profile away from the root surface. We shall use this formula to evaluate the contribution of mass flow, by difference from the solution (for flux, not c) from that in equation (4.7). In no case is c_a, the concentration at the root surface, separable into diffusive and mass-flow components.

Uptake rates at the root surface are responsive to the concentration at the root surface, c_a. We need to incorporate an accurate model of uptake to get a complete, consistent solution for both c_a and uptake rate j_a. An essentially universal form for nutrient uptake (review: Tinker and Nye 2000) is the Michaelis–Menten form,

$$j_a = V_{max}\frac{c_a}{c_a + K_m} \qquad (4.9)$$

Here, j_a is the nutrient-uptake rate v expressed specifically per unit area; V_{max} is the saturated rate; and K_m is the Michaelis–Menten constant, the concentration at which uptake is at half-maximal rate. One can include a back-leakage term, but this is often very small. We may substitute equation (4.9) into equation (4.7) to obtain an equation solely in terms of the concentration c_a. We may solve the combined equation, a quadratic in c_a, for the value of c_a. We can then substitute this value of c_a into equation (4.9) to obtain the estimate of the nutrient-uptake rate.

We have simulated uptake and the role of mass flow for a variety of test cases, ranging over:

1. Relatively low bulk concentrations (Mojave Desert; data from Schaeffer et al. 2003; Titus et al. 2002) to high concentrations (taken as 10 times higher); over this modest range, one sees the clear onset of differences in control of uptake by environment and physiology (Figure 4.1)
2. Soil diffusivities from high (2×10^{-10} m² s⁻¹; cf. Tinker and Nye 2000) to low (1/5 of former; at much lower diffusivities, uptake is less important)
3. Zero to high transpiration rates, equivalent to water velocities at the root surface from 0 through 5×10^{-10} m s⁻¹ to 1.1×10^{-7} m s⁻¹)

The results are shown in Table 4.1. Note that the increment in nutrient uptake when mass flow is present is only 0.01% to 1.23%. Mass flow appears large only as the crude measure of concentration multiplied by water velocity. This is the measure commonly cited (e.g., Marshner et al. 1991; Kage 1997). However, this measure omits the interference term (suppression of diffusion) that essentially cancels the crude increment exactly. We offer that mass flow need not be considered in estimating nutrient uptake in general. That is, the rate of nutrient uptake is essentially identical in the presence or absence of mass flow. Nutrient concentration and soil diffusivity remain the controlling factors. The conclusions apply to freely diffusing nutrients such as nitrate, and even more strongly to nutrients such as phosphate that are significantly adsorbed on soil constituents. We have omitted consideration of a few phenomena that contribute modestly to uptake. One is mass flow into young root tips, apical to the maturation zone. In our most favorable case, it represents 7% of diffusive uptake, but this is to be weighted by the low fraction of root in this condition, and we may still neglect it (Brady et al. 1993). Uptake by root tips also occurs before the diffusive gradient develops, such that mass flow is again important. Yanai (1994) shows that this generates a modest correction (always less than 20% in her conditions). These and several other factors will be discussed in a later publication.

Our results appear superficially at variance with some experimental indications that the nighttime transpiration is negatively correlated with nutrient status, and enhanced by N amendments (Bucci et al. 2006). The relation that they found may be correlational rather than causal, via some mechanism that acts adaptively for a physiological or ecological activity not directly related to nutrition. Bucci et al. (2006) noted that N and P fertilization changed the full hydraulic architecture of the plants.

4.3 INTEGRATING OTHER ROOT SYSTEM CHARACTERISTICS

Improving our knowledge of plant N uptake and subsequent development of predictive models will ultimately depend on a keen understanding of root system characteristics that collectively control N uptake. A number of important characteristics distinguish nitrogen availability in native vs. managed ecosystems. Unlike crops where fertilizer application minimizes N limitation, native ecosystems are often nitrogen limited. Nitrogen availability in natural ecosystems also differs from that of agroecosystems in terms of temporal and spatial heterogeneity. For example, early-season flushes of N mineralization and pulses of increased N availability after a rain event add an element of uncertainty to N availability that is less dramatic in typical field crop systems, particularly those under irrigation.

These differences have undoubtedly resulted in the evolution of a number of root system characteristics that would not necessarily be favored in crop production practices of today. For example, the degree of mycorrhizal infection in native plants is far more profuse than that observed in agronomic species. Integrating the relative contribution of mycorrhizal colonization in N-uptake models is a challenging task. While a large number of studies indicate a positive effect on N uptake due to mycorrhizal associations, the enhancement is highly species-specific (BassiriRad et al. 2001; BassiriRad 2006) and varies with the stages of plant development. Previ-

TABLE 4.1
Modeled Concentrations of Nitrate and Uptake Rates under Five Conditions of Physiological Status and Soil Water Content

Water Status Time of Day Remark	High Day	High a Night	High a Night Low v_a	Low Day	Low Night	High b Day Extreme	High c Day Crop
Parameters							
D (m² s⁻¹)	2e–10	2e–10	2e–10	4e–11	4e–11	2e–10	2e–10
v_a (m s⁻¹)	1.01e–8	2.02e–9	5.05e–10	2.02e–9	5.05e–10	3.24e–8	1.08e–7
k (unitless)	2.53e–3	5.05e–4	1.26e–4	5.05e–4	1.26e–4	8.10e–3	0.027
With Mass Flow							
c_a (mol m⁻³)	1.01	0.997	0.995	0.997	0.995	0.921	9.845
j_a (mol m⁻² s⁻¹)	4.91e–9	4.91e–9	4.91e–9	4.91e–9	4.90e–9	1.07e–8	1.28e–6
Zero Mass Flow							
c_a^0 (mol m⁻³)	0.995	0.995	0.995	0.995	0.995	0.0890	9.042
j_a^0 (mol m⁻² s⁻¹)	4.90e–9	4.90e–9	4.90e–9	4.90e–9	4.90e–9	1.06e–8	1.28e–6
Relative Increment[d]							
$(j_a - j_a^0)/j_a^0$	0.0005	0.0001	0.0000	0.0001	0.0000	0.0123	0.0004
Crude Increment[e]							
$c_a v_a/j_a^0$	2.0756	2.0583	2.0550	2.0583	2.0550	0.278	0.831
Suppression[f]							
$(j_a - j_a^0 - v_a c_a)/j_a^0$	–2.0752	–2.0581	–2.0548	–2.0581	–2.0548	–0.266	–0.831

Note: All parameters are as described in text; only variations in the water uptake velocity at the root surface, v_a, and the nitrate diffusion coefficient, D, are considered here, in the first five cases. Cases are described by plant and soil water status (high, low) and time of day. In all entries, "e–n" indicates 10^{-n}, the power of 10 as a multiplier.

[a] The second and third cases use two different estimates of nocturnal transpiration.

[b] The sixth case is an extreme (high) estimate of uptake by *Larrea tridentata*; see text for conditions.

[c] The seventh case is crop plant under conditions favorable to very high mass flow.

[d] "Increment" is the fractional increase in uptake as the difference between cases with and without mass flow.

[e] "Crude increment" is the (misleading, common) estimate of the fractional increment, using (water flow) × concentration. It is expressed as a fraction of uptake in the absence of mass flow, and can exceed unity.

[f] "Suppression" is the decrement in uptake from suppression of diffusion, again relative to uptake without mass flow.

ously, we attempted to integrate the role of mycorrhizal colonization in a similar but simpler version of the functional balance model presented here (BassiriRad et al. 2001). There, we concluded that effects of mycorrhizal fungi on plant N uptake must be viewed in the context of both enhancing plant N status as well as the potential effects on carbon balance. Clearly, more mechanistic understanding of the biology of mycorrhizal fungi needs to emerge before we can quantitatively assess their leverage on plant N uptake via the modeling exercises recommended here.

Another root system characteristic that may have a large amount of leverage on determining plant N uptake is root longevity or life span. Most crop plants are annuals, where root turnover may not confer a significant advantage for N uptake. On the other hand, roots of many native species such as sugar maple may live as long as a year or longer (Hendrick and Pregitzer 1993). While there is some recognition of the importance of root longevity to N uptake (Bloomfield et al. 1996; Eissenstat and Yanai 1997), there is little effort in integrating root life span into models of plant N uptake. We suggest that such integration will improve models of N uptake. Finally, much has been described about root architecture and topological differences among native species, and how such differences may have evolved in response to soil N status (Fitter 1991). We suggest that root architecture plays a role in N uptake of plants. However, we are far from being able to determine the extent of that role. It is argued that root systems' branching patterns fall into two categories: herringbone and dichotomous (Fitter 1987, 1991). Fitter (1991) argued that herringbone topology gives rise to a larger specific root length, and thus is more effective than the dichotomous rooting branching pattern in N uptake from infertile sites. However, a number of studies show that such a relationship is not universal (Pregitzer et al. 2002; Einsmann et al. 1999).

4.4 CONCLUDING REMARKS

Development of robust models of N uptake requires greater efforts to integrate factors that control availability of N at the root surface together with the root system characteristics that regulate its absorption into the plant. In this chapter we offer road maps by which mechanistic models of soil N availability can be linked to physiological plant growth models to accomplish that effort. Soil properties that control N transport to the root surface have long been recognized and are well characterized. We offer selected insights as to how such information can be integrated into uptake models.

In particular, we suggested that the relative importance of mass flow to overall N concentration at the root surface in previous studies may have been overestimated. We argued that internal cycling of N via mineralization is a major contributor to N availability in native systems. Efforts to link physiological growth models (e.g., functional balance models) with mechanistic models of N mineralization should significantly improve the predictability of N uptake by native plants. We highlighted the need to characterize and incorporate atmospheric deposition as well as soil organic N sources into models of uptake for improved predictability. We also suggested that adjustments in kinetics of root N uptake have a relatively large effect on plant performance under varying N supply, but efforts to assign optimal values for V_{max} remain incomplete. It is also concluded that, while increased biomass allocation to roots has beneficial outcomes for N uptake, plant growth is adversely affected, so that the optimal r value appears to be around 1; concurrent improvement of water uptake may drive the allocation patterns. Finally, we encourage studies that further examine the role of root system characteristics that have rarely been incorporated into uptake models, such as mycorrhizal colonization, longevity, and architecture.

To avoid making modeling efforts too cumbersome to parameterize for so many inputs, empirical studies must prove that changes in a given root system characteristic infer a significant change in plant performance in response to varying N availability. Our conclusions do rely heavily on models; they must undergo both modeling refinement and empirical testing. This highlights a major utility of modeling, that of framing hypotheses for such testing. We admit that empirical tests are challenging. They must also be analyzed critically. For example, the attribution to mass flow of a significant role in controlling uptake rates is misleading in the absence of accounting for mass flow's role in flattening the diffusional gradient, leaving the total uptake rate unchanged.

ACKNOWLEDGMENT

This work was partly supported by funding from the National Science Foundation.

REFERENCES

Aber, J. D., and C. T. Driscoll. 1997. Effects of land use, climate variation and N deposition on N cycling and C storage in northern hardwood forests. *Global Biogeochemical Cycles* 11: 639–648.

Aber, J. D., K. J. Nadelhoffer, P. Steudler, and J. M. Melillo. 1989. Nitrogen saturation in northern forest ecosystems. *Bioscience* 39: 378–386.

Abuzinadah, R. A., and D. J. Read. 1986a. The role of proteins in the nitrogen nutrition of ectomycorrhizal plants, 1: Utilization of peptides and proteins by ectomycorrhizal fungi. *New Phytol.* 103: 481–493.

Abuzinadah, R. A., and D. J. Read. 1986b. The role of proteins in the nitrogen nutrition of ectomycorrhizal plants, 3: Protein utilization by *Betula picea* and *Pinus* in mycorrhizal association with *Hebeloma crustuliniforme*. *New Phytol.* 103: 507–514.

Abuzinadah, R. A., and D. J. Read. 1988. Amino acids as nitrogen sources for ectomycorrhizal fungi: Utilisation of individual amino acids. *Trans. British Mycological Soc.* 91: 473–479.

Agren, G. I., and E. Bosatta. 1987. Theoretical analysis of the long term dynamics of carbon and nitrogen in soils. *Ecology* 68: 1181–1189.

Amelung, W., K. Zhang, K. W. Flach, and W. Zech. 1999. Amino sugar in native grassland soils along a chemosequence in North America. *Soil Sci. Soc. Am. J.* 63: 86–92.

Aslam, M., R. L. Travis, and D. W. Rains. 2001. Differential effect of amino acids on nitrate uptake and reduction system in barley roots. *Plant Sci.* 160: 219–228.

BassiriRad, H. 2006. Root characteristics and control of plant nitrogen uptake. *J. Crop Improvement* 15: 25–52.

BassiriRad, H., V. P. Gutschick, and J. Lussenhop. 2001. Root system adjustments: regulation of plant nutrient uptake and growth responses to elevated CO_2. *Oecologia* 126: 305–320.

Benbi, D. K., and J. Richter. 2002. A critical review of some approaches to modeling nitrogen mineralization. *Biol. Fertil. Soils* 35: 168–183.

Bennett, J. N., and C. E. Prescott. 2004. Organic and inorganic nitrogen nutrition of western red cedar, western hemlock and salal in mineral N-limited cedar-hemlock. *Oecologia* 141: 468–476.

Bloomfield, J., K. Vogt, and P. M. Wargo. 1996. Tree root turnover and senescence: Root structure and sites of ion uptake. In *Plant roots: The hidden half*, eds. Y. Waisel, A. Eshel, and U. Kafkafi, 363–381. 2nd ed. New York: Marcel Dekker.

Brady D. J., P. J. Gregory, and I. R. P. Fillery. 1993. The contribution of different regions of the seminal roots of wheat to uptake of nitrate from soil. *Plant Soil* 155: 155–158.

Bucci S. J., F. G. Scholz, G. G. Goldstein, F. C. Meinzer, and A. G. Franco. 2006. Nutrients limitations removal by fertilizations decreases nocturnal water loss in savanna trees. Abstracts, 90th annual meeting, Ecological Society of America, Montreal, p. 179.

Chapin, F. S., N. Fercher, K. Kielland, K. R. Everett, and A. E. Linkins. 1988. Productivity and nutrient cycling of Alaskan tundra: Enhancement by flowing soil water. *Ecology* 69: 693–702.

Chapin, F. S., L. Moilainen, and K. Kielland. 1993. Preferential use of organic acid N by a non-mycorrhizal arctic sedge. *Nature* 361: 150–153.

Chen, C., and Z. Xu. 2006. On the nature and ecological functions of soil soluble organic nitrogen (SON) in forest ecosystems. *J. Soil Sediments* 6: 63–66.

Darrah, P. R., D. L. Jones, G. J. D. Kirk, and T. Roose. 2006. Modelling the rhizosphere: A review of methods for "upscaling" to the whole-plant scale. *Eur. J. Soil Sci.* 57: 13–25.

Einsmann, J. C., R. H. Jones, M. Pu, and R. J. Mitchell. 1999. Nutrient foraging traits in 10 co-occurring plant species of contrasting life forms. *J. Ecol.* 87: 609–619.

Eissenstat, D. M., and R. D. Yanai. 1997. The ecology of root lifespan. *Adv. Ecol. Res.* 27: 1–16.

Falkengren-Greu, U. 1995. Interspecies differences in preference of ammonium and nitrate in vascular plants. *Oecologia* 102: 305–311.

Fan, W., J. C. Randolph, and J. L. Ehman. 1998. Regional estimation of nitrogen mineralization in forest ecosystems using geographic information systems. *Ecological Appl.* 8: 734–747.

Farquhar G. D., S. von Caemmerer, and J. A. Berry. 1980. A biochemical model of photosynthetic CO_2 assimilation in leaves of C_3 species. *Planta* 149: 78–90.

Fernandes, M. S., and R. O. P. Rossiello. 1995. Mineral nitrogen in plant physiology and plant nutrition. *Crit. Rev. Plant Sci.* 14: 111–148.

Fisk, M. C., and S. Schmidt. 1995. Nitrogen mineralization and microbial biomass nitrogen dynamics in three alpine tundra communities. *Soil Sci. Soc. Am. J.* 59: 1036–1043.

Fitter, A. 1987. An architectural approach to the comparative ecology of plant root systems. *New Phytologist* 106: 61–77.

Fitter, A. 1991. Characteristics and functions of root systems. In *Plant roots: The hidden half*, ed. Y. Waisel, A. Eshel, and U. Kafkafi, 3–25. New York: Marcel Dekker.

Galloway, J. N., and E. B. Cowling. 2002. Reactive nitrogen and the world: 200 years of change. *Ambio* 31 (2): 64–71.

Galloway, J. N., W. H. Schlesinger, H. Levy, I. A. Michaels, and J. L. Schmoor. 1995. Nitrogen fixation: Anthropogenic enhancement, environmental response. *Global Biogeochem. Cycles* 9: 235–252.

Gardner, W. R. 1960. Dynamic aspects of water availability to plants. *Soil Sci.* 89: 63–73.

Gaunt, J. L., S. P. Sohi, H. Yang, N. Mahieu, and J. R. M. Arah. 2001. A procedure for isolating soil organic matter fractions suitable for modeling. In *Sustainable management of soil organic matter*, ed. R. M. Rees et al. 83–90. Wallingford, Oxfordshire: CABI.

Gebauer, G., H. Rehder, and B. Wollenweber. 1988. Nitrate, nitrate reduction and organic nitrogen in plants from different ecological and taxonomic groups of Central Europe. *Oecologia* 75: 371–385.

Glass, A. D. M. 2005. Homeostatic processes for the optimization of nutrient absorption: Physiology and molecular biology. In *Nutrient acquisition by plants: An ecological perspective*, ed. H. BassiriRad, 117–145. Berlin: Springer.

Goodale, C. L., and J. D. Aber. 2001. The long-term effects of land-use history on nitrogen cycling in northern hardwood forests. *Ecol. Appl.* 11: 253–267.

Groffman, P. M., and J. M. Tiedje. 1989. Denitrification in north temperate forest soils: Spatial and temporal patterns at the landscape and seasonal scales. *Soil Biol. Biochem.* 21: 613–620.

Gutschick, V. P. 1993. Nutrient-limited growth rates: Roles of nutrient-use efficiency and of adaptations to increase uptake rate. *J. Exp. Bot.* 44: 41–51.

Gutschick, V. P. 2006. Plant acclimation to elevated CO_2: From simple regularities to biogeographic chaos. *Ecological Modelling* 200: 433–451.

Gutschick, V. P., and L. E. Kay. 1995. Nutrient-limited growth rates: Quantitative benefits of stress responses and some aspects of regulation. *J. Exp. Bot.* 46: 995–1009.

Gutschick, V. P. and J. C. Pushnik. 2005. Internal regulation of nutrient uptake by relative growth rate and nutrient-use efficiency. In *Nutrient Acquisition by Plants: An Ecological Perspective.* ed. H. BassiRad. 181: 63–88. Heidelberg: Springer.

Hendrick, R. L., and K. S. Pregitzer. 1993. Patterns of fine root mortality in two sugar maple forests. *Nature* 361: 59–61.

Henry, H. A. L., and R. L. Jefferies. 2002. Free amino acid, ammonium and nitrate concentration in soil solutions of a grazed coastal marsh in relation to plant growth. *Plant Cell. Environ.* 25: 665–675.

Henry, H. A. L., and R. L. Jefferies. 2003. Interactions in the uptake of amino acids, ammonium and nitrate ions in the Arctic salt marsh grass, *Puccinellia phryganodes. Plant Cell Environ.* 26: 419–428.

Holland, E. A., B. H. Braswell, J. Sulzman, and J. F. Lamarque. 2005. Nitrogen deposition onto the United States and western Europe: Synthesis of observations and models. *Ecol. Appl.* 15: 38–57.

Holmes, W. E., and D. R. Zak. 1994. Soil microbial biomass dynamics and net nitrogen mineralization in northern hardwood ecosystems. *Soil Sci. Soc. Am. J.* 58: 238–243.

Jones, D. L., J. R. Healey, V. B. Willett, J. F. Farrar, and A. Hodge. 2005. Dissolved organic nitrogen uptake by plants: An important N uptake pathway? *Soil Biol. Biochem.* 37: 413–423.

Jones, D. L., D. Shannon, D. V. Murphy, and J. Farrar. 2004. Role of dissolved organic nitrogen (DON) in soil N cycling in grassland soils. *Soil Biol. Biochem.* 36: 749–756.

Joshi, A. B., D. R. Vann, and A. H. Johnson. 2006. Litter quality and climate decouple nitrogen mineralization and productivity in Chilean temperate rainforest. *Soil Sci. Soc. Am. J.* 70: 153–162.

Kage, H. 1997. Zur relativen bedeutung von massenfluss und diffusion beim nitrattransport sur eitschrift für pflanzenernährung und bodenkunde. *Zeitschrift für Pflanzeneinahrung und Bedenleunde* 160: 171–178.

Keeney, D. R., and J. M. Bremner. 1964. Effect of cultivation on nitrogen distribution in soils. *Soil Sci. Soc. Am. Proc.* 28: 653–656.

Kielland, K. 1990. Processes controlling nitrogen release and turnover in Arctic tundras. Dissertation. University of Alaska, Fairbanks.

Klipp, E., and R. Heinrich. 1994. Evolutionary optimization of enzyme-kinetic parameters: Effect of constraints. *J. Theor. Biol.* 171: 309–323.

Leadley, P. W., J. F. Reynold, and F. S. Chapin. 1997. A model of nitrogen uptake by *Eryophorum vaginatum* roots in the field: Ecological implications. *Ecological Monographs* 67: 1–22.

Lipson, D., and R. K. Monson. 1998. Plant-microbe competition for soil amino acids in the alpine tundra: Effects of freeze-thaw and dry-rewet events. *Oecologia* 113: 406–414.

Lipson, D., and T. Näsholm. 2001. The unexpected versatility of plants: Organic nitrogen use and availability in terrestrial ecosystems. *Oecologia* 128: 305–316.

Lovett, G. M. 1994. Atmospheric deposition of nutrients and pollutants in North America: An ecological perspective. *Ecol. Appl.* 4: 629–650.

Lynch, J. P., and K. Brown. 2001. Topsoil foraging: An architectural adaptation of plants to low phosphorus availability. *Plant Soil* 237: 225–237.

Magill, A. H., J. D. Aber, J. J. Hendricks, R. Bowden, J. M. Melillo, and P. Steudler. 1997. Biogeochemical response of forest ecosystems to simulated chronic nitrogen deposition. *Ecol. Appl.* 7: 402–415.

Marshner, H., M. Häussling, and E. George. 1991. Ammonium and nitrate uptake rates and rhizosphere pH in nonmycorrhizal roots of Norway spruce (*Picea abies* (L.) Karst). *Trees Structure Function* 5: 14–21.

McNulty, S. G., J. D. Aber, and R. D. Boone. 1991. Spatial changes in forest floor and foliar chemistry of spruce-fir forests across New England. *Biogeochemistry* 14: 13–29.

Nadelhoffer, K. J., M. R. Downs, B. Fry, A. Magill, and J. D. Aber. 1999. Controls on N retention and exports in a fertilized forested watershed. *Environ. Monitoring Assessment* 55: 187–210.

NADP. 1997. National Atmospheric Deposition Program. http://sws.uiuc.edu/nadp.

Näsholm, T., A. Ekbladm, R. Nordin, M. Giesler, M. Högberg, and P. Högberg. 1998. Boreal forest plants take up organic N. *Nature* 392: 914–916.

Näsholm, T., K. Huss-Danell, and P. Högberg. 2000. Uptake of organic nitrogen in the field by four agriculturally important plant species. *Ecology* 81: 1155–1161.

Näsholm, T., and J. Persson. 2001. Plant acquisition of organic N in boreal forests. *Physiol. Plant.* 111: 419–426.

Nye, P. H., and J. A. Spiers. 1964. Simultaneous diffusion and mass flow to plant roots. *8th International Congress on Soil Science (Bucharest)* 11: 535–542.

Nye, P. H., and P. B. Tinker. 1977. *Solute movement in the soil-root system.* Oxford, England: Blackwell Scientific.

Owen, A. G., and D. L. Jones. 2001. Competition for amino acids between wheat roots and rhizosphere microorganisms and the role of amino acids in plant N acquisition. *Soil Biol. Biochem.* 33: 651–657.

Padgett, P. E., and R. T. Leonard. 1993. Regulation of nitrate uptake by amino acids in maize cell suspension culture and intact roots. *Plant Soil* 155: 159–161.

Pregitzer, K. S., J. L. DeForest, A. J. Buron, M. F. Allen, R. W. Ruess, and R. L. Hendrick. 2002. Fine root architecture of nine North American trees. *Ecological Monographs* 72: 293–309.

Raab, T. K., D. A. Lipson, and R. M. Monson. 1996. Non-mycorrhizal uptake of amino acids by roots of the alpine *Kobresia myosuroides*: Implications for the alpine N cycle. *Oecologia* 108: 488–494.

Raab, T. K., D. A. Lipson, and R. M. Monson. 1999. Soil amino acid utilization among species of the Cyperaceae plant and soil processes. *Ecology* 80: 2408–2419.

Rastetter, E. B., B. L. Kwiatkowski, and R. B. McKane. 2005. A stable isotope simulator that can be coupled to existing mass balance models. *Ecol. Appl.* 15: 1772–1782.

Retch, P. B., D. F. Grigal, J. A. Aber, and S. T. Gower. 1997. Nitrogen mineralization and productivity in 50 hardwood and conifer stands on diverse soils. *Ecology* 78: 335–347.

Sallih, Z., and M. Pansu. 1993. Modelling of soil carbon forms after organic amendments under controlled conditions. *Soil Biol. Biochem.* 25: 1755–1762.

Schaeffer, S. M., S. A. Billings, and R. D. Evans. 2003. Responses of soil nitrogen dynamics in a Mojave Desert ecosystem to manipulations in soil carbon and nitrogen availability. *Oecologia* 134: 547–553.

Schimel, J. P., and J. Bennett. 2004. Nitrogen mineralization: Challenges of a changing paradigm. *Ecology* 85: 591–602.

Schlesinger, W. H. 1997. *Biogeochemistry: An analysis of global change*, 383–385. San Diego: Academic Press.

Schomberg, H. H., and M. L. Cabrera. 2001. Modeling in situ N mineralization in conservation tillage fields: Comparison of two versions of the CERES nitrogen submodel. *Ecological Modelling* 145: 1–15.

Schulze, E. D. 1989. Air pollution and forest decline in a spruce (*Picea abies*) forest. *Science* 244: 776–783.

Senwo, Z. N., and M. A. Tabatabai. 1998. Amino acid composition of soil organic matter. *Biol. Fertil. Soils* 26: 235–242.

Skeffington, R. A. 1990. Accelerated nitrogen inputs: A new problem or a new perspective? *Plant Soil* 128: 1–11.

Tietema, A., B. A. Emmett, P. Gundersen, O. J. Kjønaas, and C. J. Koopmans. 1998. The fate of ^{15}N- labelled nitrogen deposition in coniferous forest ecosystems. *For. Ecol. Manage.* 101: 19–27.

Tilman, D. 1993. Species richness of experimental productivity gradient: How important is colonization limitation? *Ecology* 74: 2179–2191

Tinker, P. B., and P. H. Nye. 2000. *Solute movement in the rhizosphere.* Oxford: Oxford University Press.

Titus, J. H., R. S. Nowak, and S. D. Smith. 2002. Soil resource heterogeneity in the Mojave Desert. *J. Arid Environ.* 52: 269–292.

Vinten, A. J. A., A. P. Whitmore, J. Bloem, R. Howard, and F. Wright. 2002. Factors affecting N immobilization/mineralization kinetics for cellulose-, glucose- and straw-amended sandy soils. *Biol. Fertil. Soils* 36: 190–199.

Vitousek, P. M., J. D. Aber, R. W. Howarth, G. E. Likens, P. A. Matson, D. W. Schindler, W. H. Schlesinger, and D. G. Tilman. 1997. Human alteration of the global nitrogen cycle: Sources and consequences. *Ecol. Appl.* 7: 737–750.

Vitousek, P. M., and R. W. Howarth. 1991. Nitrogen limitation on land and in the sea: How can it occur? *Biogeochemistry* 13: 87–115.

Wallenda, T., and D. J. Read. 1999. Kinetics of amino acid uptake by ectomycorrhizal roots. *Plant Cell Environ.* 22: 179–187.

Wang, W. J., C. J. Smith, and D. Chen. 2004. Predicting soil nitrogen mineralization dynamics with a modified double exponential model. *Soil Sci. Soc. Am. J.* 68: 1256–1265.

Wilson, E. J., T. C. E. Wells, T. H. Sparks. 1995. Are calcareous grasslands in the U.K. under threat from nitrogen deposition? An experimental determination of critical load. *J. Ecol.* 83: 823–832.

Wiltshire, G. H. 1973. Response of grasses to nitrogen source. *J. Appl. Ecol.* 10: 429–435.

Xu, X., H. Ouyang, Y. Kuzyakov, A. Richter, and W. Wanek. 2006. Significance of organic nitrogen acquisition for dominant plant species in an alpine meadow on the Tibet plateau. *Plant Soil* 285: 223–231.

Yanai, R. D. 1994. A steady-state model of nutrient uptake accounting for newly grown roots. *Soil Sci. Soc. Am. J.* 58: 1562–1571.

Yu, Z., Q. Zhang, T. E. C. Kraus, R. A. Dahlgren, C. Anastasio, and R. J. Zasoski. 2002. Contribution of amino compounds to dissolved organic nitrogen in forest soils. *Biogeochemistry* 61: 173–198.

Zak, D. R., D. F. Grigel, and L. F. Ohmann. 1993. Kinetics of microbial respiration and nitrogen mineralization in Great Lakes forests. *Soil Sci. Soc. Am. J.* 57: 1100–1106.

Zerihun, A., V. P. Gutschick, and H. BassiriRad. 2000. Compensatory roles of nitrogen uptake and photosynthetic N-use efficiency in determining plant growth response to elevated CO_2: Evaluation using a functional balance model. *Ann. Bot.* 86: 723–730.

APPENDIX 4.1   OPTIMIZATION OF ROOT:SHOOT
RATIO FOR WATER RELATIONS

At any given root:shoot ratio, r, there is a potential rate of water uptake, U, equal to the root mass, m_r, multiplied by the uptake rate per mass of root, v_w. This must match the whole-plant rate of transpiration, E, which equals the leaf area of the plant, a_p, multiplied by the transpiration rate per unit leaf area, E_{La}. This can only occur if E_{La} is adjusted to the sustainable rate by an appropriate magnitude of leaf conductance for water vapor, g_{bs}. (The subscript bs refers to the combination of boundary-layer and stomatal conductances.) The relation of E_{La} to the conductance g_{bs}, the vapor-pressure deficit D (leaf-to-air difference in partial pressure of water vapor), and total air pressure P is simple:

$$E_{La} = g_{bs}\frac{D}{P} \tag{A4.1}$$

The plant's leaf area is simply the dry mass of leaves, m_L, divided by the mass per leaf area, m_{La} (inverse of specific leaf area). In turn, the mass of leaves can be expressed as the total shoot mass, m_s, multiplied by the fraction of shoot mass as leaves, α_L. Thus, we have

$$E = \frac{\alpha_L m_s}{m_{La}} g_{bs}\frac{D}{P} = U = m_r v_w \tag{A4.2}$$

Assuming that these parameters are essentially stable at a stage in growth (particularly in early growth, where all leaves experience similar environments), we can solve for the required value of conductance that balances water uptake and water use:

$$g_{bs} = \frac{m_r}{m_s}\frac{v_w m_{La}}{\alpha_L (D/P)} \tag{A4.3}$$

This value of the conductance constrains the rate of photosynthetic CO_2 uptake by the whole plant, A_p, and thus the growth rate, GR, which is A_p multiplied by a biosynthetic efficiency, β (as grams dry matter per gram of photosynthate produced or, if A_p is in mol_{CO2} per area per time, as grams dry matter per mol CO_2). The whole-plant rate, A_p, is, like water use, a product of whole-plant leaf area with photosynthetic rate per unit leaf area, A_{La}. The latter rate responds to conductance g_{bs} in well-known ways: Its magnitude as a rate of transport of CO_2 across the conductance must equal its magnitude as the enzyme-kinetic rate (Farquhar et al. 1980):

$$A_{La} = g_{bs}'\frac{(C_a - C_i)}{P} = V_{c,max}\frac{(C_i - \Gamma)}{(C_i + K_{CO})} \tag{A4.4}$$

Here, the conductance for CO_2, g_{bs} with a prime, is closely related to that for water vapor (about two-thirds of the latter, depending upon the relative importance of boundary-layer and stomatal conductances); C_a and C_i are, respectively, the CO_2 partial pressures in ambient air and inside the leaf (more precisely, at the site of carboxylation, such that there is an extra resistance in liquid-phase transport inside the leaf, ignored here; see Gutschick, 2006); $V_{c,max}$ is the maximum rate of carboxylation or CO_2 fixation; Γ is the CO_2 compensation partial pressure; and K_{CO} is the effective Michaelis constant for CO_2 binding to the Rubisco enzyme in the presence of O_2. The latter two parameters are functions only of leaf temperature and the mixing ratio of O_2 to CO_2.

The relation above can be rearranged to a quadratic equation for C_i, given the magnitudes of all the parameters g_{bs}', C_a, P, $V_{c,max}$, Γ, and K_{CO}. These values are readily obtained for plants growing in normal air and at known temperatures, and with known rates of photosynthesis in unstressed conditions. Once C_i is known, equation (A4.4) can be used to compute the leaf rate of photosynthesis, A_{La}. Finally, we compute the growth rate, as the relative growth rate, RGR, or GR divided by total plant mass, m_p. RGR is relatively stable in various growth periods, and that is critical for competitive growth:

$$\text{RGR} = \beta \frac{A_p}{m_p} = \frac{\beta \alpha_L m_s A_{La}}{m_{La} m_p} = \frac{\beta \alpha_L}{(1+r)} \frac{A}{m_{La}} \tag{A4.5}$$

Here we have used the relation that $m_s/m_p = m_s/(m_s + m_r)$, which becomes $1/(1 + r)$ upon dividing numerator and denominator by m_s.

The key trade-off in optimizing r is that increasing r will increase water uptake, U, and allow a higher leaf conductance, A_{La}, but it decreases the fraction of the plant that does photosynthesis. At high conductance (high r), A_{La} saturates, while the factor $1/(1 + r)$ cuts into whole-plant photosynthesis.

One may ask if this argument shows the optimum value of r to be near observed values. Let us consider a young herbaceous plant, with $r = 0.5$, a leaf fraction $\alpha_L = 0.5$, an unstressed leaf conductance $g_{bs}' = 0.25$ mol_{CO2} m^{-2} s^{-1}, and leaf photosynthetic capacity $V_{c,max} = 100$ μmol m^{-2} s^{-1} (achieved rate near 20 in same units). We take the environmental descriptors as $P = 10^5$ Pa (about sea-level pressure), $C_a = 38$ Pa (thus, the current CO_2 mixing ratio), and a leaf temperature near 25°C, which puts Γ near 4 Pa and K_{CO} near 90 Pa. The water-balance arguments here then yield $C_i = 26.2$ Pa and $A_{La} = 19.1$ μmol m^{-2} s^{-1}, both very representative of such plants. Varying the value of r changes RGR, as shown in Figure A4.1. Two points are important. First, the potential uptake rate of roots, v_w, can be notably higher when soil water status is high; r could be lower than posited here. The plant's developmental controls enforce a reserve capacity. Second, the model does not indicate the mode of signaling for adjustments in r when soil water status in the long term is different from our reference conditions; functional balance per se does not yield a clear signal variable. Nonetheless, functional balance for water relations is consistent with realistic root: shoot ratios, while functional balance for nutrients is not thus consistent.

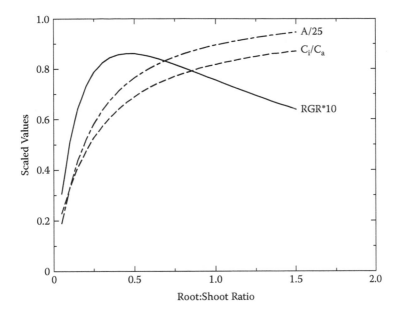

FIGURE A4.1 Occurrence of a root:shoot ratio, r, that optimizes relative growth rate (RGR) in the model of water balance between uptake and transpiration, as presented in the text. Note the continued rise, at higher r, of leaf internal CO_2 partial pressure (C_i, as ratio to ambient value C_a) and leaf photosynthetic rate (A, as ratio to limiting value of about 25 μmol m^{-2} s^{-1}).

5 Dissolved Organic Nitrogen and Mechanisms of Its Uptake by Plants in Agricultural Systems

David L. Jones, John F. Farrar, Andrew J. Macdonald, Sarah J. Kemmitt, and Daniel V. Murphy

CONTENTS

5.1   Introduction 95
5.2   DON Concentrations in Soil 96
5.3   Microbial Regulation of DON in Soil 103
5.4   Plant Production of DON 107
5.5   Plant Uptake of DON 109
      5.5.1   Plant Membranes Are Leaky to Amino Acids 110
      5.5.2   Plant Membranes Have DON Transporters That Are
              Energized 110
      5.5.3   Extracellular Enzymes? 113
      5.5.4   Evidence for Plant Uptake of DON 113
      5.5.5   Do Plants Growing in Nonsterile Soil Take Up DON? 114
      5.5.6   Plant Regulation of Amino Acid Uptake 115
5.6   Concluding Remarks 116
References 117

5.1   INTRODUCTION

The study of soluble nitrogen (N) in soil has received much attention due to N being the nutrient limiting primary production in most ecosystems. Over recent decades, there has been increasing interest in sustainable agriculture due to environmental and social problems associated with conventional agricultural systems (Kristensen et al. 1995). Loss of N from the soil–plant system via leaching of nitrate (NO_3^-) and gaseous N emission has been a matter of increasing concern, particularly in high-production systems (Addiscott et al. 1991). This loss partially stems from the increased use and low cost of inorganic fertilizers and the tendency for farmers to apply excess so that crop growth and yield are not limited by insufficient N availability (Addiscott et al. 1991). Much of the research geared toward understanding the N cycle, and how management options affect it, has failed to examine one of the potentially most important pools of N in the system, dissolved organic N (DON) (Murphy et al. 2000). As DON forms the critical link between solid organic N and the subsequent production of ammonium (NH_4^+) and nitrate, it is surprising that it is remains relatively understudied, particularly in relation to low-input agricultural systems. If our ultimate goal is to create sustainable agricultural systems, it is clear that we need to understand all the main pools of the soil–plant N cycle to achieve this. Typically, it is difficult to balance the inputs and outputs of N from agricultural systems due to inaccuracies in estimating pool sizes and fluxes and our lack of mechanistic understanding of the N cycle. For example, the leaching of DON from agricultural systems possibly contributes to the often "unaccounted for" fraction in farm N budgets (Jarvis et al. 1996; Leach et al. 2005).

Although a statistical relationship can be established between agricultural N use and rises in surface and groundwater NO_3^- concentrations and gaseous N emissions, the relationship is not simple (Addiscott et al. 1991; Goulding and Poulton 1992). This supports our view that the understanding of N cycling is satisfactory but remains incomplete. This is reinforced by recent articles showing that plants can directly take up and assimilate DON from soil, thereby effectively bypassing the need for organic N to be mineralized to NH_4^+ and NO_3^- (Jones et al. 2005a, 2005c). The potential importance of DON as a major source of N for plants, particularly in low-input systems, has highlighted our need to reexamine ecosystems N flows (van Breemen 2002). Here we review the primary factors that regulate DON concentrations and fluxes in agricultural environments.

5.2   DON CONCENTRATIONS IN SOIL

DON can enter soil from a variety of routes (Table 5.1). Each of these DON input pathways can be expected to vary markedly in both the composition and concentration of DON, depending upon the nature of the agroecosystem (e.g., organic versus conventional or low input versus high input). Similarly, each pathway can also be expected to vary in its temporal dynamics (e.g., seasonality, time of day) and its spatial input pattern (e.g., rhizosphere versus bulk soil, topsoil versus subsoil).

In most environments, the total soil organic N pool is always much larger than the mineral N pool, with the accumulation of soil organic N closely following that of soil organic matter (Stevenson 1982; Haynes 1986). For example, across a wide range of UK agricultural environments, Jarvis et al. (1996) found that the average mineral N content of soil was 76 kg N ha^{-1} compared with of an organic N content of 7000 kg ha^{-1}. Many studies have shown that the organic N in agricultural soils is predominantly composed of protein, nucleic acids (RNA, DNA), amino acids, amino sugars, and humic materials (Stevenson 1982; Xu et al. 2003; Knicker 2000). However, it must be noted that many of these studies do not distinguish between the organic N compounds held in the dead organic matter and those in living microbial cells, although in many cases the microbial N pool only composes a small fraction of the total organic N (20–500 kg N ha^{-1}) (Friedel and Scheller 2002). Consequently, quantitative interpretation of these studies with respect to the significance of specific compounds in

TABLE 5.1

Summary of the Main Inputs of Dissolved Organic N into Soil

Source	Major Attributes	References
Rainfall	Typically low concentrations of DON (<1 mg NL^{-1}; DON comprises ca. 10%–30% of the total N	Cornell et al. 2003
Irrigation water	Typically low concentrations of DON (0.1–5 mg NL^{-1}	Christou et al. 2005
Canopy throughfall	Typically low concentrations of DON (0.1–10 mg NL^{-1}; concentrations higher from trees than crops	Michalzik and Stadler 2005
Rhizodeposition	Extremely variable release rates depending upon species and environmental conditions	Jones et al. 2004
Microbial exudates	Known to occur but rarely quantified; ecological significance largely unknown in soil environments	—
Livestock urine	Concentrations of DON very high (100–5000 mg NL^{-1}; dominated by urea	Lucas and Jones 2006
Livestock feces	Concentrations high (10–500 mg NL^{-1} but dependent upon age and type of feces	Lucas and Jones 2006
Human-made fertilizers	Concentrations extremely high but localized (e.g., around urea granules)	Shah et al. 2004
Green manures and composts	Concentrations are often high, but highly dependent upon source	Saviozzi et al. 2006; de Guardia et al. 2002
Industrial waste	Can be high in some wastes spread to land (e.g., abattoir waste) and low in others (e.g., paper waste)	—
Biosolids	Typically, the concentrations are high, but less than the amount of inorganic N present	Insam and Merschak 1997

soil organic N cycling remains difficult. Their physical distribution and coexistence of the individual compounds within the soil matrix also remains unknown.

From these studies in which the soil organic matter (SOM) is hydrolyzed into its constituent units, it appears that amino acids and amino sugars in a polymeric state (peptides, protein, glycoproteins, chitin, etc.) typically constitute 30%–40% of the soil organic N (Stevenson 1982). Further, from an analysis of plant and microbial residues, it can be deduced that these compounds also represent the major inputs of N into the soil (Reddy et al. 2003). Accordingly, they also represent the biggest source of inorganic N released by mineralization. The remaining 60%–70% of soil organic N is composed largely of humic substances whose chemical structure is poorly understood but is known to be of high molecular weight and relatively recalcitrant. While large quantities of proteins are present in soil, the types of individual proteins present remain understudied, but they can be expected to contain many different catalytic (i.e., enzymes) and structural proteins in various stages of hydrolysis. It is likely that recent advances in proteomics technology will help to resolve this

TABLE 5.2

Soil Mineral N (Min–N; NO_3^- + NH_4^+) and Dissolved Organic N (DON) Contents 0–75 cm (kg N ha^{-1}) in Plots with Different Long-Term Manuring on the Broadbalk Wheat Experiment (Rothamsted, U.K.)

		Plot No. and Manuring Regime [a]			
		2.1 N FYM N2	5 Nil N PK	8 N3 PK	15 N5 PK
Sampling Date	Soil N Content				
Dec. 1999	Min–N	85	32	89	41
	DON	49	35	33	26
Mar. 2000	Min–N	67	22	25	20
	DON	64	30	21	27

[a] Fertilizer N was applied as NH_4NO_3 in spring. Since 1885, plot 2.1 has received 35 t ha^{-1} of farmyard manure (FYM) in autumn; additional fertilizer applications (96 kg N ha^{-1}) began in 1968. Plots 5 and 8 have received no additional N and 144 kg N ha^{-1}, respectively, since 1852. Plot 15 has received 240 kg N ha^{-1} since 1985. Plots received P and K as triple superphosphate and K_2SO_4, respectively.

(Schulze 2005). Most of the proteins in soil remain relatively insoluble due to their binding to humic substances present in soil organic matter or their high degree of hydrophobicity. Recent reports have also suggested that soil organic N may be dominated by substances such as glomalin, which is a glycoprotein secreted by arbuscular mycorrhizae (Rillig 2004; Rosier et al. 2006). The relationship between the components of solid organic N and those present in the soil's DON fraction has yet to be investigated in a systematic or comprehensive manner. More work is therefore required to understand the linkage between these solid and liquid pools, particularly in the light of recent SOM fractionation schemes (Cambardella and Elliott 1993; Feigl et al. 1995; Mathers et al. 2000).

The DON pool in soil represents only a small fraction of the total soil organic N pool and is often similar in size to that of the mineral N pool, typically ranging from 10 to 50 kg N ha^{-1} in arable agricultural soils, and up to 200 kg N ha^{-1} in highly organic grassland soils (Table 5.2) (Murphy et al. 2000). From the limited data available, the pool size of DON tends to be more constant than that of mineral N, which is known to be highly spatially and temporally dependent (Murphy et al. 2000; Willett et al. 2004). A study by Christou et al. (2005) has shown that DON is present in substantial quantities in all the dominant agricultural ecosystems in Europe (Figure 5.1). In this study across 94 individual sites, it was shown that the concentration of dissolved organic carbon (DOC) in soil was highly correlated with that of DON and that the DOC-to-DON ratio was not significantly affected by agroecosystem type. Typically, across all land-use types, DON constituted 57% ± 8% of the total dissolved N (TDN) pool and generally followed the order of:

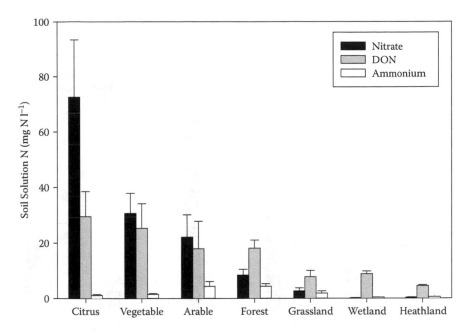

FIGURE 5.1 Dissolved organic nitrogen in agroecosystems in Europe. (From Christou, M. et al. 2005. *Soil Biol. Biochem.* 37: 1560–1563. With permission.)

citrus > vegetable > arable = forest > grassland = wetland > heathland

As a proportion of the TDN, DON was unsurprisingly more important in low-input agricultural ecosystems, where inorganic fertilizer additions were low (Christou et al. 2005). In relative contrast to DON, the amount of dissolved inorganic N (DIN) varied widely with land use, with intensive agricultural systems being dominated by NO_3^- and low-input systems dominated by NH_4^+.

Most of the uncertainty surrounding our poor understanding of the significance of DON in soil environments is due to our lack of knowledge about which organic N compounds are present, what their concentration is in soil solution, and the length of time they reside in soil. It seems paradoxical that our knowledge of the NO_3^- and NH_4^+ pool is regarded as critical to our understanding of N losses and plant productivity in both agricultural and natural ecosystems, yet in comparison, our knowledge of DON cycling is very limited. Although some of the constituents of DON remain unidentified, and are probably complex humic materials, many of the natural low-molecular-weight components (amino acids, amino sugars, nucleic acids, chlorophyll, phospholipids) and their precursors (proteins, peptides, chitin) are well known. While many of these components have been quantified in solid organic matter (Stevenson 1982; see Figure 5.2), few have been quantified in soil solution. The one exception to this is free amino acids, which typically constitute 0.5%–10% of the total DON pool and have a total concentration in soil solution typically ranging from 0.1 to 10 μM (ca. 0.01 to 0.5 kg N ha^{-1}) (Hannam and Prescott 2003; Kawahigashi et al. 2003; Henry and Jefferies 2002; Jones et al. 2005c). The signature of individual amino acids in soil solution is highly soil dependent and is known to

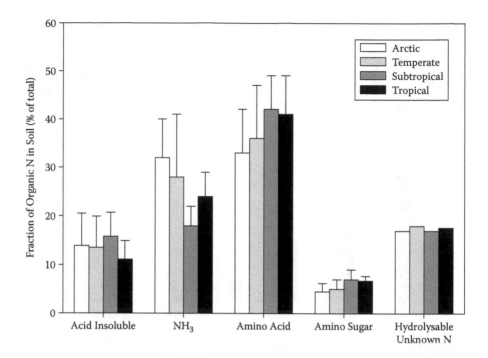

FIGURE 5.2 Nitrogen distribution in soils from widely different climatic zones. The soils were subject to acid hydrolysis to break the organic N into different fractions. (From Stevenson, F. J. 1982. *Nitrogen in agricultural soils*. Madison, WI: American Society of Agronomy. With permission.)

vary significantly over time and space and with management strategy (Kawahigashi et al. 2003). However, our mechanistic understanding of free amino acid cycling in soil still remains poor. Nonproteinaceous amino acids are also of significance, particularly in the rhizosphere, where, for example, their excretion by graminaceous plant roots aids the uptake of Fe and possibly other micronutrients from soil (Gerke 2000; Martens and Suarez 2003).

In contrast to amino acids, the major input of amino sugars into soil can be expected to be due to the turnover of soil fungal cell walls and, to a lesser extent, mesofauna. Although large amounts of amino sugars can be found in soil organic matter (Stevenson 1982; Niemann and Scheffer 1999; Amelung et al. 1999), the concentrations of free amino sugars such as galactosamine and glucosamine in soil solutions are reported to be very low (0.01–10 μM) (Michalzik and Matzner 1999; Roberts et al. 2007), representing approximately 0.001% of the total soil amino sugar pool. One reason for these low concentrations in solution may be the insolubility of the amino sugar polymers (e.g., chitin, chitosan, ergosterol) and their rapid utilization by the soil microbial community (Roberts et al. 2007). Whether smaller units formed during the biodegradation of the amino sugar polymers possess greater solubility remains understudied (Qin et al. 2003).

It is clear from studies on soil enzyme activity that many extracellular proteins and glycoproteins may be present in soil solution due to release from living or dead (lysed) microorganisms and plant roots (Richardson et al. 2001; Matsuyama et al. 1999). The functional diversity, molecular weight, and concentration of these proteins in soil solution remain largely unknown. Further, the relative extent to which these enzymes are present in an active or inactive form (e.g., after complexation with metals, minerals, or humic substances) in soil solution remains unknown for many agroecosystems (Allison 2006). One methodological problem is the lack of appropriate probes for measuring total protein contents of soil solution due to the cross-reaction of most commercial dyes with humic substances, free amino acids, or other substances (e.g., Bradford reagent) (Crossman et al. 2000; Whiffen et al. 2007). Although protein-specific assays have been developed (Herman et al. 2002), further work is still required to develop a simple and reliable assay for quantifying total or individual proteins in soil solution. Studies by Yu et al. (2002), using acid hydrolysis of soil solution DON, have indicated that approximately 50% of the amino acids in soil solution are present in a polymeric state (peptides, proteins), although whether these represent active proteins (i.e., enzymes, signal messengers, etc.) or simply partial degradation products of proteins remains unknown. While other studies have analyzed proteins in soil, the harsh extraction procedures suggest that many of these originate from other soil pools (e.g., microbial pool), and studies of this nature should be treated with caution (Murase et al. 2003). Considering the significance of peptides and other DON molecules in cell signaling and microbial community functioning, rhizosphere ecology and interplant competition (e.g., antimicrobial ε-poly-L-lysine or trifolitoxin; rhizobacteria growth-promoting opines), additional work is required to further characterize the dynamics and composition of the polymeric amino acid pool (Nishikawa and Ogawa 2002; Mansouri et al. 2002; Martin et al. 2003; Yoshida et al. 2001; Liu et al. 2000; Nielsen et al. 1999; Robleto et al. 1998).

The identification of antibiotic peptides in soil is also very important for healthcare and, in particular, in the development of drugs to control infection by human pathogens such as multidrug-resistant *Staphylococcus aureus* (Lee et al. 1997; Marinelli et al. 1996). Peptides and other DON compounds have also proved useful in human drug delivery (Kweon and Lim 2003); while in contrast some soil peptide toxins can be the cause of disease (e.g., melioidosis) (Haase et al. 1997). Similarly, the concentration of many other DON compounds in soil remains unknown, and efforts should be made to quantify these (e.g., purines, pyrimidines, polyamines). This will become increasingly easier due to the advent of proteomics and metabolomics technologies, which are capable of coping with the often low concentrations of DON found in soil solution (Zhu et al. 2003). Other DON compounds of anthropogenic origin (e.g., pesticides) may also be present in soil. However, it is not our intention to review the cycling of these compounds in soil as they are dealt with in detail elsewhere (Cheng 1991; Winton and Weber 1996; Sarmah et al. 2004). Many naturally produced (e.g., peptides) or added DON compounds (e.g., penicillamine) may also be important in bioremediation and the production of bioflocculants for wastewater treatment (Fischer 2002; Frankenberger and Arshad 2001; Deng et al. 2003; Fujita et al. 2000). In summary, although we have a detailed knowledge about some DON compounds in soil, this probably represents a very small fraction of the spectrum of N-containing compounds in soil solution.

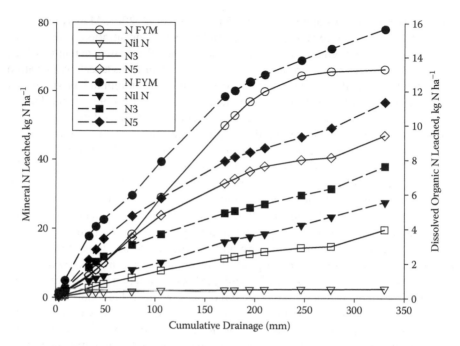

FIGURE 5.3 Cumulative N leached as Min–N (NH_4^+ + NO_3^-, open symbols) and DON (filled symbols), estimated using porous ceramic cups inserted at 65 cm in selected plots under continuous winter wheat on the Broadbalk Wheat Experiment, Rothamsted, U.K., in the winter of 1999–2000. See Table 5.2 for explanation of treatments.

The existence of soluble organic forms of N in drainage waters from agricultural land has been known for many years. Recent work indicates that, under some circumstances, substantial losses of DON from agricultural land may occur (Figure 5.3) (Siemens and Kaupenjohann 2002). The extent to which DON can be leached from the soil profile is influenced by its chemical composition. Dissolved organic matter leaching from forest floor litter is dominantly composed of decay-resistant organic acids formed by partial decomposition of plant, microbial, and animal tissues (Currie et al. 1996). In organic-rich layers, such as leaf litter or peat, hydrophobic and hydrophilic acids occur in equal concentrations. However, hydrophobic compounds are retained by adsorption and aggregation in mineral horizons, and the relative concentration of hydrophilic acids increases in drainage from upland soils as the thickness of the mineral horizons increases (Tipping et al. 1999). Dissolved organic matter in arable soils has a higher proportion of hydrophobic compounds compared with extracts from grassland and forest soils (Raber et al. 1998), indicating a greater propensity for adsorption.

Ligand exchange between the dissolved organic matter and carboxyl (Kaiser et al. 1997) or hydroxyl groups (Shen 1999) on the surface of soil minerals is thought to be an important mechanism for sorption, although a number of other mechanisms have also been proposed (Chiou et al. 1983). Adsorption isotherms for interactions between DOC and arable agricultural soils can be represented by linear isotherms, where the different adsorption capacities of different soils are related to the clay content (Riffaldi et al. 1998). Shen (1999) found the sorption capacity of the soil to be influenced by pH (maximal at pH 4–5 and decreasing with pH > 5), clay content (through its relationship with amount of mineral surface), the ionic strength of the soil solution, and the ion species present in soil solution (e.g., Ca^{2+} binding to mineral sites or complexing with dissolved organic matter). Thus management practices in agricultural systems, such as manure application and liming, will change the sorption capacity of soil and the resulting composition of the dissolved organic matter in solution.

5.3 MICROBIAL REGULATION OF DON IN SOIL

Analysis of total soil hydrolysates show that the organic N pool is not dominated by any one individual compound (Stevenson 1982). This suggests that most of the DON in soil is probably relatively bioavailable in the short-to-medium term (i.e., <1 d to <10 y). This contrasts with organic P, where phytate accumulates due to its high chemical reactivity and low bioavailability (Taranto et al. 2000; Turner et al. 2005). Some DON in soil will be physically protected from microbial attack by becoming trapped within other organic macromolecules (e.g., proteins within lignin) and mineral particles, while others will remain unavailable due to chemical protection. Indeed, chemical and physical protection is probably largely synonymous in many agricultural soils (e.g., bound in colloid-rich aggregates).

Extracellular enzymes are largely responsible for the breakdown of organic N polymers into their constitutive units (e.g., amino acids, purines, pyrimidines, amino sugars) (Burns and Dick 2002). Some abiotic ultraviolet-induced cleavage of DON can also be expected to occur at the soil surface, although this is likely to be of minimal significance in the overall flux due to the limited vertical light penetration into soil (e.g., depth of 10–200 µm) (Ciani et al. 2005). As organic N inputs into most agricultural systems are dominated by proteins, proteolytic enzymes (e.g., proteases and peptidases) are quantitatively probably the most important enzymes involved in soil organic N cycling. These enzymes hydrolyze proteins, releasing individual amino acids and peptides into solution. Based upon evidence gained from pure microbial cultures, it is likely that protein hydrolysis will be carried out by a diverse range of soil microorganisms, and that a large degree of functional redundancy exists within the population. Further, it is likely that, while these enzymes are constitutively expressed in some organisms, they will be capable of up-regulation in others, depending upon the appearance of a suitable substrate.

The signals in soil that up- or down-regulate enzyme activity in microbial cells (e.g., proteases) are largely unknown and are likely to be complex (Ruiz and Silhavy 2005). Although we know that protease activity is dependent upon many factors in soil, it is often impossible to divorce the direct effects (e.g., on gene expression or excretion rates within individual microbial cells) from indirect effects on the activity, size, and structure of the soil microbial community (e.g., climate) (Garcia et al. 1994; Marinari et al. 2006; Melero et al. 2006). A wide array of microbial proteases have been characterized at a molecular and physiological levels. Their classification is based upon which peptide linkages they break (Paul and Clark 1989). For example, enzymes such as pronase and subtilisin carry out terminal-amino-acid-chain removal, while others such as collagenase carry out cleavage at several locations within proteins (Sasagawa et al. 1993). In addition, not all proteins can be expected to denature/break down at the same rate due to differences in their chemical composition and molecular weight, which control their structure, solubility, sorption/desorption reactions, and reactivity with other compounds in soil solution (e.g., affinity for metals and humic substances). For example, some fibrous proteins with many cross-links (e.g., keratin) are relatively resistant to bacterial attack because of disulfide bonds between cysteine molecules, but are not resistant to breakdown by actinomycetes and fungi (e.g., *Streptomyces* and *Penicillium*) (Paul and Clark 1989; George et al. 1999).

Once broken down into low-molecular-weight (LMW) units, many DON components may be taken up directly by the soil microbial community or mineralized extracellularly, with the N absorbed as NH_4^+ (Hadas et al. 1992). Current evidence suggests that extracellular deamination of amino acids is not a major mineralization pathway, with most being taken up intact into microbial cells. From an energy standpoint, this is probably more efficient, as the NH_4^+ would need to be converted back to an amino acid (e.g., glutamate) after importation into the cell. In bacteria, exogenous amino acids are used for protein synthesis in preference to the endogenously formed amino acids, and they often suppress the production of the latter (Zalokar 1961).

There is no single pattern by which the biosynthetic pathways of all 20 amino acids are regulated. Even in prokaryotes, the pattern of regulation of a single amino acid biosynthetic pathway in one bacterium may be different from that in another bacterium, even though the pathway itself may be identical in the two organisms (Umbarger 1981). That <1% of the soil microorganisms have been cultured and identified highlights the fact that we know so little about the metabolic diversity of DON metabolism (Tiedje et al. 1999). A study by Vieublé Gonod et al. (2006), however, has suggested that a wide array of organisms are capable of using amino acids as a sole source of carbon (C) and N, suggesting that a high degree of functional redundancy exist in soil for some parts of the DON mineralization pathway. As most organisms in soil are C limited (Paul and Clark 1989; Raynaud et al. 2006), it has been suggested that the uptake of amino acids from soil is not regulated by N availability, but rather by C availability (Jones et al. 2004). This is supported by studies showing that the addition of large quantities of inorganic N do not suppress amino acid mineralization in soil (Jones and Shannon 1999). Catabolism of amino acids therefore serves two functions for the soil microbial community: firstly as a source of energy and C for cells that cannot attack complex carbohydrates or lipids or when the usual energy sources are not in sufficient supply; and secondly to supply N when an adequate supply of N is not available (Umbarger 1981).

Of the amino acids taken up by soil microorganisms, approximately 70% are used for anabolic processes (i.e., new biomass) and 30% are used directly in catabolic processes (i.e., respiration) (Jones et al. 2005b). The partitioning of amino acid-C into these two pathways seems to be largely independent of soil type, agronomic regime, temperature, and N availability (Jones et al. 2005b). When used in catabolic processes, amino acids supply an excess amount of N, leading ultimately to the excretion of NH_4^+ into the soil solution. A rapid excretion of NH_4^+ can be readily observed when the soil microbial community is supplied with the amino acid arginine ($C_6H_{14}N_4O_2$), which possesses an extremely low C-to-N ratio (Jones and Kielland 2002; Kemmitt et al. 2006). Umbarger (1978) provides a review of the many regulators of amino acid biosynthesis in microorganisms and whether their synthesis pathways are controlled by C or N. In some cases, the amino acid catabolic pathway may have the physiological role either of yielding energy or C, with N as a by-product, or of yielding N, with the C and energy as by-products, and regulation is accordingly adapted to the physiological role. Some amino acid catabolic enzymes are apparently not subject to either a C or an N control. It is clear from the differences and complexity in amino acid signaling and biosynthetic pathways between eukaryotes and prokaryotes that we still have a lot to learn about the role of amino acids in regulating the ecology of an often diverse microbial activity community in soil (Kohlhaw 2003).

Amino acids are taken up from soil using membrane proteins capable of transporting the amino acids against the concentration gradient. The transport is an active process and uses a respiration-driven electrochemical potential gradient set up across the membrane (Krulwich et al. 1977). Uptake may occur via a single transporter that is capable of transporting many different amino acids (Glover et al. 1975). However, most microorganisms possess multiple transport systems capable of taking up individual or groups of amino acids. Normally, they possess low-affinity permeases and high-affinity carriers designed for taking up amino acids from high (>1 mM) and low (<1 mM) external concentrations, respectively. The high-affinity transporters are normally saturable and obey Michaelis–Menten kinetics, while the permeases are generally not saturable and can be described by a first-order kinetic model (Vinolas et al. 2001; Gupta and Howard 1971; Kay and Gronlund 1971; Tang and Howard 1973; Hoban and Lyric 1977). These transport systems have the following characteristics:

1. They vary in the specificity of recognition of amino acids (Anraku 1980).
2. They require energy and are subject to temperature and pH optima (characteristics of active transport systems).
3. They may or may not be inhibited by the presence of certain amino acids (D'Ambrosio et al. 1973).

Many of these transport proteins from commonly found soil microorganisms have been well characterized at both the molecular and physiological level. Transport systems with narrow specificity may be active in the presence of nutrients, but under conditions of starvation, an active transport system with very broad specificity may be used (Tang and Howard 1973). Characterization of the kinetic properties of amino acid uptake by the entire soil microbial community suggests that microbial uptake does not become saturated even at high substrate addition rates (>10 mM), and that the uptake rate is amino acid specific. Amino acids such as glutamate, which forms the central hub of N cycling in microorganisms, seems to be taken up and utilized faster than lysine, which lies at the end of an amino acid synthesis pathway (Jones and Hodge 1999). In addition, the amino acid transporters have similar kinetic properties to those of plant roots, suggesting that plant–microbial competition for N resources in soil will be intense (Jones et al. 2004). This is not surprising considering the high degree of gene sequence homology between plant and microbial transport proteins.

In a range of agronomic systems, the turnover of amino acids in soil has been shown to be extremely rapid, with a half-life of approximately 0.5–2 h in topsoils and 3–12 h in subsoils (Jones 1999; Jones et al. 2005b, 2005c). When amino acids are added to the soil, there is no apparent lag phase in their use by soil microorganisms (Jones and Hodge 1999). This has led to the suggestion that the transport systems are constitutively expressed, reflecting the fact that amino acids are continually released into the soil solution from protease-mediated SOM hydrolysis. The rapid rate of amino acid turnover also suggests that the soil solution amino acid pool in soil solution turns over nearly 10,000 times per year. The rapid capture of amino acids by soil microorganisms also explains why their concentration in soil solution is often extremely low, with total free amino acid concentrations rarely rising above 100 µM, and individual amino acids rarely above 10 µM (Jones et al. 2005b). This rapid utilization of amino acids also partially explains why only small amounts of terrestrially derived amino acids are found in marine and fresh waters, as most will be stripped during passage to the water bodies (Lara et al. 1998; Kawasaki and Benner, 2006). Based upon the rapid uptake and mineralization of amino acids in soil, Jones et al. (2004) suggested that the primary block in the N mineralization pathway of neutral pH grasslands was the conversion of protein to amino acids, and not the conversion of either amino acids to NH_4^+ or NH_4^+ to NO_3^-. In later experiments, we also showed that the conversion of peptides to amino acids was nearly as fast as the conversion of amino acids to NH_4^+.

The rapid utilization of peptides is also helped by peptide-specific transporters in the microbial community (Walker and Altman 2005). The presence of these transporters negates the need for the peptides to be broken down firstly into amino acids before becoming accessible to the microbial community. The peptides being taken up are typically two to six amino acids in length, and uptake into the cell is mediated by specific membrane transporters similar to those used for taking up amino acids. These peptides may have a role in cell signaling and may yield more energy than the uptake of the equivalent free amino acids (Leach and Snell 1960). This result suggests that the limiting step in the soil N cycle is therefore the protease-mediated cleavage of protein into peptides. In acidic agricultural soils, however, the rate of nitrification may also be limited by the low pH, resulting in an additional block in the conversion of NH_4^+ to NO_3^- (Kemmitt et al. 2006).

As many proteins entering soil are protected within plant and animal tissues, it is still unclear whether their physical protection is more rate limiting than the availability of suitable catalytic enzymes capable of breaking them down. As apoptosis has now been shown to occur in both plant and microbial systems (Hoeberichts and Woltering 2003; Kuehn and Phillips 2005), it is also likely that some degree of protein degradation will occur even before the arrival of microbial enzymes. In this instance, leakage of degradation products (e.g., peptides and amino acids) through cell walls is likely to occur, leading to the creation of a necrosphere, akin to that of the rhizosphere around living roots.

Amino sugars may also represent a significant source of DON to the soil microbial community. As plants do not contain appreciable quantities of amino sugars, these primarily enter soil due to the extracellular cleavage of microbial polymers such as muramic acid and chitin. The uptake of free amino sugars (e.g., glucosamine) into microbial cells is also similar in nature to that of amino acids. Evidence suggests that the capacity for uptake of amino sugars is less than that for amino acids, and that the transport systems of the soil microbial community of grasslands become saturated at relatively low soil concentrations (<1 mM) (Roberts et al. 2007). These kinetic properties may be what would be predicted based upon the very low concentrations typically found in soil solution (<1 μM). As the turnover of the amino sugars is also relatively rapid in soil (half-life 1–6 h), it also suggests that the rate-limiting step in amino sugar-derived N mineralization is the conversion of polymers into monomeric units and not the subsequent mineralization of the monomers to NH_4^+.

The use of DON by soil microorganisms is likely to be significantly affected by the spatial location of microorganisms within the soil. In particular, rates of DON cycling in the rhizosphere may be much greater than in the bulk soil due to a greater microbial biomass, activity, and rate of turnover (Herman et al. 2006). Mycorrhizal fungi associated with roots may also enhance DON mineralization by exuding extracellular proteases and carrying out ammonification (Leake and Read 1990a, 1990b). These are discussed more fully in section 5.5.4 of this chapter and in other reviews on the rhizosphere (Jones et al. 2004).

While most information exists on the processing of identifiable DON compounds in soil, less work has been undertaken on the biodegradability of the unidentified components. However, from one perspective, the focus on the identifiable components may be justified in terms of the total flux of DON through the soil. While identifiable DON compounds are only present in soil solution in very small amounts, the flux through this pool is extremely rapid. In contrast, the unidentifiable components of DON are proportionally more important in pool size, but are relatively recalcitrant, indicating that the flux through this pool is relatively low (van Hees et al. 2005).

5.4 PLANT PRODUCTION OF DON

Plants represent a major input of DON into agricultural systems. This can occur in the form of throughfall (during the passage of rain or irrigation water through the crop canopy) and, more importantly, from roots as a result of root exudation and root turnover (Jones et al. 2004). DON also enters the soil when aboveground crop residues are incorporated back into the soil. In the case of green manures (e.g., alfalfa shoots, clover shoots), a large amount of DON may be returned to the soil, while for cereal residues at harvest (e.g., straw, stubble), the amounts of DON entering the soil are probably much less. Further, based upon our experience, the DON from cereal residues is generally more recalcitrant than that originating from green manures.

The factors regulating the loss of organic N from plants into soil have been well documented over the last 50 years (Rovira 1969; Curl and Trueglove 1986). Root exudates contain a wide range of compounds, many of which contain N. Typically, 1%–5% of a plant's net fixed C is lost into the soil by root exudation, of which approximately 1%–10% is present as DON. Due to the large concentration gradient that exists between the root cytoplasm and the soil solution, many LMW compounds are lost across the plasma membrane purely by diffusion (e.g., amino acids). Due to the size of the pores in the cell wall, only compounds of relatively small size are lost (e.g., <5000 MW). The plant is thought to exert little control over this outward flow (Farrar et al. 2003; Jones et al. 2005a). Exudation occurs throughout the root system but is mainly localized to root tips and the points of lateral root emergence, although the patterns of release are amino acid specific (Curl and Trueglove 1986; Jaeger et al. 1999).

Passive losses by root exudation are high in areas of cellular expansion due to:

1. Membrane expansion and vesicle fusion, which inevitably cause some cytoplasmic loss.
2. Apoplastic unloading and loss of amino acids due to the lack of symplastic connections between cells (e.g., plasmodesmata).
3. The high concentration of LMW compounds in actively growing root regions (Jones and Darrah 1994; Farrar et al. 2003).
4. The lack of an exodermal and endodermal barrier.
5. The loss of mucilage and border cells.
6. The high surface area-to-volume ratio of the root cap region in comparison to the mature root zone (Hawes et al. 2002).

Alongside sugars and organic acids, amino acids are one of the central groups of compounds in cellular metabolism, and their concentration in root cells is typically in the region of 1–10 mM, which is approximately 100- to 1000-fold higher than exists in the soil solution. The rate of amino acid efflux is also dependent upon many factors, including those regulated by the plant (plant species, ontogenetic stage, degree of disease, nutrient status, circadian rhythm), soil (nutrient status, presence of toxic metals and chemicals, soil biodiversity, soil physical factors), and the wider environment (e.g., temperature, atmospheric CO_2 level, grazing, pesticides, light intensity, etc.) (Phillips et al. 2004, 2006; Wang et al. 2006).

In addition to the LMW components of root exudates, some high-molecular-weight (HMW) components are also lost, including proteins (e.g., phosphatases, invertase). Due to the limited pore size of the cell wall and the inducible nature of release (e.g., phosphatases release under P deficiency), it is thought that these enzymes are actively excreted into the rhizosphere. Jones and Darrah (1994) showed that the rate of amino acid loss from maize roots was significantly greater than the rate of protein loss. Although there have been many attempts to quantify the rate of DON loss in root exudates, many of these studies remain difficult to interpret due to the consideration of DON flux in the rhizosphere as being solely a unidirectional flux (i.e., loss from the plant into the soil). However, as discussed in section 5.5, roots are also capable of taking up DON from the soil, and therefore DON flow in the rhizosphere is bidirectional. Most studies performed to date have ignored the influx component, and consequently only a net flux has been measured (Jones and Darrah 1994). These root-exudation experiments are typically performed in sterile solution culture in the absence of soil. Under these experimental growth conditions, the rates of root DON exudation may be significantly underestimated (Jones and Darrah 1993). This is because the root recaptures many of the amino acids exuded into the external solution. Therefore, what most experiments of this nature measure is the equilibrium point at which the rate of efflux equals the rate of influx. As the influx is very efficient, the concentration of zero net loss is very low (1–10 µM). This contrasts with a realistic soil system, where the rhizosphere microbial community continually removes amino acids from the soil solution and prevents root recapture. Further, these sterile, solution-based exudate-collection systems may underestimate net efflux because:

1. There is a lack of solid support, which normally increases exudation (Mozafar 1991).
2. The nutrient solutions rarely reflect those in soil.
3. There are no root pathogens present (e.g., fungi and nematodes) (Rengel 2002; Bardgett et al. 1999).

Consequently, it is difficult to extrapolate results from laboratory sterile-solution cultures to the field environment. More attempts need to be made, therefore, to quantify DON fluxes in soil.

One approach has been the use of DON biosensors that allow the visualization of DON efflux zones in the rhizosphere (Jaeger et al. 1999). Although semiquantitative, this technique is a significant advance in the *in situ* determination of DON fluxes, although it still needs to be tested on more plant species under a range of environmental conditions and for a diverse range of DON compounds. It is also apparent from recent studies that soil microorganisms may be able to manipulate the rates of DON release, although this still needs to be validated in a soil environment (Phillips et al. 2004).

In summary, it is clear that living roots release DON, but the quantitative importance of this at the field scale remains largely unknown in comparison with other DON inputs (e.g., from root turnover and microbial exudation and turnover).

5.5   PLANT UPTAKE OF DON

In the 1980s, it was widely assumed that the sole sources of N that plants could utilize were nitrate and ammonium. Thus the first editions of Marschner (1995), Beevers (1976), and Bray (1983) on N metabolism do not index "amino acid uptake" or "organic N uptake." Larcher's (2002) excellent text gave no role to organic N in plant N acquisition and just nine years ago, Forde and Clarkson (1999) reviewed nitrate and ammonium uptake as if they were the only significant sources of N to plants. This is strange, since the apparently ubiquitous ability of plant cells to transport amino acids and, indeed, small peptides has long been recognized (Luttge and Higinbotham 1979), and the use by plant cells of amino acids coming from legume nodules has long been appreciated (Stewart 1966). Earlier, an encyclopedic review by McKee (1962) had brought together much evidence that a wide range of organic N sources, notably a range of amino acids and urea, could support plant growth. The paradigm has changed. Now we must believe that roots are capable of taking up LMW organic N compounds from the soil, and the question has moved to asking: How much plant N is acquired by uptake of DON? To answer this requires knowledge of the bidirectional fluxes of DON at the soil–root interface (i.e., the net balance between DON influx and efflux).

5.5.1 PLANT MEMBRANES ARE LEAKY TO AMINO ACIDS

All plant membranes are leaky, so it is not surprising that amino acids and other nitrogenous solutes are lost passively down a concentration gradient from a range of cell types (Reinhold and Kaplan 1984). Amino acid loss from intact living roots is well established (Cakmak and Marschner 1988; Jones and Darrah 1993, 1994). Additionally, a wide range of polypeptides are exuded from roots of wheat (Basu et al. 1994). The concentration gradient driving passive efflux may be up to 1000-fold (i.e., the difference in concentration between the cytoplasm and the soil solution). Due to the size of this gradient, passive diffusion of organic N compounds into roots will not usually occur, despite of the permeability of membranes; rather, a net loss by exudation will occur unless the compounds are recaptured or the net influx of organics exceeds the rate of passive efflux. We argue below that urea may be an interesting exception.

5.5.2 PLANT MEMBRANES HAVE DON TRANSPORTERS THAT ARE ENERGIZED

The range of organic N compounds that plants can acquire from soil is defined by the transporters the plants have for them. Plant membrane transporters for amino acids and other organic N compounds have been functionally characterized, and the genes encoding them have been identified. The transporters generally have a high-affinity system that follows Michaelis–Menten kinetics, and they sometimes have dual kinetics with an additional low-affinity system that may or may not be saturable. Typically, the high-affinity system is a proton-coupled symporter, which is energized by association with membrane-bound ATPase (Bush 1993). However, because amino acids are the main N molecules within plants, some or most of these transporters could be largely involved with internal amino acid transport, and not with acquisition of exogenous DON.

Competitive uptake experiments were used to deduce the presence of four transporters in plants: one for basic, one for acidic, and two for neutral amino acids (Bush 1993). The carriers are promiscuous within and to some extent between these categories. Some examples follow. The range of amino acids capable of transport is frequently wide and may include alanine, leucine, glutamine, glutamate, valine, threonine, proline, phenylalanine, isoleucine, and arginine. The latter two had biphasic kinetics (Bush 1993). Arginine, aspartate, histidine, leucine, and phenylalanine share a common carrier (McDaniel et al. 1982). Wheat, barley, and mustard roots transport a range of amino acids, some by a common carrier (Wright 1962). *Lemna* has a transporter for neutral amino acids (Jung and Luttge 1980). Barley has lysine, methionine, and proline transporters with a K_m of 1.6, 20, and 3.9 µM, respectively, and a V_{max} on the order of 1 µmol DW^{-1} h^{-1}. All show multiphasic kinetics (Soldal and Nissen 1978).* A review of studies of this type lists a range of K_m from 4 µM to unsaturable (Reinhold and Kaplan 1984). Although the exact nature of the low-affinity system has been the subject of much debate (biphasic versus multiphasic; same or different carrier; saturable or nonsaturable, etc.), because it operates at a

---

* V_{max} and K_m are kinetic parameters in the Michaelis–Menten formulation of uptake rate.

concentration range well above that of DON in soil, these arguments are of little relevance here.

The application of molecular techniques has transformed our understanding of amino acid transporters in plant and microbial systems. There are a multitude of carriers with different properties (Fischer et al. 1998; Williams and Miller 2001). The genes fall into two main superfamilies: the amino acid/polyamine/organocation (APC) superfamily, and the amino acid/auxin permease (AAAP) superfamily. Some examples follow. AtCAT1 is in the APC superfamily and codes for a high-affinity transporter for basic amino acids, but it does not seem to be localized suitably to be involved in amino acid uptake (Williams and Miller 2001). A number of plant genes are known from the AAAP family. AtAAP1-6 are six amino acid permeases in *Arabidopsis*; more are known from other species. AtAAP1, 2, 4, and 6 have affinity for neutral and acidic amino acids, while AtAAP3 and AtAAP5 are not charge specific. AtAAP3 and AtAAP6 are found in roots (Hirner et al. 1998). RcAAP3 is expressed in roots of castor bean and transports neutral and basic amino acids (Neelam et al. 1999). AtLHT1 appears to be expressed in lower, young leaves and on root surfaces, where it transports histidine and lysine but not arginine (Frommer et al. 1995; Chen and Bush 1997). We know rather little about the tissue-expression patterns and possible functions in roots of most of the genes identified to date (Williams and Miller 2001; Frommer et al. 1994). One major limitation is that the molecular studies have generally not been concerned with the question of whether plants take up DON from soil. It is clear that we are not yet in a position to

1. Match the classes of transporters defined by classical kinetic techniques to those identified by molecular techniques.
2. Know whether all the transporters identified to date can be localized at potential uptake sites in roots.
3. Be sure that we have identified all the genes encoding transporters concerned with uptake.

Plants also contain transporters for a range of other organic N compounds apart from amino acids. Peptide transporters are known to exist and are often associated with protein-storage tissue (Song et al. 1996; Steiner et al. 1994; Waterworth and Bray 2006). One such transporter, AtPTR2, is expressed in roots (Williams and Miller 2001) and could provide a low rate of peptide uptake. Transporters for purine and its derivatives have been characterized at the gene level, but it is not known if any are involved in purine acquisition from soil (Williams and Miller 2001). Polyamines such as putrescine, spermidine, and cadaverine can also be taken up by membrane transport systems (DiTomaso et al. 1992), but the K_m of the transporters (120 µM) may be too high to be of relevance to the real world. Evidence from Nassar et al. (2003), however, has suggested that plant growth is promoted by the presence of a putrescine producing soil actinomycete (*Streptomyces griseoluteus*), but whether this was a direct (e.g., polyamine uptake by roots) or indirect effect (e.g., on other rhizosphere microorganisms) remains unknown. Some polyamines might also be excreted from roots as a mechanism to regulate fungal pathogen growth in the rhizosphere (Chibucos and Morris 2006).

The evidence for the important role of polyamines in cell signaling in plants and microbial communities is slowly becoming evident, and this area in particular highlights our ignorance of many root–soil interface processes (Thuler et al. 2003; Zeid and Shedeed 2006; Young and Chen 1997). Links between N_2 fixation and DON uptake by plants remain uninvestigated, and this warrants further investigation.

All of the above N-containing compounds are typically present at much higher concentrations inside the plant than outside in the soil. One exception to this is urea, which can be present at relatively high concentration in the soil (50–100 mM) (Lucas and Jones 2006), for example locally from animal urine inputs (Shand et al. 2002). At such high external concentrations, it may be able to enter plants passively. The aquaporin or MIP family of proteins transports water and also small neutral solutes (Gaspar et al. 2003). Plasma-membrane MIPs from tobacco confer low rates of permeability for urea of 1.0×10^{-6} cm s^{-1} by facilitated diffusion (Gerbeau et al. 1999). Since urease is a ubiquitous plant enzyme, rapid metabolism of the incoming urea to NH_4^+ may maintain a concentration gradient favoring urea entry (Benchemsibekkari and Pizelle 1992; Watson and Miller 1996). Our tentative suggestion that passive uptake of urea by this route might have a role in plant N nutrition remains to be tested. However, we have shown in sterile culture that *Brassica napus* plants can use urea as a sole source of N when supplied at high concentrations (10 mM). Further, urea can be used as a sole source of N when it is supplied to leaves via foliar sprays and when no N is supplied to the roots (Nicoulaud and Bloom 1998; Tan et al. 2000). Studies in sterile, hydroponically grown maize and wheat have indicated that roots are not capable of taking up exogenously applied amino sugars (e.g., glucosamine) (Jones et al. 2005a). Urea can be taken up unchanged by roots and supports the growth of sterile plants, including *Lemna*, *Brassica*, *Plantago*, and *Zea mays* (McKee 1962). A thorough study of plant growth on urea by Gerendas and Sattelmacher (1997a, 1997b) and Gerendas et al. (1998) emphasized the importance of nickel to activate urease, and showed how plants growing on urea as the sole N source are dependent on an active urease as well as, implicitly, a functional urea permease. The uptake mechanism of nitrate, urea, and ammonium by roots of intact plants (*Oryza sativa* L.) was studied using ^{15}N as a tracer to establish an absorption kinetic model and the influence of external concentration on uptake (Anti et al. 2001). In the treatment with nitrate, the relative uptake rate decreased with increasing plant age, while, in contrast, urea uptake was higher in older plants. For comparison, the uptake of ammonium remained relatively constant with age. This exemplified the complexity of generalizing from experiments performed on seedlings to mature plants.

The transporters detailed above offer the potential for uptake of specific DON fractions from soil. For that potential to be realized, the genes must be expressed in root cells that have access to the external substrate and the proteins incorporated into the plasma membrane rather than, say, the tonoplast. Thus we urgently need information on transporter localization in roots. This can potentially be achieved by localizing either the protein, the mRNA, or the function. Function is effectively localized to the appropriate part of the root if uptake can be shown, as we discuss in the following section.

5.5.3 Extracellular Enzymes?

Plants could gain DON by either or both of two strategies:

1. Direct uptake of just those molecules for which they have root membrane transporters
2. Secreting enzymes to convert more complex forms of DON (or insoluble N) into simpler compounds that they can take up

The abundant evidence for the first strategy has been dealt with, but what about the extracellular proteases that are produced by many plant pathogenic and mycorrhizal fungi? Such a strategy is feasible, since all plants produce proteases internally (Vierstra 1996; Wrobel and Jones 1992) and carnivorous plants secrete them (Schulze et al. 1999; Stephenson and Hogan 2006), as do many cells in tissue culture (Stano et al. 1998). Surprisingly perhaps, there is little evidence for the functional loss of proteases from plant roots. It may be time to reexamine the idea that limited secretion of proteases, perhaps induced by a combination of N deficiency and localized source of complex N, is possible. Early evidence from *Brassica napus* plants grown in sterile culture indicated that they could not support growth when supplied with the animal protein, bovine serum albumin, as a sole N source. This contrasts with work of Park et al. (2002), who showed that *Agrobacterium rhizogenes*-transformed hairy roots of pokeweed (*Phytolacca americana*) secreted proteases into the cell wall space. They hypothesized that this was part of an antifungal defense mechanism. It is possible, however, that the proteases secreted by pokeweed target the cleavage of specific pathogen proteins and are not used as part of a strategy for root N acquisition. Urea could also be subject to extracellular lysis to NH_3, which could then be taken up via plant NH_4^+ transporters. Although there is a great deal of evidence for free urease in soils, there is no evidence that roots secrete urease into the apoplast or soil, suggesting that plants take up urea intact.

5.5.4 Evidence for Plant Uptake of DON

Plant growth can be supported by a variety of amino acids supplied singly or in mixtures, or by urea (McKee 1962). Thus these compounds must be taken up or metabolized to mineral N outside the root before plants can acquire the N. Careful work in hydroponics, combined with demonstration that there are transporters at the root–soil interface, provides convincing evidence for direct uptake of amino acids and other forms of DON. The soil solution and apoplast in roots form one continuous diffusion pathway (Canny 1995). The surface of cortical cells and of part of the outside of the stele are in diffusive contact with the soil solution and are subject to transpiration-driven mass flow. Thus delivery of organic N to uptake sites located throughout the epidermis and root cortex is unlikely to be a problem in well-designed hydroponic-based experiments; uptake from soil is another matter.

First we ask: Can plant roots take up organic N? The early review of McKee (1962) suggests that alanine, asparagine, and histidine are commonly taken up. In such studies, however, it is possible that extracellular mineralization/deamination to NH_4^+ may have occurred prior to uptake, possibly by microbes living in the external solution and on the rhizoplane when roots are not maintained under sterile conditions (Jones et al. 2005a). The presence of mycorrhizal fungi adds substantial complications, as there is no doubt that these fungi are adept at utilizing DON (Abuzinadah and Read 1986; Finlay et al. 1992). Axenic cultures of duckweed have been shown to take up leucine, alanine, glycine, isoleucine, and valine (Borstlap 1977), but this may occur both by the leaves as well as the roots. Roots of sterile castor bean accumulate a range of amino acids by active uptake involving at least two independent carriers (Schobert and Komor 1987), while pea root tips actively take up phenylalanine and tyrosine (Watson and Fowden 1975). A careful study has demonstrated direct amino acid uptake unequivocally. *Lolium perenne* grown in sterile culture (and thus nonmycorrhizal) and supplied with ^{15}N-labeled glycine took up at least 80% of it intact. After uptake, the bulk of the amino acid was metabolized to serine by serine: glyoxylate aminotransferase (Thornton 2001).

The outcome of these studies is clear: Roots of a range of species can take up a number of amino acids and urea, and by implication these species have the appropriate transporters in their roots. Next we ask: Can plants grow using organic N alone? McKee (1962) cites several examples, some of which for urea are mentioned above. Ghosh and Burris (1950) grew initially sterile clover, tomato, and tobacco on a range of amino acids: 12 supported the growth of clover, 14 of tomato, and 11 of tobacco. Wheat grows aseptically on alanine, glycine, and urea (Newton 1957) and slowly on glutamate (Zaragoza-Castells, Jones, and Farrar, unpublished). Barley grows and produces grain on about 20 different amino acids; grain yields were highest with arginine and glutamine and, indeed, with these amino acids, they produced a higher yield than with either nitrate or ammonium (Mori et al. 1977). The kinetics of amino acid uptake suggest that castor bean could support significant growth from amino acids as a sole N source (Schobert and Komor 1987).

5.5.5 Do Plants Growing in Nonsterile Soil Take Up DON?

It is absolutely clear that a wide range of higher plants can take up a variety of forms of organic N in sufficient quantity to wholly or partly support their growth. The extent to which this ability is of importance to plants growing in the field is not yet fully known (Jones et al. 2005a). As this review has shown, if an amino acid contacts a root at the site of an appropriate transporter, it will be taken up. The question is, therefore, not one of ability, but of opportunity. Can roots compete with soil bacteria and fungi for organic N? Is there a sufficient concentration and flux of DON at the root surface to permit a significant rate of uptake?

Here, potential artifacts and problems of interpretation are acute because of the potential problems of microbial or mycorrhizal uptake and transformation, and because of the dilution of applied material into poorly understood pools in the soil (e.g., exchangeable and free soil solution pools) (Jones et al. 2005a). The heterogeneity of nutrient distribution in soil, and especially in the rhizosphere, is known but is only beginning to be addressed (Farrar et al. 2003), and this has huge implications for access of root-uptake systems to different DON fractions. The following criteria need to be met before we can accept that plants do compete successfully and directly for amino acids or urea in soil:

1. The molecule must be taken up without change so we know it has not been metabolized by soil microbes; the simplest method is to use compounds isotopically labeled with both C and N (i.e., ^{14}C, ^{13}C, ^{15}N).
2. The molecule must be taken up quickly so we know it has not been passed to the plant following uptake by mycorrhizae.
3. The concentration of tracer added to the soil in tracer experiments should not greatly raise the concentration normally seen at the uptake sites so we know that the same transporter kinetics are involved.
4. Ideally, the specific activity of radiolabeled substrate after dilution in soil should be known to evaluate the rate of uptake by the plant.
5. The half-life of radiolabeled substrate in soil needs to be matched to the duration of the experiment.

Wheat supplied with glycine labeled in both C and N (to avoid artifacts due to mineralization) acquired glycine-N and took up about 20% of the glycine intact. (Presumably, the other 80% was taken up as nitrate after mineralization.) As the roots were nonmycorrhizal, the wheat must have taken up the glycine directly (Näsholm and Persson 2001). The glycine was acquired in competition with soil microbes, but it was not possible to assess its likely importance for the N budget of the wheat crop (Näsholm and Persson 2001). To date, most experiments performed in soil have indicated that agricultural plants are poor competitors for amino acids in comparison with the soil microbial community. However, significant differences in competitive ability does exist between plant species (Owen and Jones 2001; Bardgett et al. 2003; Dunn et al. 2006). Due to the differing flux rates of the different N solutes in soil, it is likely that a mathematical modeling approach will be required to determine the significance of DON uptake to a plant's N budget.

5.5.6 PLANT REGULATION OF AMINO ACID UPTAKE

Roots can take up amino acids by active transport using a number of transporter proteins. Regulation of the rate of uptake could involve changing (a) the number of each type of transporter in the membrane by modulating gene expression, or (b) the properties of existing transporters. However, the information available on the relative importance of each of these regulation mechanisms is lacking. Because there are nonspecific transporters that recognize a range of amino acids, competition for uptake sites will occur (Watson and Fowden 1975; Schobert and Komor 1987). Ammonium inhibits amino acid uptake in melon root tips, perhaps by competition at a carrier (Watson and Fowden 1975). Conversely, nitrate uptake by maize roots is strongly inhibited by including the presence of several amino acids—aspartate, arginine, glutamine, and glycine—in the external medium (Padgett and Leonard 1993, 1996). Regulation at the gene level is scarcely understood, as amino acid uptake is sensitive to internal N and C status, but the nature of this regulation is not known. The gene AtAAP1 may be upregulated by sucrose or nitrate, and the AtAAP1 promoter has a nitrate-response element. There is strong evidence for the role of shoot-derived signals regulating N acquisition and a supposition that these may consist of one or more amino acids moving in the phloem (Forde and Clarkson 1999). Amino acids recycled in the phloem are strong candidates for downregulating nitrate uptake, but whether they also regulate DON uptake is unknown (Muller and Touraine 1992). The rapidity of this regulatory effect suggests a direct effect on uptake of the amino acids, but this warrants further study. Internal amino acids may also inhibit nitrate uptake, although this has been questioned (Tillard et al. 1998). Thornton (2001) showed that when sterile *Lolium perenne* was supplied with glucose or sucrose for 24 h, uptake of glycine increased. Although this was interpreted as showing the energy dependence of glycine uptake, the long timescale of these experiments means that it could equally well indicate sugar-mediated expression of the glycine transporter gene. It is clear, however, that despite early reports that amino acid transporters in roots were constitutively expressed in maize seedling roots (Jones and Darrah 1994), there may be many internal factors regulating DON acquisition by roots.

5.6 CONCLUDING REMARKS

It is clear from the discussion in this chapter that DON represents an extremely important component of the soil N cycle, providing a vital link between the transformation of soil organic matter into inorganic N in agricultural systems. However, it is also clear that its importance goes far beyond its role as simply a means of supplying N to the soil microbial community and, subsequently, plants. Its role in cell signaling in both plant and microbial communities has offered fascinating insights into the complex role that DON plays in processes operating in agricultural soils. It is likely that in future years a better understanding of DON flow in soil will allow us to manipulate these processes to design more-sustainable agricultural systems. It is vital, therefore, that we continue to delve deeper into the mysteries of DON cycling in soil, unraveling the complex interconnecting pathways between the individual compounds, their sources and sinks, and the factors regulating their concentrations in soil. Further, recent evidence suggesting that some components of the DON may provide plants with a source of N may prove to be of great importance in low-input systems. Additional experimental work is certainly needed to ascertain this. What is apparent, however, is that we also need to dispel some misconceptions about DON. For example, the fact that N additions in conventional agricultural systems tend to be dominated by inorganic fertilizers does not mean that we need to discount DON. From a range of leaching studies, it is now clear that DON represents a significant loss pathway that may have different downstream impacts than those of inorganic N.

Another key finding of this review is that there is a dichotomy between large-scale ecosystem-level studies investigating N budgets and those investigating small-scale processes involving DON (e.g., in the rhizosphere). At the system level, we have tended to treat DON as a single N pool, while at the small spatial scale we have tended to separate the DON pool into a myriad of compounds. Further, in many studies on the uptake of DON by plants and soil microorganisms, we have tended to focus on the bioavailability of individual compounds under limited conditions, and therefore extrapolation to the ecosystem level has proved difficult (Jones et al. 2005a). Thus there is a need to undertake more compound-specific DON work in the large-scale system studies similar to that undertaken in forest-based studies.

It is also noticeable from this review that modeling, normally omnipresent in agricultural N reviews, is noticeable by its absence. We are now at the stage where our process-level understanding of DON fluxes in soil is sufficient to allow its incorporation into predictive N-cycling models. Certainly, if we are to improve the robustness of N models to accurately describe low-input or organic agroecosystems, we will need to incorporate DON as a major model component. How this will be parameterized can be expected to require careful thought due to the complex mixture and reactivity of the compounds that constitute DON.

REFERENCES

Abuzinadah, R. A., and D. J. Read. 1986. The role of proteins in the nitrogen nutrition of ecto-mycorrhizal plants, 3: Protein-utilization by *Betula*, *Picea* and *Pinus* in mycorrhizal association with *Hebeloma crustuliniforme*. *New Phytol.* 103: 507–514.

Addiscott, T. M., A. P. Whitmore, and D. S. Powlson. 1991. *Farming, fertilizers and the nitrate problem.* Wallingford, Oxon, U.K.: CAB International.

Allison, S. D. 2006. Soil minerals and humic acids alter enzyme stability: Implications for ecosystem processes. *Biogeochemistry* 81: 361–373.

Amelung, W., X. Zhang, K. W. Flach, and W. Zech. 1999. Amino sugars in native grassland soils along a climosequence in North America. *Soil Sci. Soc. Am. J.* 63: 86–92.

Anraku, Y. 1980. Transport and utilization of amino acids by bacteria. In *Microorganisms and nitrogen sources*, ed. J. W. Payne, 281–300. New York: John Wiley & Sons.

Anti, A. B., J. Mortatti, P. C. O. Trivelin, and J. A. Bendassolli. 2001. Radicular uptake kinetics of $^{15}NO_3$, $CO(^{15}NH_2)_2$, and $^{15}NH_4^+$ in whole rice plants. *J. Plant Nutr.* 24: 1695–1710.

Bardgett, R. D., R. Cook, G. W. Yeates, and C. S. Denton. 1999. The influence of nematodes on below-ground processes in grassland ecosystems. *Plant Soil* 212: 23–33.

Bardgett, R. D., T. C. Streeter, and R. Bol. 2003. Soil microbes compete effectively with plants for organic-nitrogen inputs to temperate grasslands. *Ecology* 84: 1277–1287.

Basu, U., A. Basu, and G. J. Taylor. 1994. Differential exudation of polypeptides by roots of aluminum-resistant and aluminum-sensitive cultivars of *Triticum aestivum* L. in response to aluminum stress. *Plant Physiol.* 106: 151–158.

Beevers, L. 1976. *Nitrogen metabolism in plants.* London: Arnold Press.

Benchemsibekkari, N., and G. Pizelle. 1992. *In vivo* urease activity in *Robinia pseudoacacia*. *Plant Physiol. Biochem.* 30: 187–192.

Borstlap, A. C. 1977. Kinetics of uptake of some neutral amino acids by *Spirodela polyrhiza*. *Acta Bot. Neerl.* 26: 115–128.

Bray, C. M. 1983. *Nitrogen metabolism in plants.* London: Longman.

Burns, R. G., and R. P. Dick. 2002. *Enzymes in the environment: Activity, ecology and applications.* New York: Marcel Dekker Inc.

Bush, D. R. 1993. Proton-coupled sugar and amino-acid transporters in plants. *Annu. Rev. Plant Physiol. Plant Mol. Biol.* 44: 513–542.

Cakmak, I., and H. Marschner. 1988. Increase in membrane-permeability and exudation in roots of zinc-deficient plants. *J. Plant Physiol.* 132: 356–361.

Cambardella, C. A., and E. T. Elliott. 1993. Methods for physical separation and characterization of soil organic-matter fractions. *Geoderma* 56: 449–457.

Canny, M. J., 1995. Apoplastic water and solute movement, new rules for an old space. *Annu. Rev. Plant Physiol. Plant Mol. Biol.* 46: 215–236.

Chen, L., and D. R. Bush. 1997. LHT1, A lysine, and histidine-specific amino acid transporter in *Arabidopsis*. *Plant Physiol.* 115: 1127–1134.

Cheng, H. H., ed. 1991. *Pesticides in the soil environment: Processes, impacts, and modelling.* Madison, WI: Soil Science Society of America.

Chibucos, M. C., and P. F. Morris. 2006. Levels of polyamines and kinetic characterization of their uptake in the soybean pathogen *Phytophthora sojae. Appl. Environ. Microbiol.* 72: 3350–3356.

Chiou, C. T., P. E. Porter, and D. W. Schmedding. 1983. Partition equilibrium of non-ionic organic compounds between soil organic matter and water. *Environ. Sci. Technol.* 17: 295–297.

Christou, M., E. J. Avramides, and D. L. Jones. 2005. Dissolved organic nitrogen in contrasting agricultural ecosystems. *Soil Biol. Biochem.* 37: 1560–1563.

Christou, M., E. J. Avramides, and D. L. Jones. 2006. Dissolved organic nitrogen dynamics in a Mediterranean vineyard soil. *Soil Biol. Biochem.* 38: 2265–2277.

Ciani, A., K. U. Goss, and R. P. Schwarzenbach. 2005. Light penetration in soil and particulate minerals. *Eur. J. Soil Sci.* 56: 561–574.

Cornell, S. E., T. D. Jickells, J. N. Cape, A. P. Rowland, and R. A. Duce. 2003. Organic nitrogen deposition on land and coastal environments: A review of methods and data. *Atmos. Environ.* 37: 2173–2191.

Crossman, D. J., K. D. Clements, and G. J. S. Cooper. 2000. Determination of protein for studies of marine herbivory: A comparison of methods. *J. Exp. Mar. Biol. Ecol.* 244: 45–65.

Curl, E. A., and B. Trueglove. 1986. *The rhizosphere.* Berlin: Springer-Verlag.

Currie, W. S., J. D. Aber, W. H. McDowell, R. D. Boone, and A. H. Magill. 1996. Vertical transport of dissolved organic C and N under long-term N amendments in pine and hardwood forests. *Biogeochemistry* 35: 471–505.

D'Ambrosio, S. M., G. I. Glover, and R. A. Jensen. 1973. Specificity of the tyrosine-phenylalanine transport system of *Bacillus subtilis. J. Bacteriol.* 115: 673–681.

de Guardia, A., S. Brunet, D. Rogeau, and G. Matejka. 2002. Fractionation and characterisation of dissolved organic matter from composting green wastes. *Bioresour. Technol.* 83: 181–187.

Deng, S. B., R. B. Bai, X. M. Hu, and Q. Luo. 2003. Characteristics of a bioflocculant produced by *Bacillus mucilaginosus* and its use in starch wastewater treatment. *Appl. Micrbiol. Biotechnol.* 60: 588–593.

DiTomaso, J. M., J. J. Hart, and L. V. Kochian. 1992. Transport kinetics and metabolism of exogenously applied putrescine in roots of intact maize seedlings. *Plant Physiol.* 98: 611–620.

Dunn, R. M., J. Mikola, R. Bol, and R. D. Bardgett. 2006. Influence of microbial activity on plant-microbial competition for organic and inorganic nitrogen. *Plant Soil* 289: 321–334.

Farrar, J., M. Hawes, D. Jones, S. Lindow. 2003. How roots control the flux of carbon to the rhizosphere. *Ecology* 84: 827–837.

Feigl, B. J., J. Melillo, and C. C. Cerri. 1995. Changes in the origin and quality of soil organic-matter after pasture introduction in Rondonia (Brazil). *Plant Soil* 175: 21–29.

Finlay, R. D., A. Frostegard, and A. M. Sonnerfeldt. 1992. Utilization of organic and inorganic nitrogen-sources by ectomycorrhizal fungi in pure culture and in symbiosis with *Pinus contorta* Dougl. Ex. Loud. *New Phytol.* 120: 105–115.

Fischer, K. 2002. Removal of heavy metals from soil components and soil by natural chelating agents, part 1: Displacement from clay minerals and peat by L-cysteine and L-penicillamine. *Water Air Soil Poll.* 137: 267–286.

Fischer, W. N., B. Andre, D. Rentsch, S. Krolkiewicz, M. Tegeder, K. Breitkreuz, W. B. and Frommer. 1998. Amino acid transport in plants. *Trends Plant Sci.* 3: 188–195.

Forde, B. G., and D. T. Clarkson. 1999. Nitrate and ammonium nutrition of plants: Physiological and molecular perspectives. *Adv. Bot. Res.* 30: 1–90.

Frankenberger, W. T., and M. Arshad. 2001. Bioremediation of selenium-contaminated sediments and water. *Biofactors* 14: 241–254.

Friedel, J. K., and E. Scheller. 2002. Composition of hydrolysable amino acids in soil organic matter and soil microbial biomass. *Soil Biol. Biochem.* 34: 315–325.

Frommer, W. B., M. Kwart, B. Hirner, W. N. Fischer, S. Hummel, and O. Ninnemann. 1994. Transporters for nitrogenous compounds in plants. *Plant Mol. Biol.* 26: 1651–1670.

Frommer, W. B., S. Hummel, M. Unseld, and O. Ninnemann. 1995. Seed and vascular expression of a high-affinity transporter for cationic amino acids in *Arabidopsis. Proc. Natl. Acad. Sci.* 92: 12036–12040.

Fujita, M., M. Ike, S. Tachibana, G. Kitada, S. M. Kim, and Z. Inoue. 2000. Characterization of a bioflocculant produced by *Citrobacter* sp. TKF04 from acetic and propionic acids. *J. Biosci. Bioeng.* 89: 40–46.

Garcia, C., T. Hernandez, and F. Costa. 1994. Microbial activity in soils under Mediterranean environmental conditions. *Soil Biol. Biochem.* 26: 1185–1191.

Gaspar, M., A. Bousser, I. Sissoeff, O. Roche, J. Hoarau, and A. Mahe. 2003. Cloning and characterization of ZmPIP1-5b, an aquaporin transporting water and urea. *Plant Sci.* 165: 21–31.

George, E., C. Stober, and B. Seith. 1999. The use of different soil nitrogen sources by young Norway spruce plants. *Trees* 13: 199–205.

Gerbeau, P., J. Guclu, P. Ripoche, and C. Maurel. 1999. Aquaporin Nt-TIPa can account for the high permeability of tobacco cell vacuolar membrane to small neutral solutes. *Plant J.* 18: 577–587.

Gerendas, J., and B. Sattelmacher. 1997a. Significance of N source (urea vs. NH_4NO_3) and Ni supply for growth, urease activity and nitrogen metabolism of zucchini (*Cucurbita pepo* convar. giromontiina). *Plant Soil* 196: 217–222.

Gerendas, J., and B. Sattelmacher. 1997b. Significance of Ni supply for growth, urease activity and the concentrations of urea, amino acids and mineral nutrients of urea-grown plants. *Plant Soil* 190: 153–162.

Gerendas, J., Z. Zhu, and B. Sattelmacher. 1998. Influence of N and Ni supply on nitrogen metabolism and urease activity in rice (*Oryza sativa* L.). *J. Exp. Bot.* 49: 1545–1554.

Gerke, J., 2000. Mathematical modelling of iron uptake by graminaceous species as affected by iron forms in soil and phytosiderophore efflux. *J. Plant Nutr.* 23: 1579–1587.

Ghosh, B. P., and R. H. Burris. 1950. Utilization of nitrogenous compounds by plants. *Soil Sci.* 70: 187–200.

Glover, G. I., S. M. D'Ambrosio, and R. A. Jensen. 1975. Versatile properties of a nonsaturatable, homogeneous transport system in *Bacillus subtilis*: Genetic, kinetic, and affinity labelling studies. *Proc. Natl. Acad. Sci.* 72: 814–818.

Goulding, K., and P. Poulton. 1992. Unwanted nitrate. *Chem. Brit.* 28: 1100–1102.

Gupta, R. K., and D. H. Howard. 1971. Comparative physiological status of the yeast and mycelial forms of *Histoplasma capsulatum:* Uptake and incorporation of L-leucine. *J. Bacteriol.* 105: 690–700.

Haase, A., J. Janzen, S. Barrett, B. Currie. 1997. Toxin production by *Burkholderia pseudomallei* strains and correlation with severity of melioidosis. *J. Med. Microbiol.* 46: 557–563.

Hadas, A., M. Sofer, J. A. E. Molina, P. Barak, and C. E. Clapp. 1992. Assimilation of nitrogen by soil microbial population: NH_4 versus organic N. *Soil Biol. Biochem.* 24: 137–143.

Hannam, K. D., and C. E. Prescott. 2003. Soluble organic nitrogen in forests and adjacent clearcuts in British Columbia, Canada. *Can. J. For. Res.* 33: 1709–1718.

Hawes, M. C., G. Bengough, G. Cassab, and G. Ponce. 2002. Root caps and rhizosphere. *J. Plant Growth Reg.* 21: 352–367.

Haynes, R. J. 1986. *Mineral nitrogen in the plant soil system*. New York: Academic Press.

Henry, H. A. L., and R. L. Jefferies. 2002. Free amino acid, ammonium and nitrate concentrations in soil solutions of a grazed coastal marsh in relation to plant growth. *Plant Cell Environ.* 25: 665–675.

Herman, D. J., K. K. Johnson, C. H. Jaeger, E. Schwartz, and M. K. Firestone. 2006. Root influence on nitrogen mineralization and nitrification in *Avena barbata* rhizosphere soil. *Soil Sci. Soc. Am. J.* 70: 1504–1511.

Herman, R. A., J. D. Wolt, and W. R. Halliday. 2002. Rapid degradation of the Cry1F insecticidal crystal protein in soil. *J. Agric. Food Chem.* 50: 7076–7078.

Hirner, B., W. N. Fischer, D. Rentsch, M. Kwart, and W. B. Frommer. 1988. Developmental control of H+/amino acid permease gene expression during seed development of *Arabidopsis*. *Plant J.* 14: 535–544.

Hoban, D. J., and R. M. Lyric. 1977. Glutamate uptake in *Thiobacillus novellas*. *Can. J. Microbiol.* 23: 271–277.

Hoeberichts, F. A., and E. J. Woltering. 2003. Multiple mediators of plant programmed cell death: Interplay of conserved cell death mechanisms and plant-specific regulators. *BioEssays* 25: 47–57.

Insam, H., and P. Merschak. 1997. Nitrogen leaching from forest soil cores after amending organic recycling products and fertilizers. *Waste Manage. Res.* 15: 277–291.

Jaeger, C. H., S. E. Lindow, S. Miller, E. Clark, and M. K. Firestone. 1999. Mapping of sugar and amino acid availability in soil around roots with bacterial sensors of sucrose and tryptophan. *Appl. Environ. Microbiol.* 65: 2685–2690.

Jarvis, S. C., E. A. Stockdale, M. A. Shepherd, and D. S. Powlson. 1996. Nitrogen mineralisation in temperate agricultural soils: Processes and measurement. *Adv. Agron.* 57: 187–235.

Jones, D. L. 1999. Amino acid biodegradation and its potential effects on organic nitrogen capture by plants. *Soil Biol. Biochem.* 31: 613–622.

Jones, D. L., and P. R. Darrah. 1993. Re-sorption of organic-compounds by roots of *Zea mays* L. and its consequences in the rhizosphere, 2: Experimental and model evidence for simultaneous exudation and re-sorption of soluble C compounds. *Plant Soil* 153: 47–59.

Jones, D. L., and P. R. Darrah. 1994. Amino-acid influx at the soil-root interface of *Zea mays* L. and its implications in the rhizosphere. *Plant Soil* 163: 1–12.

Jones, D. L., J. R. Healey, V. B. Willett, J. F. Farrar, and A. Hodge. 2005a. Dissolved organic nitrogen uptake by plants: An important N uptake pathway? *Soil Biol. Biochem.* 37: 413–423.

Jones, D. L., and A. Hodge. 1999. Biodegradation kinetics and sorption reactions of three differently charged amino acids in soil and their effects on plant organic nitrogen availability. *Soil Biol. Biochem.* 31: 1331–1342.

Jones, D. L., A. Hodge, and Y. Kuzyakov. 2004. Plant and mycorrhizal regulation of rhizodeposition. *New Phytol.* 163: 459–480.

Jones, D. L., S. J. Kemmitt, D. Wright, S. P. Cuttle, R. Bol, and A. C. Edwards. 2005b. Rapid intrinsic rates of amino acid biodegradation in soils are unaffected by agricultural management strategy. *Soil Biol. Biochem.* 37: 1267–1275.

Jones, D. L., and K. Kielland. 2002. Soil amino acid turnover dominates the nitrogen flux in permafrost-dominated taiga forest soils. *Soil Biol. Biochem.* 34: 209–219.

Jones, D. L., and D. Shannon. 1999. Mineralization of amino acids applied to soils: Impact of soil sieving, storage, and inorganic nitrogen additions. *Soil Sci. Soc. Am. J.* 63: 1199–1206.

Jones, D. L., D. Shannon, T. Junvee-Fortune, and J. F. Farrar. 2005c. Plant capture of free amino acids is maximized under high soil amino acid concentrations. *Soil Biol. Biochem.* 37: 179–181.

Jung, K. D., and U. Luttge. 1980. Amino-acid uptake by *Lemna gibba* by a mechanism with affinity to neutral L-amino and D-amino acids. *Planta* 150: 230–235.

Kaiser, K., G. Guggenberger, L. Haumaier, and W. Zech. 1997. Dissolved organic matter sorption on subsoils and minerals studied by ^{13}C-NMR and DRIFT spectroscopy. *Eur. J. Soil Sci.* 48: 301–310.

Kawahigashi, M., H. Sumida, and K. Yamamoto. 2003. Seasonal changes in organic compounds in soil solutions obtained from volcanic ash soils under different land uses. *Geoderma* 113: 381–396.

Kawasaki, N., and R. Benner. 2006. Bacterial release of dissolved organic matter during cell growth and decline: Molecular origin and composition. *Limnol. Oceanogr.* 51: 2170–2180.

Kay, W. W., and A. F. Gronlund. 1971. Transport of aromatic amino acids by *Pseudomonas aeruginosa. J. Bacteriol.* 105: 1039–1046.

Kemmitt, S. J., D. Wright, K. W. T. Goulding, and D. L. Jones. 2006. pH regulation of carbon and nitrogen dynamics in two agricultural soils. *Soil Biol. Biochem.* 38: 898–911.

Knicker, H. 2000. Biogenic nitrogen in soils as revealed by solid-state carbon-13 and nitrogen-15 nuclear magnetic resonance spectroscopy. *J. Environ. Qual.* 29: 715–723.

Kohlhaw, G. B. 2003. Leucine biosynthesis in fungi: Entering metabolism through the back door. *Microbiol. Mol. Biol. Rev.* 67: 1–15.

Kristensen, L., C. Stopes, P. Kolster, and A. Granstedt. 1995. Nitrogen leaching in ecological agriculture: Summary and recommendations. *Biol. Agric. Hort.* 11: 331–340.

Krulwich, T. A., R. Blanco, and P. A. McBride. 1977. Amino-acid transport in whole cells and membrane-vesicles of *Arthrobacter pyridinolis. Arch. Biochem. Biophys.* 178: 108–117.

Kuehn, G. D., and G. C. Phillips. 2005. Role of polyamines in apoptosis and other recent advances in plant polyamines. *Crit. Rev. Plant Sci.* 24: 123–130.

Kweon, D. K., and S. T. Lim. 2003. Preparation and characteristics of a water-soluble chitosan-heparin complex. *J. Appl. Polymer. Sci.* 87: 1784–1789.

Lara, R. J., V. Rachold, G. Kattner, H. W. Hubberten, G. Guggenberger, A. Skoog, and D. N. Thomas. 1998. Dissolved organic matter and nutrients in the Lena River, Siberian Arctic: Characteristics and distribution. *Mar. Chem.* 59: 301–309.

Larcher, W. 2002. *Physiological plant ecology: Ecophysiology and stress physiology of functional groups.* Berlin: Springer Verlag.

Leach, F. R., and E. E. Snell. 1960. The absorption of glycine and alanine and their peptides by *Lactobacillus casei. J. Biol. Chem.* 235: 3523–3531.

Leach, K. A., K. D. Allingham, J. S. Conway, K. W. T. Goulding, and D. J. Hatch. 2005. Nitrogen management for profitable farming with minimal environmental impact: The challenge for mixed farms in the Cotswold Hills, England. *Int. J. Agric. Sustainabil.* 2: 21–32.

Leake, J. R., and D. J. Read. 1990a. Proteinase activity in mycorrhizal fungi, 1: The effect of extracellular pH on the production and activity of proteinase by ericoid endophytes from soils of contrasted pH. *New Phytol.* 115: 243–250.

Leake, J. R., and D. J. Read. 1990b. Proteinase activity in mycorrhizal fungi, 2: The effects of mineral and organic nitrogen-sources on induction of extracellular proteinase in *Hymenoscyphus ericae* (Read) Kort and Kernan. *New Phytol.* 116: 123–128.

Lee, H. W., J. W. Choi, H. W. Kim, D. P. Han, W. S. Shin, and D. H. Yi. 1997. A peptide antibiotic AMRSA1 active against multidrug-resistant *Staphylococcus aureus* produced by *Streptomyces* sp. HW-003. *J. Microbiol. Biotechnol.* 7: 402–408.

Liu, Y. F., I. C. Luo, C. Y. Xu, F. C. Ren, C. Peng, C. Y. Wu, and J. D. Zhao. 2000. Purification, characterization, and molecular cloning of the gene of a seed-specific antimicrobial protein from pokeweed. *Plant Physiol.* 122: 1015–1024.

Lucas, S. D., and D. L. Jones. 2006. Biodegradation of estrone and 17 beta-estradiol in grassland soils amended with animal wastes. *Soil Biol. Biochem.* 38: 2803–2815.

Luttge, U. and N. Higinbotham. 1979. *Transport in Plants.* Heidelberg: Springer-Verlag.

Mansouri, H, A. Petit, P. Oger, and Y. Dessaux. 2002. Engineered rhizosphere: The trophic bias generated by opine-producing plants is independent of the opine type, the soil origin, and the plant species. *Appl. Environ. Microbiol.* 68: 2562–2566.

Marinari, S., R. Mancinelli, E. Carnpiglia, and S. Grego. 2006. Chemical and biological indicators of soil quality in organic and conventional farming systems in Central Italy. *Ecol. Indic.* 6: 701–711.

Marinelli, F., L. Gastaldo, G. Toppo, and C. Quarta. 1996. Antibiotic GE37468A: A new inhibitor of bacterial protein synthesis, 3: Strain and fermentation study. *J. Antibiotic.* 49: 880–885.

Marschner, H. 1995. *Mineral nutrition of higher plants*. New York: Academic Press.

Martens, D. A., and D. L. Suarez. 2003. Soil methylation-demethylation pathways for metabolism of plant-derived selenoamino acids. In *Sym. Series 835, Biogeochemistry of Environmentally Important Trace Elements*, ed. Y. Cai and O. C. Braids, 355–369. Washington DC: American Chemical Society.

Martin, N. I., H. J. Hu, M. M. Moake, J. J. Churey, R. Whittal, R. W. Worobo, and J. C. Vederas. 2003. Isolation, structural characterization, and properties of mattacin (Polymyxin M), a cyclic peptide antibiotic produced by *Paenibacillus kobensis. J. Biol. Chem.* 278: 13124–13132.

Mathers, N. J., X. A. Mao, Z. H. Xu, P. G. Saffigna, S. J. Berners-Price, and M. C. S. Perera. 2000. Recent advances in the application of ^{13}C and N-15 NMR spectroscopy to soil organic matter studies. *Aust. J. Soil Res.* 38: 769–787.

Matsuyama, T., H. Satoh, Y. Yamada, and T. Hashimoto. 1999. A maize glycine-rich protein is synthesized in the lateral root cap and accumulates in the mucilage. *Plant Physiol.* 120: 665–674.

McDaniel, C. N., R. K. Holterman, R. F. Bone, and P. M. Wozniak. 1982. Amino-acid-transport in suspension-cultured plant cells, 3: Common carrier system for the uptake of L-arginine, L-aspartic acid, L-histidine, L-leucine, and L-phenylalanine. *Plant Physiol.* 69: 246–249.

McKee, H. S. 1962. *Nitrogen metabolism in plants*. Oxford: Clarendon Press.

Melero, S., J. C. R. Porras, J. F. Herencia, and E. Madejon. 2006. Chemical and biochemical properties in a silty loam soil under conventional and organic management. *Soil Till. Res.* 90: 162–170.

Michalzik, B., and E. Matzner. 1999. Dynamics of dissolved organic nitrogen and carbon in a Central European Norway spruce ecosystem. *Eur. J. Soil Sci.* 50: 579–590.

Michalzik, B., and B. Stadler. 2005. Importance of canopy herbivores to dissolved and particulate organic matter fluxes to the forest floor. *Geoderma* 127: 227–236.

Mori, S., N. Nishizawa, H. Uchino, and Y. Nishimura. 1977. Utilization of organic nitrogen as the sole source of nitrogen for barley. *J. Sci. Soil Manure* 48: 612–617.

Mozafar, A. 1991. Contact with ballotini (glass spheres) stimulates exudation of iron reducing and iron chelating substances from barley roots. *Plant Soil* 130: 105–108.

Muller, B., and B. Touraine. 1992. Inhibition of NO_3^- uptake by various phloem-translocated amino-acids in soybean seedlings. *J. Exp. Bot.* 43: 617–623.

Murase, A., M. Yoneda, R. Ueno, and K. Yonebayashi. 2003. Isolation of extracellular protein from greenhouse soil. *Soil Biol. Biochem.* 35: 733–736.

Murphy, D. V., A. J. Macdonald, E. A. Stockdale, K. W. T. Goulding, S. Fortune, J. L. Gaunt, P. R. Poulton, J. A. Wakefield, C. P. Webster, and W. S. Wilmer. 2000. Soluble organic nitrogen in agricultural soils. *Biol. Fert. Soils* 30: 374–387.

Näsholm, T., and J. Persson. 2001. Plant acquisition of organic nitrogen in boreal forests. *Physiol. Plant.* 111: 419–426.

Nassar, A. H., K. A. El-Tarabily, and K. Sivasithamparam. 2003. Growth promotion of bean (*Phaseolus vulgaris* L.) by a polyamine-producing isolate of *Streptomyces griseoluteus. Plant Growth Reg.* 40: 97–106.

Neelam, A., A. C. Marvier, J. L. Hall, and L. E. Williams. 1999. Functional characterization and expression analysis of the amino acid permease RcAAP3 from castor bean. *Plant Physiol.* 120: 1049–1056.

Newton, W. 1957. The utilization of single organic nitrogen compounds by wheat seedlings and by *Phytophthora parasitica*. *Can. J. Bot.* 34: 445–448.

Nicoulaud, B. A. L., and A. J. Bloom. 1998. Nickel supplements improve growth when foliar urea is the sole nitrogen source for tomato. *J. Am. Soc. Hort. Sci.* 123: 556–559.

Nielsen, T. H., C. Christophersen, U. Anthoni, and J. Sorensen. 1999. Viscosinamide, a new cyclic depsipeptide with surfactant and antifungal properties produced by *Pseudomonas fluorescens* DR54. *J. Appl. Microbiol.* 87: 80–90.

Niemann, A., and B. Scheffer. 1999. Polymeric aminoacid- and aminosugar-compounds in two fen soils from northwestern Germany. *Agribiol. Res.* 52: 127–136.

Nishikawa, M., and K. Ogawa. 2002. Distribution of microbes producing antimicrobial epsilon-poly-L-lysine polymers in soil microflora determined by a novel method. *Appl. Environ. Microbial.* 68: 3575–3581.

Owen, A. G., and D. L. Jones. 2001. Competition for amino acids between wheat roots and rhizosphere microorganisms and the role of amino acids in plant N acquisition. *Soil Biol. Biochem.* 33: 651–657.

Padgett, P. E., and R. T. Leonard. 1993. Regulation of nitrate uptake by amino-acids in maize cell-suspension culture and intact roots. *Plant Soil* 156: 159–162.

Padgett, P. E., and R. T. Leonard. 1996. Free amino acid levels and the regulation of nitrate uptake in maize cell suspension cultures. *J. Exp. Bot.* 47: 871–883.

Park, S. W., C. B. Lawrence, J. C. Linden, and J. M. Vivanco. 2002. Isolation and characterization of a novel ribosome-inactivating protein from root cultures of pokeweed and its mechanism of secretion from roots. *Plant Physiol.* 130: 164–178.

Paul, E. A., and F. E. Clark. 1989. *Soil microbiology and biochemistry*. New York: Academic Press.

Phillips, D. A., T. C. Fox, M. D. King, T. V. Bhuvaneswari, and L. R.Teuber. 2004. Microbial products trigger amino acid exudation from plant roots. *Plant Physiol.* 136: 2887–2894.

Phillips, D. A., T. C. Fox, and J. Six. 2006. Root exudation (net efflux of amino acids) may increase rhizodeposition under elevated CO_2. *Global Change Biol.* 12: 561–567.

Qin, C. Q., Y. M. Du, L. T. Zong, F. A. Zeng, Y. Liu, and B. Zhou. 2003. Effect of hemicellulase on the molecular weight and structure of chitosan. *Polym. Degrad. Stab.* 80: 435–441.

Raber, B., I. Kogel-Knabner, C. Stein, and D. Klem. 1998. Partitioning of polycyclic aromatic hydrocarbons to dissolved organic matter from different soils. *Chemosphere* 36: 79–97.

Raynaud, X., J. C. Lata, and P. W. Leadley. 2006. Soil microbial loop and nutrient uptake by plants: A test using a coupled C:N model of plant-microbial interactions. *Plant Soil* 287: 95–116.

Reddy, K. S., M. Singh, A. K. Tripathi, M. Singh, and M. N. Saha. 2003. Changes in amount of organic and inorganic fractions of nitrogen in an Eutrochrept soil after long-term cropping with different fertilizer and organic manure inputs. *J. Plant Nutr. Soil Sci.* 166: 232–238.

Reinhold, L., and A. Kaplan. 1984. Membrane transport of sugars and amino acids. *Annu. Rev. Plant Physiol. Plant Mol. Biol.* 35: 45–83.

Rengel, Z. 2002. Genetic control of root exudation. *Plant Soil* 245: 59–70.

Richardson, A. E., P. A. Hadobas, and J. E. Hayes. 2001. Extracellular secretion of *Aspergillus phytase* from *Arabidopsis* roots enables plants to obtain phosphorus from phytate. *Plant J.* 25: 641–649.

Riffaldi, R., R. Levi-Minzi, A. Saviozzi, and A. Benetti. 1998. Adsorption on soil of dissolved organic carbon from farmyard manure. *Agric. Ecosys. Environ.* 69: 113–119.

Rillig, M. C. 2004. Arbuscular mycorrhizae and terrestrial ecosystem processes. *Ecol. Lett.* 7: 740–754.

Roberts, P., R. Bol, and D. L. Jones. 2007. Free amino sugar reactions in soil in relation to soil carbon and nitrogen cycling. *Soil Biol. Biochem.* 39: 3081–3092.

Robleto, E. A., K. Kmiecik, E. S. Oplinger, J. Nienhuis, and E. W. Triplett. 1998. Trifolitoxin production increases nodulation competitiveness of *Rhizobium etli* CE3 under agricultural conditions. *Appl. Environ. Microbiol.* 64: 2630–2633.

Rosier, C. L., A. T. Hoye, and M. C. Rillig. 2006. Glomalin-related soil protein: Assessment of current detection and quantification tools. *Soil Biol. Biochem.* 38: 2205–2211.

Rovira, A. D. 1969. Plant root exudates. *Bot. Rev.* 35: 35–57.

Ruiz, N., and T. J. Silhavy. 2005. Sensing external stress: Watchdogs of the *Escherichia coli* cell envelope. *Curr. Op. Microbiol.* 8: 122–126.

Sarmah, A. K., K. Muller, and R. Ahmad. 2004. Fate and behaviour of pesticides in the agro-ecosystem: A review with a New Zealand perspective. *Aust. J. Soil Res.* 42: 125–154.

Sasagawa, Y., Y. Kamio, Y. Matsubara, Y. Matsubara, K. Suzuki, H. Kojima, and K. Izaki. 1993. Purification and properties of collagenase from *Cytophaga* sp. 143-1 strain. *Biosci. Biotechnol. Biochem.* 57: 1894–1989.

Saviozzi, A., R. Cardelli, P. N'kou, R. Levi-Minzi, and R. Riffaldi. 2006. Soil biological activity as influenced by green waste compost and cattle manure. *Compost Sci. Util.* 14: 54–58.

Schobert, C., and E. Komor. 1987. Amino-acid-uptake by *Ricinus-communis* roots: Characterization and physiological significance. *Plant Cell Environ.* 10: 493–500.

Schulze, W. X. 2005. Protein analysis in dissolved organic matter: What proteins from organic debris, soil leachate and surface water can tell us—a perspective. *Biogeosciences* 2: 75–86.

Schulze, W., W. B. Frommer, and J. M. Ward. 1999. Transporters for ammonium, amino acids and peptides are expressed in pitchers of the carnivorous plant *Nepenthes. Plant J.* 17: 637–646.

Shah, S. B., M. L. Wolfe, and J. T. Borggaard. 2004. Simulating the fate of subsurface-banded urea. *Nutr. Cycl. Agroecosys.* 70: 47–66.

Shand, C. A., B. L. Williams, L. A. Dawson, S. Smith, and M. E. Young. 2002. Sheep urine affects soil solution nutrient composition and roots: differences between field and sward box soils and the effects of synthetic and natural sheep urine. *Soil Biol. Biochem.* 34: 163–171.

Shen, Y. H. 1999. Sorption of natural dissolved organic matter on soil. *Chemosphere* 38: 1505–1515.

Siemens, J., and M. Kaupenjohann. 2002. Contribution of dissolved organic nitrogen to N leaching from four German agricultural soils. *J. Plant Nutr. Soil Sci.* 165: 675–681.

Soldal, T., and P. Nissen. 1978. Multiphasic uptake of amino acids by barley roots. *Physiol. Plant.* 43: 181–188.

Song, W., H. Y. Steiner, L. Zhang, F. Naider, G. Stacey, and J. M. Becker. 1996. Cloning of a second *Arabidopsis* peptide transport gene. *Plant Physiol.* 110: 171–178.

Stano, J., P. Kovacs, I. Safarik, D. Kakoniova, and M. Safarikova. 1998. A simple procedure for the detection of plant extracellular proteolytic enzymes. *Biol. Plant.* 40: 475–477.

Steiner, H. Y., W. Song, L. Zhang, F. Naider, J. M. Becker, and G. Stacey. 1994. An *Arabidopsis* peptide transporter is a member of a new class of membrane-transport proteins. *Plant Cell* 6: 1289–1299.

Stephenson, P., and J. Hogan. 2006. Cloning and characterization of a ribonuclease, a cysteine proteinase, and an aspartic proteinase from pitchers of the carnivorous plant *Nepenthes ventricosa blanco. Int. J. Plant Sci.* 167: 239–248.

Stevenson, F. J. 1982. *Nitrogen in agricultural soils*. Madison, WI: American Society of Agronomy.

Stewart, W. D. 1966. *Nitrogen fixation in plants*. New York: Oxford University Press.

Tan, X. W., H. Ikeda, and M. Oda. 2000. Effects of nickel concentration in the nutrient solution on the nitrogen assimilation and growth of tomato seedlings in hydroponic culture supplied with urea or nitrate as the sole nitrogen source. *Scientia Hort.* 84: 265–273.

Tang, S. L., and D. H. Howard. 1973. Uptake and utilization of glutamic acid by *Cryptococcus albidus*. *J. Bacteriol.* 115: 98–106.

Taranto, M. T., M. A. Adams, and P. J. Polglase. 2000. Sequential fractionation and characterisation (^{31}P-NMR) of phosphorus-amended soils in *Banksia integrifolia* (L.f.) woodland and adjacent pasture. *Soil Biol. Biochem.* 32: 169–177.

Thornton, B. 2001. Uptake of glycine by non-mycorrhizal *Lolium perenne*. *J. Exp. Bot.* 52: 1315–1322.

Thuler, D. S., E. I. S. Floh, W. Handro, and H. R. Barbosa. 2003. *Beijerinckia derxii* releases plant growth regulators and amino acids in synthetic media independent of nitrogenase activity. *J. Appl. Microbiol.* 95: 799–806.

Tiedje, J. M., S. Asuming-Brempong, K. Nusslein, T. L. Marsh, and S. J. Flynn. 1999. Opening the black box of soil microbial diversity. *Appl. Soil Ecol.* 13: 109–122.

Tillard, P., L. Passama, and A. Gojon. 1998. Are phloem amino acids involved in the shoot to root control of NO$_3^-$ uptake in *Ricinus communis* plants? *J. Exp. Bot.* 49: 1371–1379.

Tipping E, C. Woof, E. Rigg, A. F. Harrison, P. Ineson, K. Taylor, D. Benham, J. Poskitt, A. P. Rowland, R. Bol, and D. D. Harkness. 1999. Climatic influences on the leaching of dissolved organic matter from upland U.K. moorland soils, investigated by a field manipulation experiment. *Environ. Int.* 25: 83–95.

Turner, B. L., E. Frossard, and D. S. Baldwin. 2005. *Organic phosphorus in the environment*. Wallingford, U.K.: CAB International.

Umbarger, H. E. 1978. Amino-acid biosynthesis and its regulation. *Annu. Rev. Biochem.* 47: 533–606.

Umbarger, H. E. 1981. Regulation of amino acid metabolism. In *Amino Acid and Sulphur Metabolism*, ed. A. Neuberger and L. L. M. van Deenen. New York: Elsevier.

van Breemen, N. 2002. Natural organic tendency. *Nature* 415: 381–382.

van Hees, P. A. W., D. L. Jones, R. Finlay, D. L. Godbold, and U. S. Lundström. 2005. The carbon we do not see: The impact of low molecular weight compounds on carbon dynamics and respiration in forest soils—a review. *Soil Biol. Biochem.* 37: 1–13.

Vierstra, R. D. 1996. Proteolysis in plants: Mechanisms and functions. *Plant Mol. Biol.* 32: 275–302.

Vieublé Gonod, L., D. L. Jones, and C. Chenu. 2006. Sorption regulates the fate of the amino acids lysine and leucine in soil aggregates. *Eur. J. Soil Sci.* 57: 320–329.

Vinolas, L. C., J. R. Healey, and D. L. Jones. 2001. Kinetics of soil microbial uptake of free amino acids. *Biol. Fert. Soils.* 33: 67–74.

Walker, J. R., and E. Altman. 2005. Biotinylation facilitates the uptake of large peptides by *Escherichia coli* and other gram-negative bacteria. *Appl. Environ. Microbiol.* 71: 1850–1855.

Wang, P., S. P. Bi, S. Wang, and Q. Y. Ding. 2006. Variation of wheat root exudates under aluminum stress. *J. Agric. Food Chem.* 54: 10040–10046.

Waterworth, W. M., and C. M. Bray. 2006. Enigma variations for peptides and their transporters in higher plants. *Ann. Bot.* 98: 1–8.

Watson, C. J., and H. Miller. 1996. Short-term effects of urea amended with the urease inhibitor N-(n-butyl) thiophosphoric triamide on perennial ryegrass. *Plant Soil* 184: 33–45.

Watson, R., and L. Fowden. 1975. The uptake of phenylalanine and tyrosine by seedling root tips. *Phytochemistry* 14: 1181–1186.

Whiffen, L. K., D. J. Midgley, and P. A. McGee. 2007. Polyphenolic compounds interfere with quantification of protein in soil extracts using the Bradford method. *Soil Biol. Biochem.* 39: 691–694.

Willett, V. B., B. A. Reynolds, P. A. Stevens, S. J. Ormerod, and D. L. Jones. 2004. Dissolved organic nitrogen regulation in freshwaters. *J. Environ. Qual.* 33: 201–209.

Williams, L. E., and A. J. Miller. 2001. Transporters responsible for the uptake and partitioning of nitrogenous solutes. *Annu. Rev. Plant Physiol. Plant Mol. Biol.* 52: 659–688.

Winton, K., and J. B. Weber. 1996. A review of field lysimeter studies to describe the environmental fate of pesticides. *Weed Technol.* 10: 202–209.

Wright, D. E. 1962. Amino acid uptake by plant roots. *Arch. Biochem. Biophys.* 97: 174–180.

Wrobel, R., and B. L. Jones. 1992. Appearance of endoproteolytic enzymes during the germination of barley. *Plant Physiol.* 100: 1508–1516.

Xu, Y. C., Q. R. Shen, and W. Ran. 2003. Content and distribution of forms of organic N in soil and particle size fractions after long-term fertilization. *Chemosphere* 50: 739–745.

Yoshida, S., S. Hiradate, T. Tsukamoto, K. Hatakeda, and A. Shirata. 2001. Antimicrobial activity of culture filtrate of *Bacillus amyloliquefaciens* RC-2 isolated from mulberry leaves. *Phytopathology* 91: 181–187.

Young, C. C., and L. F. Chen. 1997. Polyamines in humic acid and their effect on radical growth of lettuce seedlings. *Plant Soil* 195: 143–149.

Yu, Z., Q. Zhang, T. E. C. Kraus, R. A. Dahlgren, C. Anastasio, and R. J. Zasoski. 2002. Contribution of amino compounds to dissolved organic nitrogen in forest soils. *Biogeochemistry* 61: 173–198.

Zalokar, M. 1961. Kinetics of amino acid uptake and protein synthesis in *Neurospora*. *Biochim. Biophys. Acta* 46: 423–432.

Zeid, I. M., and Z. A. Shedeed. 2006. Response of alfalfa to putrescine treatment under drought stress. *Biol. Plant.* 50: 635–640.

Zhu, H., Bilgin, M., and Snyder, M. 2003. Proteomics. *Annu. Rev. Biochem.* 72: 783–812.

6 Water and Nitrogen Uptake and Responses in Models of Wheat, Potatoes, and Maize

P. D. Jamieson, R. F. Zyskowski,
F. Y. Li, and M. A. Semenov

CONTENTS

6.1 Introduction 127
6.2 The Models 129
6.3 Phenology 130
6.4 Canopy Development 131
6.5 Biomass Accumulation and Partitioning 132
6.6 Nitrogen Dynamics 133
6.7 Water Use and Responses to Water Shortage 135
6.8 Roots and Soil Properties 136
6.9 Interactions between Shortages of N and Water 137
6.10 Experimental Verification 137
6.11 Concluding Remarks 141
Acknowledgments 143
References 143

6.1 INTRODUCTION

Much of the effort of early crop simulation models was aimed at predicting the timing of phenological events, the accumulation of biomass, and its partitioning to harvest in the absence of major stresses (e.g., Weir et al. 1984). Restrictions in growth due to shortages of water or nitrogen (N) were added to these models via reduction factors based on the ratios of supply to demand (water) or current whole-plant concentration to a supposed optimum concentration of N (e.g., Porter 1993; Ritchie and Otter 1985). The ratios were used to calculate stress indices for water and N that were applied in a very similar fashion. Many current models carry these as part of their history, and their users apply them uncritically.

Plants use water and N quite differently. Although water is an essential and substantial part of growing plant tissue, most of the water drawn up from soil is lost to the atmosphere on the same day and may represent substantially more mass per unit area than the standing dry biomass of the crop. For example, a crop of standing biomass 5 t ha^{-1} may transpire 50 t ha^{-1} of water in a single day. The same crop at harvest may have a standing biomass of around 20 t ha^{-1}, but may have transpired around 5000 t ha^{-1} of water between emergence and maturity. In contrast, most N taken up by a crop is retained as an active part of photosynthetic tissue, as part of the plant structure, and in labile storage. In comparison to water, the amounts are tiny: the 20-t ha^{-1} mature crop will have taken up and retained only about 0.2–0.4 t ha^{-1} of N. Water is the transport medium for nutrients and assimilation products in crops, and is the major part of their cooling system. Nitrogen is one of the nutrients that gets transported but is also a vital part of the system that gathers light to turn CO_2 into biomass. The mechanism of response to water shortage is that stomata close to conserve water; this restricts CO_2 uptake, and growth slows. Ultimately, the crop will shed the energy load by leaf curling and changed leaf angle, by shedding leaves, and then by desiccation. In contrast, shortage of N will be reflected by the inability of the crop to generate green area, by the diminution of RuBisCO (ribulose-1,5-bisphosphate carboxylase/oxygenase) in leaves, and finally by the premature loss of green area as N is remobilized for use in other tissue. So growth is slowed primarily because of reduced light interception, and then (Grindley 1997) by reduced light-use efficiency (LUE). Given the differing roles of water and N in plants, it is unlikely that respective shortages will affect plants in the same way.

The calculation of water demand as a function of solar radiation, temperature, etc., is straightforward, well developed, and tested (Jamieson 1999). It involves good physical theory and a minimum of empiricism. There tends to be more empiricism in calculating maximum supply rate because, even though the physics is well understood (Jamieson and Ewert 1999), the complexity of root penetration and distribution makes this a less tractable problem. A stress-index approach based on a supply–demand ratio is appropriate and widely applied (Jamieson 1999; Porter 1993; Ritchie and Otter 1985). The major modeling challenge is the description of the interacting soil and root processes that affect the maximum rate at which water can be supplied. There are minor empiricisms that have to do with the mismatch of model time steps with the timescale of evaporative processes, but the main empiricisms are the ones that relate to the effect of the shortage of water on growth processes, i.e., how stress indices are calculated from the supply–demand ratio to affect leaf area expansion and senescence, or photosynthetic processes.

In contrast, the calculation of the "critical" N concentration (N_c) that maximizes production at any time is entirely empirical. N_c is not constant; it decreases non-linearly as the crop grows (Greenwood et al. 1990). A great deal of effort has gone into establishing nitrogen dilution curves for a range of crops (Colnenne et al. 1998; Plénet and Lemaire 1999). Any model that uses the curve then needs to define an upper and lower limit of N concentration to account for luxury consumption or N shortage, and to define a concentration ratio that can be used to calculate growth limitations. The model may look quite simple, but its great disadvantage is that it is entirely empirical and the reasons for "stress" effects are not clear.

For wheat, Jamieson and Semenov (2000) addressed the questions above by dividing crop biomass into three compartments. These were:

1. Structural, characterized by low N concentration
2. Green area, characterized by high N concentration assumed to be constant when expressed per unit area
3. Storage—that component associated with luxury N consumption

When expressed this way, the reason for the decline in N concentration with increasing biomass becomes clear. At the beginning, most tissue is in leaves and has high N concentration. As the crop grows, an increasing proportion is low N structural material such as enclosed pseudostem and true stem. The model can be used to *derive* a critical N curve, i.e., when N supply is just sufficient to support structure and green area with no N in storage. Luxury N consumption is allowed for as storage, in proportion to the biomass, and the upper limit is reached when storage is full. When N is in short supply, the main effect is either the reduction in expansion of green area (leading to a change in the ratio of high to low N tissue) or premature senescence to provide N for structure or growing grain. Thus, N shortages restrict growth through effects on light interception. The only empirical requirements are estimates of specific leaf N, the concentration of N in structural tissue, and a measure of the N storage capacity of tissue and grain. These can all be measured directly, and none requires time-dependent data to be fitted. Nitrogen stress factors are unnecessary. Potentially, it is possible to account for N dilution in green tissue through its effect on LUE (Muchow and Sinclair 1994), but in practice Jamieson and Semenov (2000) found this an unnecessary complication. More recently, Sinclair et al. (2003) have shown a model for soybean using a similar approach for N dynamics that better explains soybean growth and yield than earlier models, and Jamieson et al. (2003) used the same principles in creating a potato model.

In this chapter we briefly describe models of wheat, potato, and maize based on the outline above. We include measures of performance based on comparisons of simulations with experimental data from research sites and in farm fields.

6.2 THE MODELS

The three models are the Sirius wheat model, as described by Jamieson and Semenov (2000), the potato model described by Jamieson et al. (2003), and a recently implemented version of the Muchow–Sinclair maize model (Muchow et al. 1990) as modified by Wilson et al. (1995) for cool climates, but expanded to include the crop N economy. All three models share code for calculating evapotranspiration, soil water and nitrate percolation, and an aggregated N mineralization model taken from the NITCROS model of Hansen and Aslyng (1984). They require daily inputs of solar radiation, precipitation, maximum and minimum temperature, in addition to a soil physical description at least to the maximum root depth. The models can be represented as

$$Y = \int_{em}^{t} \xi A Q \, dt \tag{6.1}$$

where Y is the yield at time t, A is the LUE, Q is the amount of light intercepted, *em* is the time of crop emergence, and ξ is a partitioning coefficient that determines how much biomass is assigned to the economically important pool at any time, i.e., grains in the case of wheat and maize, and tubers for potatoes. Each model contains submodels for phenology (so that timing of important events can be simulated), canopy development and light interception, LUE, and partitioning of biomass. In all three models, the carbon economy and biomass accumulation is controlled by the N economy. Nitrogen is the focus of this chapter, and most attention will be given to the models of N economy. However, brief descriptions of other parts of the models will be given as a necessary context.

6.3 PHENOLOGY

Phenological development determines the duration of the crop in real time. Because most processes are temperature dependent, calculations are made in thermal time—the time integral of a temperature-response function above a base temperature. Because plants are made up of phytomers, ultimately the duration of a crop from emergence to maturity will depend on the rate of phytomer production, the number of phytomers produced on a main stem, and how long the senescence process takes.

The simplest phenological description is for potatoes. The major assumption is that the thermal time (base 0°C) from emergence to complete senescence is constant, provided there is no shortage of water and N. The only event of note is tuber initiation, until which time all biomass is assigned to haulm and leaf growth and to maximum canopy size, and after which all new biomass is assigned to tubers. For the cultivar Russet Burbank, the thermal lifetime is 2000°C days from emergence, tuber initiation is 300°C days after emergence, and maximum canopy size occurs midway through the crop's lifetime. Thermal time from planting to emergence is also assumed constant at 300°C days, although we recognize that this may be influenced by the physiological age of seed tubers (Jamieson et al. 2006).

The major influence on the thermal duration of a maize crop is the final leaf number (FLN). Our implementation of the model assumes that FLN is fixed for any cultivar and is uninfluenced by day length or any other factor. After the final leaf has appeared, then there are cultivar-specific constant thermal time intervals from last leaf to silking, and from silking to start of grain fill. Thermal time from sowing to emergence is also constant. During the leaf expansion phase, thermal time is calculated from a segmented linear model of thermal time accumulation from 0 to 34°C, with a change at 18°C (Wilson et al. 1995). Above 18°C, the model is identical to the original Muchow–Sinclair model, where thermal time is calculated over a base temperature of 8°C. The number of fully expanded leaves is calculated from an exponential function of thermal time from emergence (Wilson et al. 1995). Shortages of N and water can accelerate maturity by reducing the duration of grain filling. Base temperature for the post-silking phases is set at 0°C after Muchow et al. (1990).

The Sirius wheat model also uses main-stem FLN as the basis for its phenology, with fixed thermal time intervals (base temperature 0°C) from sowing to emergence, flag leaf ligule appearance to anthesis, anthesis to the beginning of linear grain fill, and grain fill duration. Wheat phenology, and therefore the model, is more complex than that of the other two crops, because wheat final leaf number responds to both day length (Brooking at al. 1995) and vernalization (Brooking and Jamieson 2002). A more complete discussion is given by Jamieson et al. (1998a).

6.4 CANOPY DEVELOPMENT

The canopy is the main engine that drives crop growth; it intercepts light and uses it to generate biomass. In this section we describe the potential canopy models, unmodified by shortages of N and water. They thus provide the unstressed envelope of canopy development, i.e., they provide the maximum simulated canopy size in response to temperature and resource allocation in the absence of water stress and with a plentiful N supply. Effects of water and N limitations are dealt with later. Canopy green area depends on the addition of new green area (new leaves and exposed sheaths or stem) and the senescence of older leaves. These are handled in different ways in the three models.

The most highly aggregated canopy model is that for potatoes. As noted above in the phenology section, canopy duration defines crop growth duration. The canopy expands for half the thermal time, and senesces for the other half. The implication is that new leaf production stops about half way through the thermal lifetime of the canopy, and thereafter only senescence is occurring. This is supported by an observed potential leaf lifetime of up to 70 days (Vos and Oyarzún 1987), representing about half the crop lifetime. Additionally, we have observed that, from the canopy midpoint, few and only small leaves are produced. Hence the canopy will stop expanding when the last leaves are produced, and will have senesced completely when they die. Both expansion and senescence are linear in thermal time. Only two canopy parameters are required: thermal duration and peak green area index (GAI). A major simplification is that no effects of population are simulated; the population is simply assumed to be "sufficient." This was based on the observation that potato population could be halved from a supposed optimum without affecting yield. This occurs because the branching nature of the potato plant allows the canopy to grow into gaps, so yield is unaffected even though tuber numbers are.

The maize canopy model is based closely on the work of Dwyer and Stewart (1986) and Muchow and Carberry (1989). The fully expanded area of each leaf is calculated as a function of the leaf number and the area of the largest leaf. Muchow and Carberry (1989) also observed that the area of all of the leaves expanding on the plant at any time was equal to the fully expanded area of the two leaves immediately above the last fully expanded leaf. So potential green area per plant is the sum of the areas of the fully expanded leaves plus the fully expanded area of the next two leaves, until the expansion of the final leaf. Green area loss is accounted for by a calculation of the fraction of area that has senesced, expressed as a function of thermal time from emergence. After the full expansion of the last leaf, only senescence occurs. GAI is calculated by assuming a standard population. Implicit in that is an assumption that area per plant and cob size will change to accommodate changes in population.

The Sirius model exists in three versions. Results shown here are from Sirius 2000, developed from Jamieson and Semenov (2000), where canopy area is highly aggregated. GAI expands as an exponential function of thermal time to an upper limit, stays constant until anthesis, and then declines as a quadratic function of thermal time from anthesis. The underlying assumption is that, from maximum GAI until anthesis, green area gains and losses are equal, and GAI declines post-anthesis because no new green area is formed. A more detailed version was developed by Lawless et al. (2005), which specified the area of leaf layers associated with main-stem leaves. That meant that leaf layers could senesce in the order they appeared as a function of their age. This canopy description was incorporated into the SiriusQuality model by Martre et al. (2003, 2006) in a version that explicitly simulates the accumulation of classes of proteins during grain growth.

6.5   BIOMASS ACCUMULATION AND PARTITIONING

All three models accumulate biomass using the concept of LUE (Monteith 1977). In both Sirius and the potato model, shoot biomass accumulation is calculated as the product of intercepted photosynthetically active radiation (PAR) and LUE, incremented daily. The LUE is calculated as a function of the balance of direct to diffuse radiation using approaches suggested by Sinclair et al. (1992) and Hammer and Wright (1994). In practical terms, this gives a mean LUE of 2.2 g MJ^{-1} for sites in the United States, Australia, and New Zealand, whereas in Europe and the United Kingdom, mean LUE is close to 2.5 g MJ^{-1} because skies are cloudier. LUE is also affected by temperature, with the peak occurring at 20°C and falling as the square of the difference in temperature from 20°C to be zero at 0°C and 40°C. In the maize model, LUE has a maximum of 3.3 g MJ^{-1} PAR when temperature is above 16°C, is zero below 8°C, and increases linearly between 8 and 16°C, above which LUE is constant. In extreme water stress, LUE is reduced (Jamieson et al. 1998c). Light interception is related to green area index (GAI) via Beer's law, with an extinction coefficient of 0.45 (Jamieson et al. 1998c) for wheat, 0.6 for potatoes (Jamieson et al. 2003), and 0.4 for maize (Muchow et al. 1990). Variations in N supply have no effect on the LUE because specific leaf N is fixed.

Biomass partitioning is via simple partitioning rules, partly linked to phenology. All three models assume a fixed specific leaf mass (mass per unit leaf area). Excess biomass is assigned to a "stem" pool, described both by biomass and N content. During stem growth, stem biomass can increase only; in the case where a biomass increment is insufficient to grow the required green area at the required specific leaf mass, then average specific leaf mass is allowed to be smaller (thinner leaves). The rules for potato are slightly different, in that stem biomass decreases from maximum canopy mass as stem carbohydrates are remobilized to tubers.

The last part of partitioning is to the harvest organs. For wheat and maize, either vegetative tissue or grain is growing at any time; vegetative growth stops when grain growth commences. The partitioning rules are a little different for the two models. In Sirius, all new biomass after the beginning of grain filling is assigned to grain, plus a proportion (25%) of the shoot biomass existing at anthesis. The latter portion is transferred linearly in thermal time so that it has all been translocated by physiological maturity. In the maize model, the rate of change of harvest index is fixed in time, and grain filling stops when the harvest index gets to 50%, unless stopped earlier because of water stress or canopy loss through reallocation of N.

In the potato model, all biomass is initially assigned to the shoots. From tuber initiation until maximum canopy, 75% of new biomass is assigned to tubers and 25% to the canopy. After maximum canopy, all new biomass is assigned to tubers, plus a transfer of biomass from stems to tubers, linear in thermal time so that most stem biomass is in the tubers at maturity.

None of the models assigns biomass or nitrogen into the fibrous root systems. This means that the LUE values take account of the slightly reduced biomass requirements, and root N is considered to be part of the unavailable soil N (Jamieson and Semenov 2000).

6.6 NITROGEN DYNAMICS

The three models divide plant N into four pools. Structural N is that associated with cell wall construction and is assumed to be immobile and a constant proportion of the biomass during growth, although the N content is allowed to decrease further during senescence in wheat (Table 6.1). Active N is associated with green area. The main simplification is that specific leaf N concentration, i.e., expressed as mass of N per unit green area, is assumed to be constant, so that any changes in the amount of active N are expressed through changes in green area (Grindley 1997). There is a storage pool for N, associated with "stem" biomass. The quotation marks are used because the exact location is both unidentified and unimportant within the context of the way the models work. The excess N may be stored in leaves as nitrate (Vos and van der Putten 1998) or as RuBisCO in excess of concentrations required to maximize photosynthesis (Irving and Robinson 2006). Finally, there is N that is translocated into the harvested organs: grains and tubers. Model parameters describing the distribution of N in the crops are given in Table 6.1.

The nitrogen dynamics of the plant can be divided into two main phases. The first of these we term the N-loading phase, when vegetative tissue is taking up N. The second phase is N-unloading, when vegetative tissue gives up much of its N content

---

TABLE 6.1

Nitrogen Contents of the Pools in the Three Models

	Wheat	Maize	Potato
Specific leaf N content (g m^{-2})	1.5	1.5	2.0
Structural N content[a] (g kg^{-1})	5.0	5.0	5.0
Stem N storage range (g kg^{-1})	0–10	0–10	0–40
Harvested organ N content[b] (g kg^{-1})	15–30	15	8–20

[a] This is represented by the minimum stem N content. In wheat, the value reduces to 0.3 g kg^{-1} after anthesis.

[b] In extreme circumstances, wheat grain N content may exceed these values.

to the harvest organs. To some extent, the phases may overlap, with soil N uptake occurring while vegetative tissue is unloading to grain or tubers.

During the N-loading phase, all three models work in similar ways. Demand for N is set to satisfy the need for N in structural components (cell walls), green area, and storage. There is a hierarchy of allocation, in the order given. In the absence of sufficient supply from the soil, the stored N is reallocated first to satisfy structural requirements, then green area. Once N storage is exhausted, N is removed from green area to satisfy structural needs. Because of the assumption that specific leaf N content is constant, N withdrawn from green area results in a reduction in GAI. An additional N pool is involved in potatoes once tubers start to grow. After tuber initiation, the minimum N concentration of tubers must also be met, and that sits in the hierarchy of allocation just above structural N, but is similar in that transport of N to tubers is one-way. When N is plentiful, extra N is allocated to tubers to increase the N content up to a maximum level, which means that as soon as N supply becomes limited, tuber growth can continue with no extra N input until the minimum N concentration is reached. No explicit reallocation of N within tubers is simulated.

In maize and wheat, vegetative and grain growth are very nearly disjoint. The crop may be treated as if it is either growing vegetative tissue or it is growing grain, which means that simulating the N-unloading process is straightforward. Nitrogen loading ceases when grain begins to grow, and N unloading commences. In potatoes, tuber growth is really part of the vegetative growth process, and for a substantial fraction of the life of the crop, tubers and canopy are growing concurrently. In addition, tuber biomass substantially exceeds the biomass of the canopy, and the harvest index approaches unity. That means that these processes are handled slightly differently in detail in the potato model during both the N-loading and -unloading phases.

A major difference between the models described here and other models is that, in the case of maize and wheat, transport of N to the harvested organs is source regulated (Jamieson and Semenov 2000; Martre et al. 2003). Any system for moving mass among locations requires a source, a sink, and a transport system. Earlier models such as CERES-Wheat and AFRCWHEAT2 assumed implicitly that N had to be transported from the shoots to satisfy the demand of grains, which implies that the transport system is controlled by the sink, through the product of grain number and the demand of individual grains. Because these models transfer carbon to the grain using similar rules, there is little room for simulating variations in grain protein content, except through a scheme that preloads N into grain (Asseng and Milroy 2006) by assigning a high N concentration to grain at its first growth increment. In contrast, Sirius assigns all new biomass and a proportion of stored biomass to the grain once grain growth commences. New biomass is created in chronological time, and retranslocation of stored assimilate is done in thermal time. That means that grain biomass accumulation is only partly temperature controlled, whereas transport of N occurs in thermal time and thus is wholly temperature controlled. Therefore, variations in temperature and irradiation result in variations in grain protein content. For instance, hot conditions have little effect on grain growth rate, but do shorten the duration, reducing yield. In contrast, the same amount of N is transported regardless of temperature, so that in hot conditions the protein concentration of the wheat will be high. This accords with observation (Triboi and Triboi-Blondel 2002; Triboi et al. 2006). The maize model is a little different in detail, but similar in principle.

In the potato model the N content of tubers is treated in very much the same way as structural N, and has a high priority. The large biomass of the tubers (10–20 t ha^{-1}), compared with that of the canopy (2–5 t ha^{-1}), and the high harvest index (>90%) of tubers mean that the sink plays a substantial role in controlling transport. This is also partly due to the fact that tubers are growing through cell division as well as cell expansion, and cell walls need protein, and the demand for N is controlled at least partly by the increasing number of cells—a measure of sink strength. The main effect of having insufficient N storage is the premature destruction of the canopy (i.e., accelerated senescence) to satisfy the N requirements of the still-growing tubers.

An important consequence of the way N is handled in these models is that the concept of N stress is unnecessary and not used. When N is short during the loading phase, then storage pools cannot be filled and green area cannot be expanded. If N is short during the unloading phase, then storage pools are emptied by reallocation and green area is lost prematurely. Because specific leaf N is constant, there are no effects of N shortages on LUE; these are all expressed through the loss of green area and hence light interception.

6.7 WATER USE AND RESPONSES TO WATER SHORTAGE

The three models use the Ritchie (1972) approach for estimating daily evapotranspiration, with separate calculations of transpiration and soil evaporation using the current GAI and potential evapotranspiration. Potential evapotranspiration is calculated according to Penman (1948; as formulated by French and Legg 1979) or Priestley and Taylor (1972), depending on whether wind-run and vapor-pressure information are available. Monteith (1965) supplied a more complete description that takes account of stomatal function. However, Jamieson et al. (1995b) showed, using a Taylor expansion, that the Ritchie (1972) formulation is equivalent to that of Monteith (1965) to a reasonable degree of approximation. Restrictions to transpiration occur when the crop root–soil combination is unable to supply water at the rate set by the available energy (Ritchie 1972). The daily supply rate is set as a fraction of the root-zone-available water in each layer (Dardanelli et al. 2004). Daily evapotranspiration is the minimum of the energy- or supply-limited rate. None of these models explicitly deals with the dynamics of water potential gradients or stomatal function, partly because the one-day time step does not match the timescale of these phenomena. Jamieson (1999) noted that, in some senses, the details are not important, because they are a response to the physics. Plants must have a control system to limit water loss when moisture cannot be supplied as fast as the energy supply demands; otherwise they will desiccate and die. Death before reproduction has no evolutionary reward.

There are inevitable consequences for growth when transpiration is restricted. Water egress and CO_2 ingress share much of the same pathway, so control of one means control of the other. Indeed, one interpretation of the thesis put forth by Tanner and Sinclair (1983) is that growth processes control stomatal resistance to keep the ratio of internal to atmospheric CO_2 concentration constant. Whatever the precise mechanism, energy and water balance considerations mean that plants will reduce their water content and heat up in response to water shortages. These are symptoms of stress.

The existence and evidence of stress makes use of stress indices appropriate for modeling drought response. Stress leads first to a rise in canopy temperature, followed by partial dehydration, sometimes leaf curling, and evidence of wilting. Longer term effects are reduction in canopy size, accelerated senescence, and reduced LUE, all leading to reduced growth rates. In the three models, the ratio of supply rate to demand rate (F_W), constrained to a maximum of unity, is used as the basis for calculating water stress indices that are used to reduce the rate of canopy expansion, accelerate senescence, reduce the LUE, and change the partitioning. In the three models here, water stress indices that affect the canopy are the most severe. The simplest in this regard is Sirius, where GAI is calculated as a function of thermal time in four phases: exponential expansion, linear expansion, constant, and ontogenetic decline (Brooking et al. 1995; Jamieson et al. 1995a). A single factor must therefore slow leaf expansion and accelerate senescence. During the leaf "expansion" phases, the potential increment of GAI each day is reduced by a factor

$$F_L = 1.5F_W - 0.5 \qquad (6.2)$$

The restriction of F_W to values between 0 (maximum stress) and 1 means that F_L has values between −0.5 and 1. Therefore, during severe drought, the canopy can contract during the "expansion" phase. Simultaneously, the maximum value the GAI can attain is reduced by the difference between the actual and potential GAI increments. At the end of the leaf expansion phase, GAI is affected by senescence only. A factor F_{LKLR} is used to remove leaf area during the subsequent phase of constant GAI and to accelerate the rate of GAI decline during senescence. F_{LKLR} is defined to have a minimum value of 1 when F_W is 0.7 or greater and to increase linearly to a maximum value of 1.5 as F_W declines to 0.4. At this level of stress, GAI is removed 50% faster than for no stress. During phase 3 ("constant" GAI), GAI may be reduced at a maximum absolute rate of 0.04 per day. In phase 4 (*post-anthesis*), the accumulation of thermal time for GAI calculation is increased by F_{LKLR} to accelerate senescence (Jamieson et al. 1998b). The stress factor for LUE is set at $2F_W$, again constrained to values between 0 and 1, so that stress must be more severe to affect LUE than GAI. For instance, F_W must be less than 0.5 to reduce LUE, but at that value, F_L is 0.25. During ear growth, at the end of which grain number is calculated, partitioning of biomass to the ear structure is reduced by the same factor as LUE (Jamieson et al. 1998c).

Similar, though not identical, functions are applied in the potato and maize models.

6.8   ROOTS AND SOIL PROPERTIES

Soil water and N transport are based on the model of Addiscott and Whitmore (1991). The soil is divided into 5-cm layers with specific properties. Values for moisture contents at saturation (θ_{sat}), the drained upper limit (θ_{dul}), and lower limit of extraction (θ_{ll}) are specified. Any water above θ_{dul} is mobile, and a proportion (k_q) drains to the next layer each day. Any water above θ_{ll} is available for extraction. Water below θ_{ll} is unavailable, but may have N dissolved in it. An increment of N is added to the soil each day from mineralization (Hansen and Aslyng 1984), calculated from total N content of the top 30 cm of soil, soil moisture content, and temperature. Denitrification is characterized as a single pulse loss just after fertilizer is applied, proportional to the amount of applied fertilizer N.

The only root system property simulated in the profile is depth. Roots extend at a fixed rate in thermal time until a particular phenological event (anthesis for wheat) or a particular depth is reached (700 mm for Russet Burbank potatoes). Resources within the root zone are then available to the crop. Uptake rate is limited to 10% of the water above the lower limit of extraction in any layer in any day (Dardanelli et al. 2004). A maximum of 5 kg N ha^{-1} day^{-1} as nitrate (NO_3^-) or ammonium (NH_4^+) may be extracted from the root zone. This limit was arbitrarily applied based on Sinclair and Amir (1992). In practice, the limit is seldom reached in wheat crops (P. Martre, personal communication). Otherwise, N uptake is independent of moisture content, but N may be removed only from the available water fraction.

6.9 INTERACTIONS BETWEEN SHORTAGES OF N AND WATER

There is no explicit interaction between shortages of N and soil water built into the models. The effects are assumed to be independent, and N uptake is assumed to be active, i.e., it has aspects of both mass flow and diffusion. However, interactions occur because the independent processes affect each other. For instance, water stress influences canopy expansion and accelerates senescence. The canopy is a major pool for N, so during water stress, N uptake is reduced because demand is reduced. That applies to other N-containing biomass pools as well. Any reduction in biomass accumulation affects the size of the N pool and therefore the demand. There are supply-side interactions as well. Nitrogen cannot be removed from dry soil. Conversely, shortages of N will affect the demand for water when energy interception is reduced because of the inability to produce GAI or through premature senescence.

6.10 EXPERIMENTAL VERIFICATION

The most mature of the models is Sirius, which has been extensively tested in many environments, and some of these tests have been published (Jamieson and Semenov 2000; Martre et al. 2006). The comparisons shown in Jamieson and Semenov (2000) were limited to yields between about 4 and 11 t ha^{-1}. Recently, the range was extended to include comparisons with data between 1 and 4 t ha^{-1} from western Australia in the low-rainfall wheat belt (Bowden et al. 1999). The comparison is consolidated in Figure 6.1. Jamieson and Semenov (2000) also showed that simulations based on the assumption that N transfer to grain is entirely source-regulated provided much more accurate estimates of grain protein content than those based on assumptions that control was from the sink. Based on the former assumption, Sirius is able to simulate grain protein content over a wide range (Armour et al. 2004) with reasonable accuracy (Figure 6.2). Even where there was a systematic overestimate of protein content, changes in grain protein content caused by variations in N fertilizer were well simulated. Those systematic differences may have been associated with the fact that all the cultivars were assumed to have the same nitrogen responses and

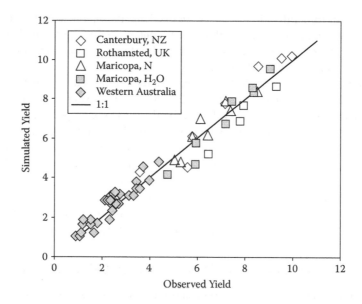

FIGURE 6.1 Comparison of yield (t ha^{-1}) simulated with the wheat model Sirius with observations. Main experimental treatments were water supply (Canterbury), water and N supply (Rothamsted), N, water supply with and without high atmospheric CO_2 concentration (Maricopa, Arizona), and N and water supply (western Australia). Correlation coefficient r is 0.98, RMSD = 0.53 t ha^{-1}, with no bias. (Adapted from Jamieson, P. D., and M. A. Semenov. 2000. *Field Crops Res.* 68: 21–29. Additional western Australian data are from Bowden, Bill, et al. 1999. Department of Agriculture and Food, Western Australia. With permission.)

pool concentrations. Nevertheless, variations in simulated yield were correlated with observations and were associated with variations in site, season, and N supply.

Allocation of N into pools associated with green area, storage, and structure resulted in very accurate simulation of the changes in GAI and biomass accumulation caused by changes in N-fertilizer supply. This is illustrated in Figures 6.3 and 6.4 for the ambient CO_2 high and low N treatments from the Maricopa free air CO_2 enhancement experiments in Arizona in 1995/1996 (Kimball et al. 1999). Simulated biomass was systematically high, but the pattern of accumulation and the differences associated with N supply were accurately simulated, including aspects of the nitrogen balance. For instance, grain N concentration and absolute N amount were well simulated (Jamieson and Semenov 2000).

The maize model, as now formulated, is very new. It has been tested so far with data from a single season for several sites where N supply was varied in experiments in farm crops (Li et al. 2006). Phenological parameters were estimated from relative maturity values so that silking dates were matched. That was the extent of "calibration" of the model. It reproduced the yield and yield variations in unirrigated production at sites with varying rainfall at a range of sowing dates (Figure 6.5). Yield was underestimated at low N-fertilizer application rates at two sites, but this was most likely caused by underestimating the soil supply of N through mineralization in the two soils. The model also produced a similar variation in crude protein in material harvested for silage, although it did systematically overestimate the protein

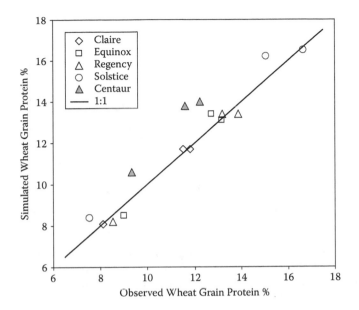

FIGURE 6.2 Comparison of Sirius model simulations of wheat grain protein with observations for five wheat cultivars in Canterbury, New Zealand. Each cultivar was sown on a different day at a different location in replicated experiments with three rates of N. Correlation coefficient r = 0.95, RMSD = 0.92 protein %, bias = 0.44 protein %. (Adapted from Armour, T., et al. *Agron. NZ* 34: 171–176. With permission.)

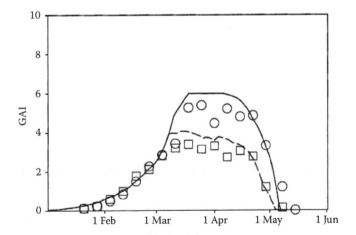

FIGURE 6.3 Time course of GAI for high (circles, continuous line) and low N (squares, broken line) ambient CO_2 treatments at Maricopa, Arizona, in 1995/1996. Symbols represent observations, and the line the simulations. (Reproduced from Jamieson, P. D., and M. A. Semenov. 2000. *Field Crops Res.* 68: 21–29. With permission.)

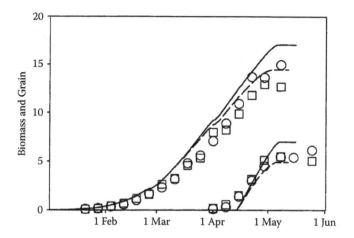

FIGURE 6.4 Observed and simulated aboveground and grain biomass accumulation (t ha^{-1}) for 1996/1997 high and low N ambient CO_2 treatments at Maricopa, Arizona. Symbols are as given in Figure 6.3. (Reproduced from Jamieson, P. D., and M. A. Semenov. 2000. *Field Crops Res.* 68: 21–29. With permission.)

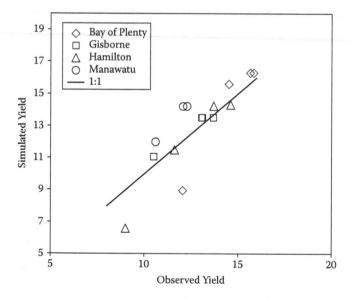

FIGURE 6.5 Comparison of simulated and measured maize grain yield (t ha^{-1}) at four sites in New Zealand. Correlation coefficient r = 0.85, RMSD = 1.4 t ha^{-1}, bias = 0.31 t ha^{-1}. (Adapted from Li, F. Y., et al. 2006. *Agron. NZ* 36: 61–70. With permission.)

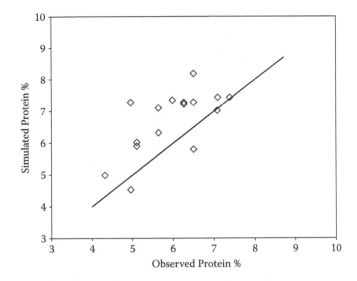

FIGURE 6.6 Comparison of simulated and observed crude biomass protein content for maize crops grown in New Zealand. Correlation coefficient r = 0.66, RMSD = 1.06 protein %, bias = 0.73 protein %. (Adapted from Li, F. Y., et al. 2006. *Agron. NZ* 36: 61–70. With permission.)

content (Figure 6.6). The values for specific leaf N concentration and stem storage were taken directly from Sirius, i.e., they were values derived from measurements on wheat. Thus the cause of the overestimate may simply be that those values are a little high. As yet we do not have time-dependent data.

Most of the validation of the potato model has been with the cultivar Russet Burbank, but through a range of environments from the cool temperate climate of South Canterbury in New Zealand to the hot dry climate of South Australia. All validation has been with irrigated crops, although they were not always irrigated sufficiently to avoid yield reductions (Jamieson et al. 2006). A comparison of simulated with observed yield shows a systematic tendency to overestimate tuber yield (Figure 6.7). However, as with the maize, yield changes with N supply were well simulated.

In all cases, accuracy of simulations was related to accurate simulation of GAI. In an experiment at Lincoln, New Zealand (Martin et al. 2001), fertilizer-N supply was varied from adequate to substantially insufficient in an irrigated Russet Burbank potato crop. Crop response to N shortage was to produce a smaller canopy (lower peak GAI) and for the canopy to senesce early. Observations and simulations are closely matched (Figure 6.8B). In response to reduced light interception, the crop simply stopped growing prematurely, although high and nil N-fertilizer treatments had similar growth rates until then. Again, the simulation approximated the behavior of biomass accumulation reasonably well (Figure 6.8A), although the prediction was for a slightly reduced growth rate through much of tuber bulking.

6.11 CONCLUDING REMARKS

Our main thesis is that, although growth and yield of crops is driven by light interception, the latter is itself controlled by the N supply because N is a vital part of

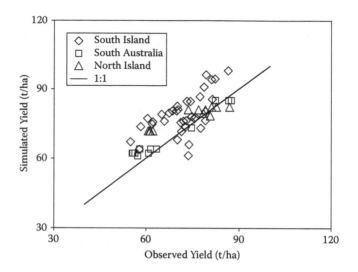

FIGURE 6.7 Comparison of simulated and observed potato yields at sites in North and South Island, New Zealand, over several seasons, and for South Australian sites in one season. Correlation coefficient r = 0.75, RMSD = 8.47 t ha⁻¹, bias = 5.48 t ha⁻¹. (New Zealand data from Jamieson, P. D., et al. 2006. *Agron. NZ* 36: 49–53. South Australian data supplied by Norbert Maier, SARDI.)

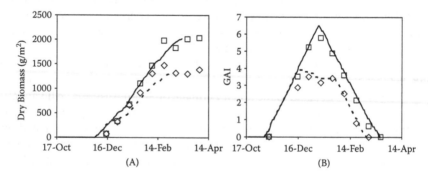

FIGURE 6.8 Time courses of biomass (A) and GAI (B) for the highest (squares, solid lines) and nil N (diamonds, broken lines) fertilizer treatments at Lincoln, New Zealand. Symbols represent observations, and lines are from the simulations. (Adapted from Martin, R. J., et al. 2001. *Agron. NZ* 31: 81–86. With permission.)

the light-collection system. Shortages of N and water should be treated differently because both the mechanisms and the timescales they operate in are different. Plant water is transitory, and enormous quantities pass through a crop compared with the standing biomass at any time. The effects of water shortages on production are through stomatal closure, reduced nutrient transport, desiccation, and heat stress. Some of this expression will be through reduced canopy size (and therefore light interception) as a method of shedding the energy load. Because of that, simulation of effects through stress indices based on the ratio of the rate of supply of soil water to the evaporative demand is very appropriate. In contrast, N comprises a very small fraction of the total biomass of a crop, and much of it is recycled during growth (Irving and Robinson 2006). The effects of shortages of N on production are expressed through diminished ability to build GAI, and then through its rapid loss as N is moved elsewhere in the crop. Simplifying assumptions, e.g., that specific leaf N and mass concentration in structure are constant, together with an upper-tissue concentration representing labile N storage, make for an easily parameterized model that is not data-hungry for calibration. Accurate simulations of biomass, GAI, and grain protein suggest that these approximations are also reasonable. No N-stress indices are necessary, so that the effects of N shortage are simulated in a manner entirely different from those of water shortages.

ACKNOWLEDGMENTS

The authors thank all participants in the experimental work reported here, in particular the New Zealand maize and potato growers whose crops were monitored to provide validation data. Potato experimental work in New Zealand was coordinated by Sarah Sinton, and the maize experimental work by Andrea Pearson. Our thanks to the late Norbert Maier, Louise Chvl, Bob Peake (SARDI), and the South East Potato Growers group for the provision of data. Funding for the New Zealand maize and potato evaluations was from the New Zealand Foundation for Arable Research, Horticulture New Zealand, the New Zealand Ministry for Agriculture and Forestry Sustainable Farming Fund, Ballance Agri-Nutrients, and the New Zealand Fertiliser Manufacturers' Research Association Inc. The Australian evaluations were funded through the Australian Department of Agriculture, Fisheries and Forestry; the National Landcare Programme; the South Australian Potato Trust; and the South East Potato Growers. Model development was funded through various programs of the New Zealand Foundation for Research, Science and Technology. Rothamsted Research receives grant-aided support from the Biotechnology and Biological Sciences Research Council of the United Kingdom.

REFERENCES

Addiscott, T. M., and A. P. Whitmore. 1991. Simulation of solute leaching in soils of differing permeabilities. *Soil Use Manage.* 7: 94–102.

Armour, T., P. D. Jamieson, and R. F. Zyskowski. 2004. Using the Sirius Wheat Calculator to manage wheat quality: The Canterbury experience. *Agron. NZ* 34: 171–176.

Asseng, S., and S. P. Milroy. 2006. Simulation of environmental and genetic effects on grain protein concentration in wheat. *Eur. J. Agron.* 25: 115–128.

Bowden, Bill, Ross Brennan, Reg Lunt, and Asseng Senthold. 1999. *Fertiliser nitrogen, applied late, needs rain to increase grain nitrogen and protein levels in wheat.* Department of Agriculture and Food, Western Australia. Crop Updates 1999: Cereals. http://www.agric.wa.gov.au/agency/pubns/cropupdate/1999/cereals/Bowden.htm.

Brooking, I. R., P. D. Jamieson, and J. R. Porter. 1995. The influence of daylength on the final leaf number in spring wheat. *Field Crops Res.* 41: 155–165.

Brooking, I. R., and P. D. Jamieson. 2002. Temperature and photoperiod response of vernalization in near-isogenic lines of wheat. *Field Crops Res.* 79: 21–38.

Colnenne, C., J. M. Meynard, R. Reau, E. Justes, and A. Merrien. 1998. Determination of a critical nitrogen dilution curve for winter oilseed rape. *Ann. Bot.* 81: 311–317.

Dardanelli, J. L., J. T. Ritchie, M. Calmon, J. M. Andriani, and D. J. Collino. 2004. An empirical model for root water uptake. *Field Crops Res.* 87: 59–71.

Dwyer, L. M., and D. W. Stewart. 1986. Leaf area development in field grown maize. *Agron. J.* 78: 334–343.

French, B. K., and B. J. Legg. 1979. Rothamsted irrigation, 1964–1976. *J. Agric. Sci., Camb.* 92: 15–37.

Greenwood, D. J., G. Lemaire, G. Gosse, P. Cruz, A. Draycott, and J. J. Neeteson. 1990. Decline in percentage N of C3 and C4 crops with increasing plant mass. *Ann. Bot.* 66: 425–436.

Grindley, D. J. C. 1997. Towards an explanation of crop nitrogen demand based on the optimisation of leaf nitrogen per unit leaf area. *J. Agric. Sci. Camb.* 128: 377–396.

Hammer, G. L., and G. C. Wright. 1994. A theoretical analysis of nitrogen and radiation effects on radiation use efficiency in peanut. *Aust. J. Agric. Res.* 45: 575–589.

Hansen, S., and H. C. Aslyng. 1984. *Nitrogen balance in crop production (simulation model NITCROS).* Copenhagen: The Royal Veterinary and Agricultural University.

Irving, L. J., and D. Robinson. 2006. A dynamic model of Rubisco turnover in cereal leaves. *New Phytol.* 169: 493–504.

Jamieson, P. D., I. R. Brooking, J. R. Porter, and D. R. Wilson. 1995a. Prediction of leaf appearance in wheat: A question of temperature. *Field Crops Res.* 41: 35–44

Jamieson, P. D., G. S. Francis, D. R. Wilson, and R. J. Martin. 1995b. Effects of water deficits on evapotranspiration from barley. *Agric. For. Meteorol.* 76: 41–58.

Jamieson, P. D., I. R. Brooking, M. A. Semenov, and J. R. Porter. 1998a. Making sense of wheat development: A critique of methodology. *Field Crops Res.* 55: 117–127.

Jamieson, P. D., J. R. Porter, J. Goudriaan, J. T. Ritchie, H. van Keulen, and W. Stol. 1998b. A comparison of the models AFRCWHEAT2, CERES-Wheat, Sirius, SUCROS2 and SWHEAT with measurements from wheat grown under drought. *Field Crops Res.* 55: 23–44.

Jamieson, P. D., M. A. Semenov, I. R. Brooking, and G. S. Francis. 1998c. Sirius: A mechanistic model of wheat response to environmental variation. *Eur. J. Agron.* 8: 161–179.

Jamieson, P. D. 1999. Drought effects on transpiration, growth and yield in crops. *J. Crop Prod.* 2: 71–83.

Jamieson, P. D., and F. Ewer. 1999. The role of roots in controlling soil water extraction during drought: An analysis by simulation. *Field Crops Res.* 60: 267–280.

Jamieson, P. D., and M. A. Semenov. 2000. Modelling nitrogen uptake and redistribution in wheat. *Field Crops Res.* 68: 21–29.

Jamieson, P. D., P. J. Stone, R. F. Zyskowski, and S. Sinton. 2003. Implementation and testing of the Potato Calculator, a decision support system for nitrogen and irrigation management. Chap. 6 in *Decision support systems in potato production: bringing models to practice*, ed. A. J. Haverkort and D. K. L. Mackerron. Wageningen: Academic Publishers.

Jamieson, P. D., R. F. Zyskowski, S. M. Sinton, H. E. Brown, and R. C. Butler. 2006. The Potato Calculator: A tool for scheduling nitrogen fertilizer applications. *Agron. NZ* 36: 49–53.

Kimball, B. A., R. L. LaMorte, Jr., P. J. Pinter, G. W. Wall, D. J. Hunsaker, F. J. Adamsen, S. W. Leavitt, T. L. Thompson, A. D. Matthias, and T. J. Brooks. 1999. Free-air CO_2 enrichment and soil nitrogen effects on energy balance and evapotranspiration of wheat. *Water Resour. Res.* 35: 1179–1190.

Lawless, C., M. A. Semenov, and P. D. Jamieson. 2005. A wheat canopy model that links canopy and phenological development. *Eur. J. Agron.* 22: 19–32.

Li, F. Y., P. D. Jamieson, and A. J. Pearson. 2006. It's AmaizeN: Developing a decision-support tool to optimise nitrogen management of maize. *Agron. NZ* 36: 61–70.

Martin, R. J., M. D. Craighead, P. D. Jamieson, and S. M. Sinton. 2001. Methods of estimating the amount of N required by a potato crop. *Agron. NZ* 31: 81–86.

Martre, P., J. R. Porter, P. D. Jamieson, and E. Triboi. 2003. Modeling grain nitrogen accumulation and protein composition to understand the sink/source regulations of nitrogen remobilization for wheat. *Plant Physiol.* 133: 1959–1967.

Martre, P., P. D. Jamieson, M. A. Semenov, R. F. Zyskowski, J. R. Porter, and E. Triboi. 2006. Modelling protein content and composition in relation to crop nitrogen dynamics for wheat. *Eur. J. Agron.* 25: 138–154.

Monteith, J. L. 1965. Evaporation and environment. *Symp. Soc. Exp. Biol.* 19: 205–234.

Monteith, J. L., 1977. Climate and the efficiency of crop production in Britain. *Phil. Trans. R. Soc. (London)* B281: 277–294.

Muchow, R. C., and P. S. Carberry. 1989. Environmental control of phenology and leaf growth in a tropically adapted maize. *Field Crops Res.* 20: 221–236.

Muchow, R. C., T. R. Sinclair, and J. M. Bennett. 1990. Temperature and solar radiation effects on potential maize yield across locations. *Agron. J.* 82: 338–343.

Muchow, R. C., and T. R. Sinclair. 1994. Nitrogen response of leaf photosynthesis and canopy radiation use efficiency in field-grown maize and sorghum. *Crop Sci.* 34: 721–727.

Penman, H. L. 1948. Natural evaporation from open water, bare soil and grass. *Proc. R. Soc. (London)* A193: 120–145.

Plénet, D., and G. Lemaire. 1999. Relationships between dynamics of nitrogen uptake and dry matter accumulation in maize crops: Determination of critical N concentration. *Plant Soil* 216: 65–82.

Porter, J. R. 1993. AFRCWHEAT2: A model of the growth and development of wheat incorporating responses to water and nitrogen. *Eur. J. Agron.* 2: 69–82.

Priestley, C. H. B., and R. J. Taylor. 1972. On the assessment of surface heat flux and evaporation using large scale parameters. *Monthly Weather Rev.* 100: 81–92.

Ritchie, J. T. 1972. Model for predicting evaporation from a row crop during incomplete cover. *Water Resour. Res.* 8: 1204–1213.

Ritchie, J. T., and S. Otter. 1985. *Description and performance of CERES-Wheat: A user oriented wheat yield model.* United States Department of Agriculture, ARS-38, 159–175.

Sinclair, T. R., and J. Amir. 1992. A model to assess nitrogen limitations on the growth and yield of spring wheat. *Field Crops Res.* 30: 63–78.

Sinclair, T. R., T. Siraiwa, and G. L. Hammer, 1992. Variation in radiation-use-efficiency with increased diffuse radiation. *Crop Sci.* 32: 1281–1284.

Sinclair, T. R., J. R. Farias, N. Neumaier, and A. L. Nepomuceno. 2003. Modeling nitrogen accumulation and use by soybean. *Field Crops Res.* 81: 149–158.

Tanner, C. B., and T. R. Sinclair. 1983. Efficient water use in crop production: Research or re-search? In *Limitations to efficient water use in crop production,* ed. H. Taylor, W. Jordan, and T. Sinclair, 1–25. Madison, Wis.: American Society of Agronomy.

Triboi, E., and A. M. Triboi-Blondel. 2002. Productivity and grain or seed composition: A new approach to an old problem. *Eur. J. Agron.* 16: 163–186.

Triboi, E., P. Martre, C. Girousse, C. Ravel, and A. M. Triboi-Blondel. 2006. Unravelling environmental and genetic relationships between grain yield and nitrogen concentration for wheat. *Eur. J. Agron.* 25: 108–118

Vos, J., and P. J. Oyarzún. 1987. Photosynthesis and stomatal conductance of potato leaves: Effects of leaf age, irradiance, and leaf water potential. *Photosynthesis Res.* 11: 253–264.

Vos, J., and P. E. L. van der Putten. 1998. Effect of nitrogen supply on leaf growth, leaf nitrogen economy and photosynthetic capacity in potato. *Field Crops Res.* 59: 63–72

Weir, A. E., P. L. Bragg, J. R. Porter, and J. H. Rayner. 1984. A winter wheat crop simulation model without water or nutrient limitations. *J. Agric. Sci., Camb.* 102: 371–382.

Wilson, D. R., R. C. Muchow, and C. J. Murgatroyd. 1995. Model analysis of temperature and solar radiation limitations to maize potential productivity in a cool climate. *Field Crops Res.* 43: 1–18.

7 Modeling Grain Protein Formation in Relation to Nitrogen Uptake and Remobilization in Rice

Yan Zhu, Hongbao Ye, Gregory S. McMaster,
Weiguo Li, and Weixing Cao

CONTENTS

7.1   Introduction 148
7.2   Materials and Methods 149
    7.2.1   Experiment Design 149
        7.2.1.1   Experiments 1 and 2: Different Varieties and Sites in
                    Three Countries 149
        7.2.1.2   Experiment 3: Different Varieties and Sites in
                    Jiangsu Province 151
        7.2.1.3   Experiments 4 and 5: Different Nitrogen Rates and
                    Irrigation Regimes 151
        7.2.1.4   Experiment 6: Different Nitrogen Rates and
                    Varieties 152
    7.2.2   Model Development 152
    7.2.3   RiceGrow Model 153
7.3   Model Description 154
    7.3.1   Pre-Anthesis Nitrogen Uptake 154
    7.3.2   Post-Anthesis Nitrogen Uptake 156
    7.3.3   Post-Anthesis Nitrogen Remobilization 157
    7.3.4   Grain Protein Accumulation 158
    7.3.5   Estimation of Genetic Parameter 160
7.4   Model Validation 160
7.5   Discussion 162
7.6   Summary and Conclusion 165
Acknowledgment 165
References 165

7.1   INTRODUCTION

Rice (*Oryza sativa* L.) is one of the most important food crops in the world and provides a substantial amount of protein in many diets. Therefore, superior grain quality of rice has become an important target in current rice production (Cheng and Zhu 1998; Huang et al. 1998; Li et al. 2005). This target has spurred interest in understanding the processes related to grain protein formation, with the goal of developing management practices to improve grain protein concentration. Crop simulation models are also being used to provide quantitative tools for crop prediction and management recommendations in modern rice production (Gao et al. 1992; Cao and Luo 2003).

During the past two decades, substantial progress has been made in rice growth modeling (Lin et al. 2003), and several simulation models, such as CERES-Rice (Ritchie et al. 1987), ORYZA2000 (Bouman et al. 2001), SIMRIW (Horie et al. 1995), RCSODS (Gao et al. 1992), and RiceGrow (Meng 2002), have been developed and tested across diverse conditions. These simulation models quantify the processes of growth and development in rice and their relationships with environmental factors and cultural practices. They can predict ecophysiological processes and yield formation in rice under varying levels of soil nutrition and water status. Applications of these growth models have generated significant benefits in yield prediction and management decision support in rice production (Lin et al. 2003). However, none of these models quantify the dynamic processes of grain quality formation in rice, which limits their use in predicting and managing grain quality.

With research advances in crop physiology, our knowledge of the processes involved in grain protein formation has greatly improved, making the likelihood of accurately simulating these processes more feasible. In rice, grain protein concentration normally accounts for 50 to 90 g/kg of grain dry weight, and the concentration is a key index to reflect grain quality, and in particular nutritional quality. Formation of protein in rice grains is closely related to plant nitrogen (N) status and is affected by varietal traits (Jiang et al. 2003; Yang et al. 2002). Leaf nitrogen status affects leaf area development and leaf photosynthesis, thus regulating grain productivity (Grindlay 1997; Hasegawa and Horie 1996; Yin et al. 2000; Sheehy et al. 2004). Accumulation of nitrogen in vegetative organs and its subsequent reallocation to reproductive organs in rice are the most important processes for grain protein formation (Norman et al. 1992; Sheehy et al. 2004). After the seedling stage when seed N is depleted, nitrogen uptake from the soil is required to meet the needs of physiological functioning, structural growth, and subsequent grain protein accumulation after anthesis. N-uptake rates vary among genotypes and environmental conditions (Dingkuhn 1996; Ying et al. 1998; Ntanos and Koutroubas 2002; Jiang et al. 2003). From jointing to booting, a large amount of nitrogenous compounds is stored as amino acids and proteins in the tissues of stems and leaves (Ntanos and Koutroubas 2002; Jiang et al. 2003). During the grain-filling period, a large amount of N is required for grain filling, and N uptake is generally much less than the demand for nitrogen by the grain. Therefore, most N demand by the grain for protein synthesis is retranslocated from the vegetative organs, especially from leaf blades, in the form of amino acids (Mae 1997). In general, greater total plant nitrogen concentration at anthesis resulted in greater nitrogen translocation to the grain, thus leading to a higher grain protein concentration (Ntanos and Koutroubas 2002). These processes of nitrogen uptake, absorption, storage, retranslocation, and synthesis of nitrogen compounds comprise the main nitrogen-flow dynamics in the rice plant (Dingkuhn 1996; Huang et al. 1998).

To date, many studies have attempted to characterize nitrogen uptake and utilization in relation to grain yield formation and protein accumulation in rice, and to establish the regulating effects of nitrogen fertilization and genotypes (e.g., Huang et al. 1998; Ying et al. 1998; Ntanos and Koutroubas 2002; Jiang et al. 2003; Sheehy et al. 2004). As a typical modeling approach in rice (Ritchie et al. 1987; Bouman et al. 2001), N concentration and protein formation in grain are determined by calculating the fraction of N partitioned into grains based on the final amount of nitrogen accumulated by the plant and adjusted by a nitrogen harvest index (NHI). This approach might give reasonable predictions of final grain protein concentration based on accumulation and distribution of dry matter and nitrogen in grains, but does not simulate the time course of nitrogen flow processes within the plant and neglects the interaction among pre-anthesis nitrogen accumulation, post-anthesis nitrogen uptake, and nitrogen remobilization in relation to final grain nitrogen accumulation. Therefore, the robustness in explaining and predicting the processes of grain protein formation under diverse conditions is reduced by this approach. In contrast, efforts to simulate wheat grain nitrogen accumulation and protein content in relation to nitrogen dynamic processes in plant have been developed (Jamieson and Semenov 2000; Asseng et al. 2002; Martre et al. 2006; Pan et al. 2006).

The objectives of the present study were to (a) quantify the dynamic relationships of plant nitrogen accumulation and remobilization to the grain as influenced by environmental and genetic factors in rice, and (b) use these relationships to develop and test a simplified process-based nitrogen flow model for predicting grain protein formation in rice. This work was intended to provide a quantitative tool to support improved management of rice production systems for desired grain protein accumulation.

7.2 MATERIALS AND METHODS

7.2.1 EXPERIMENT DESIGN

Six field experiments were conducted from 2001 to 2003 involving different sites, varieties, nitrogen rates, and irrigation regimes over three years, as summarized in Table 7.1. Daily climatic data, including maximum and minimum air temperature, precipitation, and radiation, were obtained from weather stations installed at each experimental site. The data from experiments 1 and 6 involved more detailed observations under different cultivars and N rates, and thus were used for building the model algorithms; whereas the data from experiments 2, 3, 4, and 5 were used for testing the model performance under different conditions.

7.2.1.1 Experiments 1 and 2: Different Varieties and Sites in Three Countries

Experiments 1 and 2 were two experiments conducted in 2001 and 2002 and consisted of different varieties and sites in three countries. In 2001 there were three

TABLE 7.1
Overview of Six Different Field Experiments

Experiment No.	Year	Site	Treatment	Sowing and Transplanting Date
1[a]	2001	Nanjing, China (32°03′N)	Varieties (9) and sites (4)	May 11 and June 15
		Kyoto, Japan (35°00′N)		April 30 and May 22
		Shimane, Japan (35°51′N)		April 17 and May 16
		Iwate, Japan (39°91′N)		April 12 and May 15
2	2002	Lijiang, China (26°00′N)	Varieties (9) and sites (5)	April 7 and May 5
		Nanjing, China (32°03′N)		May 11 and June 15
		Shimane, Japan (35°51′N)		April 17 and May 16
		Iwate, Japan (39°91′N)		April 12 and May 15
		Changmai, Thailand (32°03′N)		June 24 and July 22
3[b]	2002	Eight sites in Jiangsu, China (31°31′–33°42′N)	Varieties (10) and sites (8)	May 10 and June 12
4	2002	Nanjing Agricultural Univ., China (32°03′N)	Nitrogen rates (5) and irrigation regimes (2)	May 10 and June 12
5	2003	Nanjing Agricultural Univ., China (32°03′N)	Nitrogen rates (5) and irrigation regimes (2)	May 10 and June 12
6	2003	Nanjing Agricultural Univ., China (32°03′N)	Nitrogen rates (4) and irrigation regimes (4)	May 11 and June 12

Note: Except where noted, all data were used in model validation. These data were used to develop model algorithms and estimate grain protein concentration at maturity for cultivar Wuxiangjing.

[a] These data were used to estimate cultivar-specific grain protein concentrations at maturity for nine varieties grown at these sites.

[b] Locations Lishui, Rudong, Baoying, and Huaiying were used to estimate cultivar-specific grain protein concentrations at maturity for 10 varieties grown at these sites, and locations Kunshan, Jingtan, Danyang, and Sihong were used for model validation.

---

sites in Japan and one in China, and in 2002 there were two sites in Japan, two in China, and one in Thailand. In both years, nine varieties (provided by Kyoto University of Japan) were grown at each site, including five Indica types (Takanari, IR72, Sankeiso, Ch86, and IR65564-44-2-2) and four Japonica types (Nipponbare, Takenari, Banten, and WAB450-1-B-P-38-HB), as listed in Table 7.2.

Both experiments were a two-factorial randomized complete block design with three replications, and plot size was 15 m². In both years, the sowing and transplanting dates were, respectively, April 12 and May 15 at Iwate, April 17 and May 16 at Shimane, April 30 and May 22 at Kyoto, May 11 and June 15 at Nanjing, April 7 and May 5 at Lijiang, and June 24 and July 22 at Changmai. The application rates of N, P_2O_5, and K_2O were each 120 kg·ha^{-1}; row and hill spacing was 30 × 15 cm², with two plants per hill. Other routine management followed local standard practices. Following harvest after maturity, grain protein concentration was determined by

TABLE 7.2

Genotype-Specific Grain Protein Concentrations at Maturity for Different Cultivars Estimated from Different Experiments in the Present Study

Genotype	Protein Concentration (g kg⁻¹)	Genotype	Protein Concentration (g kg⁻¹)
Banten	74.5	Shanyou 63	79.2
CH86	63.4	Sidao 10	62.8
Changyou 1	64.6	Takanari	68.1
Fengyouxiangzhan	74.5	Tekenari	69.7
Guanglinxiangjing	66.5	WAB450-1-B-P-38-HB	80.4
IR65564-44-2-2	79.6	Wuxiangjing	65.1
IR72	61.4	Wuyujing	64.3
Lianjing 3	70.5	Wuyujing 3	65.0
Nipponbare	68.2	Yangdao 6	70.4
Sankeiso	69.9	Zaofeng 9	64.5

measuring grain N concentration according to the NY 147-88 rice quality protocol established by the Ministry of Agriculture of China (1988), and then using a conversion coefficient of 5.95 for estimating protein concentration in the grain.

7.2.1.2 Experiment 3: Different Varieties and Sites in Jiangsu Province

Experiment 3 was conducted in 2002 with different varieties and locations within the Jiangsu province of China. There were eight sites and ten varieties, including five Indica types (Changyou 1, Sidao 10, Yangdao 6, Fengyouxiangzhan, and Shanyou 63) and five Japonica types (Wuyujing 7, Wuyujing 3, Lianjing 3, Guanglinxiangjing, and Zaofeng 9), as listed in Table 7.2.

Each site was a randomized complete block design with three replications, and plot size was 20 m². The sowing and transplanting dates were May 10 and June 12, respectively, and row and hill spacing was 30×15 cm², with two plants per hill. The application rates of N, P_2O_5, and K_2O were each 150 kg·ha⁻¹, and other routine management followed local standard practices. Grain protein concentration was determined as described for experiments 1 and 2.

7.2.1.3 Experiments 4 and 5: Different Nitrogen Rates and Irrigation Regimes

Experiments 4 and 5 were conducted at the Nanjing Agricultural University Experiment Station in 2002 and 2003. Treatments were factorial combinations of five nitrogen application rates (0, 75, 150, 225, 300 kg N·ha⁻¹) and two irrigation regimes (shallow water irrigation and intermittent irrigation) using the Japonica variety Wuxiangjing 9.

The experiment was a randomized complete block design with three replications, and plot size was 20 m². In both years, the sowing and transplanting dates were

May 10 and June 12, respectively, and row and hill spacing was 30×15 cm^2, with 2 plants per hill. Application rates of P_2O_5 and K_2O were each 150 kg·ha^{-1}, and other routine management followed standard local practices. Grain protein concentration was determined as described for experiments 1 and 2.

7.2.1.4   Experiment 6: Different Nitrogen Rates and Varieties

Experiment 6 was conducted at the Nanjing Agricultural University Experiment Station in 2003. Treatments involved four nitrogen application rates (0, 90, 180, 270 kg N·ha^{-1}) and four Japonica varieties (Wuxiangjing 9, Nipponbare, Yangdao 6, and Takanari), as listed in Table 7.2.

The experiment was a randomized complete block design with three replications, and plot size was 20 m^2. The sowing and transplanting dates were May 11 and June 12, respectively, and row and hill spacing was 30×15 cm^2, with two plants per hill. The application rates of P_2O_5 and K_2O were all 150 kg·ha^{-1}, and other routine management followed the local standard practices.

Plants were sampled at the growth stages or periods of tillering, panicle initiation, heading, grain filling, and maturity, with sample size of at least 10 plants, and panicles were sampled every six days after anthesis. Green leaf blade area was measured with the CI-203 (CID, Vancouver, WA) area meter. Aboveground population dry matter was determined after oven-drying at 70°C to constant weight. Nitrogen concentration in total aboveground biomass at each sampling period was determined by the semi-micro Kjeldahl method. Grain protein concentration was determined as described for experiments 1 and 2.

7.2.2   Model Development

It is assumed that the nitrogen absorbed by the rice plant is mainly used for nonstructural functioning, structural growth and storage before anthesis, and grain protein accumulation after anthesis. A large amount of nitrogenous compounds in the form of amino acids or proteins is stored in the stem and leaves before anthesis. After anthesis, this material is remobilized as amino acids to the grain for protein synthesis. This forms a dynamic nitrogen flow from the soil to the grain involving the processes of uptake, storage, remobilization, and synthesis of nitrogen compounds. Thus, our model concentrates on quantifying the simplified processes of nitrogen uptake, translocation and utilization in the whole plant, and protein accumulation and concentration in the grains.

Data from experiments 1 and 6 and recent literature on plant nitrogen flow and grain protein formation (Huang et al. 1998; Ying et al. 1998; Ntanos and Koutroubas 2002; Jiang et al. 2003; Piao et al. 2003) were analyzed to develop the conceptual model of the dynamic processes of nitrogen uptake and remobilization. Based on our rice growth model (RiceGrow; Zhuang 2001; Hu 2002; Meng 2002), and using physiological development time (PDT) as a general timescale of plant development (Cao and Moss 1997; Meng et al. 2003), the fundamental functions and algorithms were formulated to describe the processes involved in plant nitrogen dynamics and grain protein accumulation in rice.

The grain protein model was programmed using Visual C^{++} language into a standard software component, and then incorporated into the RiceGrow simulation

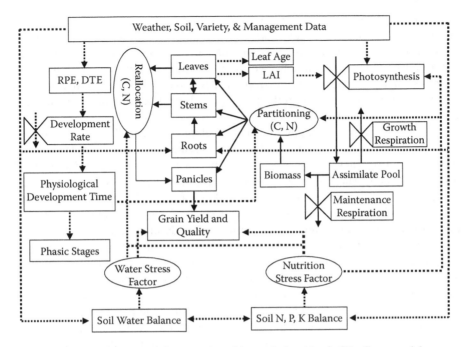

FIGURE 7.1 Basic structural framework and interrelationships in RiceGrow model.

model (Meng 2002), which is briefly described below. If the grain protein model is run as a stand-alone component, required inputs can be estimated from the Rice-Grow model or, if known, entered directly into the program by the user.

Finally, data from experiments 2, 3 (sites Kunshan, Danyang, Jingtan, and Sihong only), 4, and 5 were used for testing the model performance under different conditions. The root mean square errors (RMSE) between observed and predicted values were calculated to evaluate model predictions (Cao and Moss 1997; Bouman et al. 2001; Cao and Luo 2003), along with 1:1 plotting of measured against predicted values. Simple linear regression of observed against predicted values was also calculated to determine the goodness of fit for model performance.

7.2.3 RiceGrow Model

The RiceGrow model is an ecophysiological process-based rice simulation model developed by the authors' laboratory (Zhuang 2001; Hu 2002; Meng 2002). The structural components and interrelationships of the entire growth model are shown schematically in Figure 7.1. Among the individual model components, the processes of growth and yield formation were quantified and integrated in relation to weather factors, soil properties, genotypic parameters, and management practices in rice production. The model includes six different submodels for predicting phasic and phenological development, morphological and organ formation, photosynthesis and biomass accumulation, partitioning and yield formation, soil water relations (drought and waterlogged), and soil nutrient (N, P, K) balance.

The RiceGrow system utilizes an object-oriented and component-based software approach and is programmed in Visual Basic and Visual C++ based on the COM standard. The implemented model system can be used for simulating the processes of growth and yield formation under various environmental conditions, production levels, and genetic parameters. As a basic simulation platform of rice growth processes, RiceGrow provides input variables for use by the grain protein submodel, including the physiological development time from the development module, aboveground dry weight and leaf area index from the organ growth module, and nitrogen and water deficit factors from the environment module.

In the RiceGrow model, physiological development time (PDT, in relative days) is established as a general and continuous timescale to predict plant development progress for different genotypes and environmental conditions (Cao and Moss 1997; Meng et al. 2003). The PDT is defined as the developmental time accumulated under an optimum environment of photoperiod and temperature. In concept, it is an integration of different physiological components involved in the development processes, and thus is a relatively universal scale for predicting development progress in different environments. Calculation of PDT is based on the dynamic interaction between the processes of photoperiod response and thermal effect as influenced by genotype and environment. For instance, the PDT of a given day can be 0.5 or 1.0, depending on the combined physiological effects of relative photoperiod response, thermal effect, and genotypic coefficient. The total PDT required to reach a given stage is a fixed value, e.g., 13 PDT (days) at panicle initiation and 28 PDT at booting in rice (Meng et al. 2003). Thus, the PDT is a physiologically accountable and unified timescale for the rice plant, and can be quantitatively estimated under various growing conditions. In the present grain protein model, PDT is used as a developmental scale for describing time-course processes in relation to N uptake and remobilization in the plant and protein accumulation in the grain.

7.3   MODEL DESCRIPTION

7.3.1   Pre-Anthesis Nitrogen Uptake

Nitrogen for the growth of new plant organs is primarily from two sources: uptake by the roots and remobilization from aging organs (Huang et al. 1998; Peng and Cassman 1998; Su et al. 2001). As observed in experiment 6, the rate of nitrogen uptake and accumulation in rice changed over time as a logistic curve, and genotypic differences became more apparent with the progress of plant development (Figure 7.2). The physiological development time (PDT) was a useful parameter to represent development progress (Cao and Moss 1997; Cao and Luo 2003; Meng et al. 2003), so the general time-course pattern of potential nitrogen-uptake rate per unit PDT before anthesis ($PNURBP_{(PDT)}$ in kg·ha^{-1}·PDT^{-1}) was described by

$$PNURB_{(PDT)} = MNURP\left(1 - \exp\left(\frac{-INURP \times PDT}{MNURP}\right)\right) \tag{7.1}$$

where PDT is physiological development time under optimal conditions (Cao and Moss 1997; Cao and Luo 2003) obtained from the phasic development module of RiceGrow; INURP is the initial nitrogen-uptake rate or accumulation rate per unit

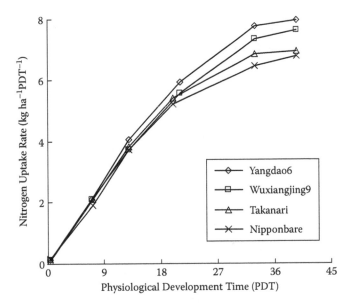

FIGURE 7.2 Relationship between nitrogen-uptake rate before anthesis and physiological development time (PDT) for different rice genotypes (experiment 6).

PDT (kg·ha⁻¹·PDT⁻¹), defined as the product of seed nitrogen concentration and sowing rate (equation 7.2); and MNURP is the maximum nitrogen-uptake rate per unit PDT (kg·ha⁻¹·PDT⁻¹) as calculated from equation (7.3).

$$INURP = PC \times SR/N2P \tag{7.2}$$

$$MNURP = NCM \times DWAP \tag{7.3}$$

In equations (7.2) and (7.3), PC (%) is the cultivar-specific grain protein concentration (i.e., genotypic parameter normally ranging from 5% to 9%); N2P is the conversion coefficient from nitrogen to protein concentration in grain and set as 5.95; SR (kg·ha⁻¹) is the sowing rate; NCM (%) is the maximum nitrogen concentration in plant, with a range of 3%–4% (Wang 1994; Sheehy et al. 1998; Jiang et al. 2003; Piao et al. 2003), and the default is set to 3.5%; and DWAP (kg·ha⁻¹·PDT⁻¹) is dry-matter increment per unit PDT under the maximum leaf area index, with the value calculated in the biomass production module of the RiceGrow model (Meng 2002; Meng et al. 2004).

Potential nitrogen-uptake rate per day before anthesis (PNURDB$_{(i)}$, kg·ha⁻¹·day⁻¹) is calculated by

$$PNURDB_{(i)} = PNURPB_{(PDT(i))} \times \Delta PDT_{(i)} \tag{7.4}$$

where $\Delta PDT_{(i)}$ is the increment of physiological development time on day i; PDT$_{(i)}$ is the accumulated physiological development time from sowing date until day i;

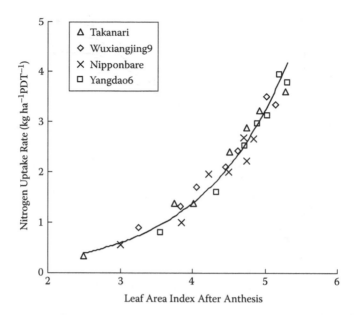

FIGURE 7.3   Relationship between nitrogen-uptake rate and leaf area index after anthesis (experiment 6).

and PNURPB$_{(PDT(i))}$ is potential nitrogen-uptake rate per unit PDT at PDT$_{(i)}$ as determined in equation (7.1).

Nitrogen uptake is affected by the nitrogen and water available in the soil, and thus the actual nitrogen-uptake rate usually is less than the potential. Actual daily N uptake per day (ANURDB$_{(i)}$) and accumulation over time (ANAB$_{(i)}$) are calculated as

$$ANURDB_{(i)} = PNURDB_{(i)} \times \min(FN_{(i)}, FW_{(i)}) \tag{7.5}$$

$$ANAB_{(i)} = \sum_{i=0}^{n} ANURDB_{(i)} \tag{7.6}$$

where ANAB$_{(i)}$ and ANURDB$_{(i)}$ are in kg N·ha^{-1} for day i; FN$_{(i)}$ and FW$_{(i)}$ are 0–1 nitrogen- and water-deficit factors on day i as calculated from the nutrient-balance module (Zhuang 2001; Zhuang et al. 2004) and the water-balance module (Hu 2002; Hu et al. 2004), respectively, with both factors normally ranging from 0.5 to 1.0.

7.3.2   POST-ANTHESIS NITROGEN UPTAKE

After anthesis, N uptake continues and contributes to grain protein in rice (Ntanos and Koutroubas 2002; Jiang et al. 2003). Our experimental data (experiment 6) indicated that nitrogen-uptake rate per unit PDT after anthesis was exponentially related to the leaf area index (LAI) during this period (Figure 7.3) and positively related to the capacity or sink size of plant nitrogen accumulation:

$$ANURPA_{(PDT)} = \Delta DW_{an} \times (NC_{an} - NCT) \times (0.02 \times \exp(0.8482 \times LAI_{(PDT)})) \ (7.7)$$

where $ANURPA_{(PDT)}$ (kg·ha^{-1}·PDT^{-1}) = actual nitrogen-uptake rate per unit PDT at PDT time after anthesis; ΔDW_{an} (kg·ha^{-1}·PDT^{-1}) = increment of aboveground dry weight (kg·ha^{-1}) at anthesis (thus using *an* as subscript) estimated in the biomass production module; NC_{an} (%) = aboveground plant nitrogen concentration at anthesis, obtained from equation (7.8); NCT (%) = minimum nitrogen concentration that could be translocated from NC_{an} and set to 0.55 NC_{an} (Wang 1994; Sheehy et al. 1998; Jiang et al. 2003); and $LAI_{(PDT)}$ = leaf area index at PDT time during grain filling, obtained from the biomass production module in RiceGrow.

$$NC_{an} = ANAB_{an}/DW_{an} \qquad (7.8)$$

$$NCT = 0.55NC_{an} \qquad (7.9)$$

where $ANAB_{an}$ (kg·ha^{-1}) = amount of plant nitrogen accumulated at anthesis, as given by equation (7.6).

The actual nitrogen-uptake rate per day on day i after anthesis ($ANURDA_{(i)}$, kg·ha^{-1}·day^{-1}) is calculated by

$$ANURDA_{(i)} = ANURPA_{(PDT(i))} \times \Delta PDT_{(i)} \qquad (7.10)$$

in which $\Delta PDT_{(i)}$ is the increment of physiological development time on day i; $PDT_{(i)}$ is the accumulated physiological development time from sowing date until day i; and $ANURPA_{(PDT(i))}$ is the actual nitrogen-uptake rate per unit PDT at $PDT_{(i)}$ after anthesis, as given by equation (7.7).

7.3.3 POST-ANTHESIS NITROGEN REMOBILIZATION

Grain growth and protein synthesis begin after anthesis. During grain filling, almost all N uptake by the plant is translocated to the grain, and nitrogenous compounds (including amino acids) located in plant organs are also remobilized to the grain when N uptake is less than the demand for N in grains (Dingkuhn 1996; Mae 1997; Huang et al. 1998; Ntanos and Koutroubas 2002). For simplification, the nitrogenous compounds in vegetative plant organs at anthesis are all considered as available nitrogen that can be remobilized and translocated to the grain.

Nitrogen translocation rate from vegetative organs to grains was used to represent the daily remobilization intensity of the transferable nitrogen stored in the plant before anthesis. Our experimental data (experiment 6) and the literature (Peng and Cassman 1998; Ying et al. 1998; Ntanos and Koutroubas 2002) indicated that the N translocation rate to grain after anthesis exhibited a dynamic time pattern of slow-fast-slow during grain filling (Figure 7.4). The translocation rate depended on the amount of plant nitrogen accumulated at anthesis and the amount of nitrogen in the plant at maturity. We calculated the actual nitrogen translocation rate per unit PDT at PDT time ($ANTRP_{(PDT)}$, kg·ha^{-1}·PDT^{-1}) by the following equation:

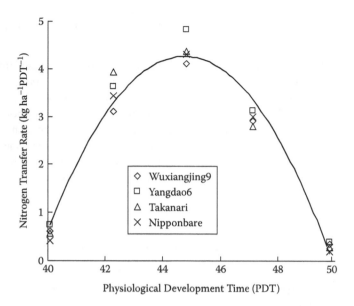

FIGURE 7.4 Relationship between nitrogen translocation rate after anthesis and physiological development time (PDT) in rice (experiment 6).

$$ANTRP_{(PDT)} = (ANAB_{an} - NCT \times DW_m)$$

$$\times (-0.0015 \times PDT^2 + 0.138 \times PDT - 3.06) \tag{7.11}$$

where $ANAB_{an}$ (kg·ha^{-1}) = amount of plant nitrogen accumulated (equation 7.6) at anthesis; NCT (%) = minimum transferable nitrogen concentration (equation 7.9); DW_m (kg·ha^{-1}) = aboveground dry weight (kg·ha^{-1}) at maturity estimated in the biomass production module; and PDT = physiological development time.

Actual nitrogen translocation rate per day on day i ($ANTRD_{(i)}$, kg·ha^{-1}·day^{-1}) is calculated as

$$ANTRD_{(i)} = ANTRP_{(PDT(i))} \times \Delta PDT_{(i)} \tag{7.12}$$

where $\Delta PDT_{(i)}$ = increment of physiological development time on day i; $PDT_{(i)}$ = accumulated physiological development time from sowing date until day i; and $ANTRP_{(PDT(i))}$ = actual nitrogen translocation rate per unit PDT at $PDT_{(i)}$, as given by equation (7.11).

7.3.4 GRAIN PROTEIN ACCUMULATION

Based on our experimental data (experiment 6) and the literature (Huang et al. 1998; Ntanos and Koutroubas 2002; Wei et al. 2002; Jiang et al. 2003), the pattern of grain protein accumulation during grain development is very similar for different cultivars. In general, protein synthesis in grain starts five days after anthesis (=40 PDT) and continues until physiological maturity (=57 PDT). As noted before, the N substrate for protein synthesis comes from N remobilization from vegetative tissue after anthesis and post-anthesis N uptake. About 70% of the total nitrogen in the grain at maturity is derived from nitrogenous compounds remobilized from vegetative organs. Remobilization is assumed to occur from anthesis (=40 PDT) to late-filling (=50 PDT). Post-anthesis nitrogen uptake from anthesis (=40 PDT) to physiological maturity (=57 PDT) accounts for the remaining 30% of the total nitrogen in the grain at maturity. In addition, protein formation is affected by air temperature (T).

Considering these relationships, the protein accumulation in the grain on day i ($GPA_{(i)}$, kg·ha^{-1}) is calculated as

$$GPA_{(i)} = 5.95 \times NAG_{(i)} \tag{7.13}$$

$$NAG_{(i)} = \begin{cases} \sum_{i=0}^{k} ((ANURDA_{(i)} + ANTRD_{(i)}) \times FT_{(i)} & 40 < PDT_{(i)} \leq 50 \\ \\ \sum_{i=k}^{n} (ANURDA_{(i)} \times FT_{(i)}) & 50 < PDT_{(i)} \leq 57 \end{cases} \tag{7.14}$$

where $NAG_{(i)}$ (kg·ha^{-1}) = amount of nitrogen accumulation in grain on day i; 5.95 = coefficient for converting grain nitrogen to protein; $ANURDA_{(i)}$ (kg·ha^{-1}·day^{-1}) = nitrogen-uptake rate per day on day i after anthesis (see equations 7.7 through 7.10); $ANTRD_{(i)}$ (kg·ha^{-1}·day^{-1}) = actual nitrogen translocation rate per day on day i as calculated in equation (7.12); and $FT_{(i)}$ = temperature impact factor on day i as calculated in equation (7.15).

Optimum temperatures for grain filling and protein formation in rice are between 24°C and 27°C, with a minimum limit of 16°C to 18°C and a maximum limit of 37°C to 40°C (Gao et al. 1992; Zhou et al. 1997; Huang et al. 1998). To reflect the dynamic relationship between temperature and protein accumulation in grain, the following sine functions are used to calculate the impact factor of temperature ($FT_{(i)}$):

$$FT_{(i)} = \begin{cases} \sin[(T_{(i)} - T_b)/(T_{ol} - T_b) \times \pi/2] & T_b \leq T_{(i)} < T_{ol} \\ 1 & T_{ol} \leq T_{(i)} \leq T_{oh} \\ \sin[(T_m - T_{(i)})/(T_m - T_{oh}) \times \pi/2] & T_{oh} < T_{(i)} \leq T_m \\ 0 & T_m < T_{(i)}, \text{ or } T_{(i)} < T_b \end{cases} \tag{7.15}$$

where $T_{(i)}$ (°C) = daily mean air temperature on day i; T_m (°C) and T_b (°C) = maximum and base temperatures for protein synthesis in grain, and set at 40°C and 16°C, respectively; and T_{ou} (°C) and T_{ol} (°C) = upper and lower limits of optimum temperature ranges for protein synthesis, and determined as 25°C and 27°C, respectively, for Indica rice, and 24°C and 26°C, respectively, for Japonica rice (Bi 1980; Zhou et al. 1997).

Protein concentration in the grain is obtained as the ratio of grain protein accumulation to grain yield from five days after anthesis (=40 PDT) to physiological maturity (=57 PDT):

$$GPC_{(i)} = GPA_{(i)}/GY_{(i)} \times 100 \qquad (7.16)$$

where $GPC_{(i)}$ (%) = grain protein concentration on day i; $GPA_{(i)}$ (kg·ha^{-1}) = amount of grain protein accumulation on day i; and $GY_{(i)}$ (kg·ha^{-1}) = grain yield accumulation on day i, as provided by the biomass partitioning module of RiceGrow.

7.3.5 ESTIMATION OF GENETIC PARAMETER

The grain protein model requires one cultivar-specific parameter m typical grain protein concentration at maturity for a given cultivar m which was estimated for the individual cultivars used in the present study, as listed in Table 7.2. The grain protein data obtained from the cultivar and site experiment in 2001 (experiment 1) were used to determine the grain protein concentrations of the varieties IR72, Ch86, Takanari, Nipponbare, Tekenari, Sankeiso, Banten, IR65564-44-2-2, and WAB450-1-B-P-38-HB, with a range of 61.4 to 80.4 g·kg^{-1}. The data from the sites of Lishui, Rudong, Baoying, and Huaiying in Experiment 3 measured the grain protein concentrations for the cultivars Sidao 10, Changyou 1, Yangdao 6, Fengyouxiangzhan, Shanyou 63, Wuyujing, Zaofeng 9, Guanglinxiangjing, Wuyujing 3, and Lianjing 3, with a range of 62.8 to 79.2 g·kg^{-1}. Based on experiment 6, the grain protein concentration for cultivar Wuxiangjing was set as 65.1 g·kg^{-1}.

7.4 MODEL VALIDATION

Experiments 2 and 3 were used to validate model simulations of grain protein concentrations for different cultivars over diverse ecological environments. Experiment 2 tested nine rice cultivars grown at five sites in Japan, China, and Thailand (Figures 7.5A1 and 7.5A2). A high goodness of fit between simulated and observed values was indicated by an r^2 of 0.8763 and RMSE of 3.48 g·kg^{-1}, while the RMSE was relatively higher for the Lijiang site and cultivars IR65564 and Takenari. Four sites (Kunshan, Jingtan, Danyang, and Sihong) in experiment 3 tested the model performance under different conditions in Jiangsu region for five Indica and five Japonica cultivars (Figures 7.5B1 and 7.5B2). For these data we also observed a good relationship between simulated and observed grain protein concentrations (r^2 = 0.8417 and RMSE = 3.12 g·kg^{-1}), although the RMSE was relatively greater for the Sihong site and cultivar Zaofeng 9. These results give us confidence that the model can reasonably predict grain protein concentration for different rice cultivars under varied ecological environments.

To further evaluate model responses to varied nitrogen supply and irrigation regimes, which influence grain protein formation, two years of data from experiments 4 and 5 were used (Figure 7.6). The simulation results for grain protein concentration showed that the predicted values were close to the observed values, with an r^2 of 0.8963. The RMSE values averaged 1.83 g·kg^{-1} for five N application rates, and 1.89 g·kg^{-1} for two irrigation regimes. Among them, the prediction error reached 2.82 g·kg^{-1} for high N rate at 300 kg·ha^{-1}, obviously greater than under lower N rates, where more consistent performance was seen with two irrigation regimes. Yet, the overall results indicated the model could reliably simulate grain protein concen-

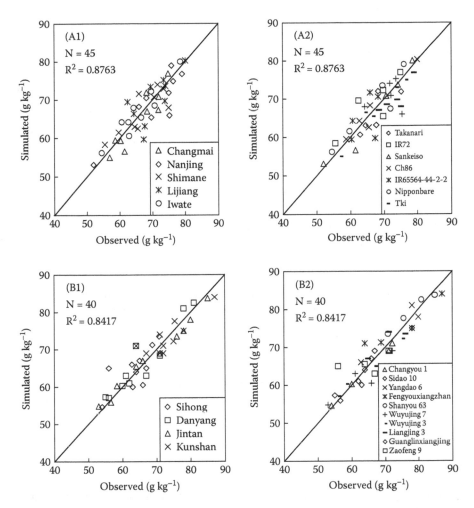

FIGURE 7.5 Comparison of simulated with measured grain protein concentrations for five sites (A1) and nine cultivars (A2) in three countries (data from experiment 2), and four sites (B1) and ten cultivars (B2) in Jiangsu Province, China (data from experiment 3). A1 and A2 or B1 and B2 are the same data, just sorted by sites and cultivars, respectively.

tration under varied conditions of nitrogen fertilization and irrigation management used in rice production systems.

In addition, the integrated model of RiceGrow exhibited good performance in predicting phenological development and dry-matter accumulation in plant organs with different rice cultivars under varied growing conditions (Meng et al. 2003, 2004). As listed in Table 7.3, the RMSE values in eight cultivars averaged 1.4, 4.5, 3.8, and 3.7 days, respectively, for prediction of emergence, panicle initiation, heading, and maturity. Prediction errors were greater for the panicle initiation stage than for other growth stages in rice. RMSE values for six cultivars under different cultural experiments averaged 41, 439, 37, and 401 kg·ha^{-1}, respectively, for leaf weight, stem weight, root weight, and grain yield (Table 7.4). The greater

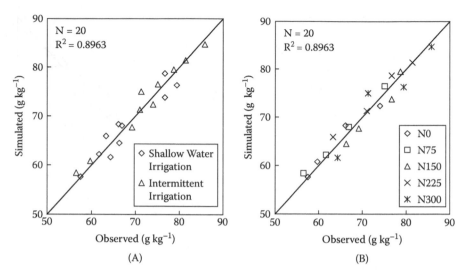

FIGURE 7.6 Comparison of simulated with measured grain protein concentrations for different irrigation regimes (A) and nitrogen rates (B). (Data from experiments 4 and 5.)

prediction errors for stem weight and grain yield than for leaf and root weights were largely due to differences in absolute weights of the organs. In a separate study with nitrogen rates varying from 0 to 405 kg N·ha^{-1} (Figure 7.7), the simulated shoot dry weights fit well the observations from the different treatments, with an r^2 of 0.9844 and RMSE of 1222.7 kg·ha^{-1}. Thus, the RiceGrow model generally simulated diverse production conditions and provided accurate inputs required by the grain protein submodel.

7.5   DISCUSSION

In this study we developed a simplified quantification of the processes of nitrogen accumulation and translocation following the physiological development time course (PDT), and created a general simulation model of grain protein formation in rice. The model was tested with data sets from field experiments comprising a broad range of sites, cultivars, fertilization levels, and irrigation rates. Validation results on grain protein concentrations indicated that overall model performance was reliable and accurate across the range of conditions tested. Yet, deviations up to nearly 10 g·kg^{-1} of predicted grain protein from measured values occurred for a few cultivars at a few sites (Figure 7.5) and under high N rates (Figure 7.6). It seems that some cultivars exhibited unstable performance under specific environmental conditions and excess N supply, which needs further evaluation and quantification. Thus, the present model is only a preliminary effort on predicting grain protein formation in rice, and remains to be tested with data from a wider range of experimental conditions. Future study is needed to determine if the constant N-to-protein conversion factor is accurate for all rice cultivars. More data are also needed to better understand and model the dynamic patterns and interrelationships of the different protein components in the rice grain, as demonstrated by Martre et al. (2006)

TABLE 7.3
RMSE for Prediction of Different Development Stages in Different Rice Cultivars

Cultivar	Emergence (day)	Panicle Initiation (day)	Heading (day)	Maturity (day)
Wuxiangjing	1.3	3.4	2.6	3.4
Yangdao 6	1.6	5.3	4.4	5.4
Koshihikari	1.2	4.1	3.3	3.2
Nipponbare	1.3	2.2	3.3	2.0
Takanari	1.6	5.1	4.0	3.9
Laolaiqing	1.4	6.4	5.2	3.4
RR109	1.2	4.1	4.1	4.4
Shanyou 63	1.4	5.2	3.7	3.7
Average value	1.4	4.5	3.8	3.7

Source: Meng (2002) and Meng et al. (2003).

Note: Data were pooled from individual experiments.

TABLE 7.4
RMSE for Prediction of Organ Dry Weights and Grain Yields in Different Rice Cultivars

Cultivar	Leaf Weight (kg·ha⁻¹)	Stem Weight (kg·ha⁻¹)	Root Weight (kg·ha⁻¹)	Grain Yield (kg·ha⁻¹)
Nipponbare	43.2	431.1	41.0	401.8
Takanari	51.1	426.5	36.9	422.3
IR72	39.3	412.4	43.2	379.8
Laolaiqing	43.6	512.3	38.5	417.8
Shanyou63	39.1	453.1	34.2	352.1
Jing9325	31.3	398.7	27.3	432.4
Average value	41.3	439.0	36.9	401.0

Source: Meng (2002) and Meng et al. (2004).

Note: Data were pooled from individual experiments.

for wheat. This will lead to a more comprehensive simulation model on the formation of various protein components in rice grains by expanding the present model of total protein accumulation.

Accurate grain protein simulation is based on reliable inputs required by the submodel. We used the RiceGrow model to provide these inputs, and it appears that sufficiently accurate inputs were provided to adequately simulate grain protein formation. The RiceGrow model has been validated for different cultivar types and management plans in the main rice-growing regions of China in other previ-

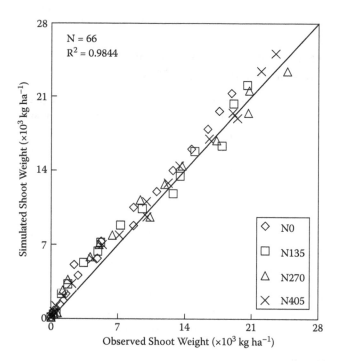

FIGURE 7.7 Comparison of simulated with observed shoot dry weights under different nitrogen rates with cultivar 9325. (Data obtained from Meng et al., 2004.)

ous studies. As detailed by Meng (2002) and Meng et al. (2003, 2004), the model exhibited reliable performance in predicting phenological development, leaf area growth, biomass production, and grain yield formation under normal cultural conditions. Further improvements are needed to enhance model predictions under high-yielding conditions as influenced by interactive nitrogen rates and water regimes. In particular, the model has shown a tendency to underpredict leaf area and biomass under high-yielding growing conditions. In addition, the processes of carbon and nitrogen assimilation and translocation should be better integrated for a coordinated simulation of starch and protein accumulation in grains. Another task is to make the model more user friendly by simplifying data input and standardizing parameterization.

Given that there is only one cultivar-specific genetic parameter (typical grain protein concentration) required by the grain protein submodel, we are encouraged that the grain protein submodel can be utilized by other rice process-based growth models such as CERES-Rice (Ritchie et al. 1987), RCSODS (Gao et al. 1992), SIM-RIW (Horie et al. 1995), and ORYZA2000 (Bouman et al. 2001) that can simulate the required inputs (e.g., LAI, aboveground dry weight, and N and water deficit factors). This provides an extension to existing rice growth models and contributes to technological tools available for rice quality management. One caveat is that the PDT model will need to be adjusted to accommodate the different phenological development approaches used in these models.

7.6   SUMMARY AND CONCLUSION

Grain protein concentration is an important quality index, and formation of grain protein largely depends on pre-anthesis nitrogen assimilation and post-anthesis nitrogen remobilization. In this study, we developed a simplified process model for simulating plant nitrogen uptake and remobilization and grain protein formation. Six field experiments involving different years, sites, varieties, nitrogen rates, and irrigation regimes provided the data for model building, genotypic parameter estimation, and model validation. Using physiological development time (PDT) as the time scale of developmental progress and cultivar-specific grain protein concentration as a genotypic parameter, the dynamic relationships of plant nitrogen accumulation and translocation with environmental and genetic factors were quantified and incorporated into the model. Pre-anthesis nitrogen-uptake rate changed with PDT in a negative exponential relationship, and post-anthesis nitrogen-uptake rate changed with LAI in an exponential equation. Post-anthesis nitrogen translocation rate depended on the plant nitrogen concentration and dry weight at anthesis and plant nitrogen concentration at maturity. Nitrogen for protein synthesis in the grain was derived from two sources: the nitrogen prestored in vegetative components before anthesis and then remobilized after anthesis, and the nitrogen uptake after anthesis. Model validation of final grain protein concentrations at maturity with the observed data exhibited reliable performance for different cultivars, ecological sites, nitrogen rates, and irrigation regimes. These results suggest that integrating the grain protein formation model with other rice growth models will aid in predicting grain protein concentration and grain protein yield of rice under various environments and genotypes. Future studies are necessary for testing and improving the performance of the model for grain protein formation under diverse conditions, while research is continuing on digitizing the whole rice production system.

ACKNOWLEDGMENT

This research was supported by the National Natural Science Foundation of China (30571092), the National Hi-tech R&D Plan of China (2006AA10Z219, 2006AA10A303), and the Hi-tech R&D Plan of Jiangsu Province (BG2004320, BG2006340).

REFERENCES

Asseng, S., A. Bar-Tal, J. W. Bowden, B. A. Keating, A. van Herwaarden, J. A. Palta, N. I. Huth, and M. E. Probert. 2002. Simulation of grain protein content in wheat. *Eur. J. Agron.* 16: 25–42.

Bi, B. 1980. Effect of low temperature on growth and development. *Plant Physiol. Commun.* 2: 15–17 (in Chinese with English abstract).

Bouman, B. A. M., M. J. Kropff, T. P. Tuong, M. C. S. Wopereis, H. F. M. ten Berge, and H. H. van Laar. 2001. *ORYZA2000: Modeling lowland rice.* Wageningen: IRRI and WURC.

Cao, W., and W. Luo. 2003. *Crop system simulation and intelligent management.* Beijing: Higher Education Press (in Chinese).

Cao, W., and D. N. Moss. 1997. Modeling phasic development in wheat: An integration of physiological components. *J. Agric. Sci.* 129: 163–172.

Cheng, F., and B. Zhu. 1998. Present research on the effect of meteoroecological factors on rice quality. *Chin. Agric. Weather* 5: 39–45 (in Chinese with English abstract).

Dingkuhn, M. 1996. Modeling concepts for the phenotypic plasticity of dry matter and nitrogen partitioning in rice. *Agric. Syst.* 52: 383–397.

Gao, L., Z. Jing, and Y. Huang. 1992. Rice cultivational simulation, optimization, and decision making system. *China Agric.* Sci. Tech: Press Beijing (in Chinese).

Grindlay, D. J. C. 1997. Towards an explanation of crop nitrogen demand based on the optimization of leaf nitrogen per unit leaf area. *J. Agric. Sci.* 12: 377–396.

Hasegawa, T., and T. Horie. 1996. Leaf nitrogen, plant age and crop dry matter production in rice. *Field Crops Res.* 47: 107–116.

Horie, T., H. Nakagawa, and H. G. S. Centeno. 1995. The rice crop simulation model SIM-RIW and its testing. In *Modelling the impact of climate change on rice in Asia*, ed. R. G. Matthews et al., 51–66. Wallingford, U.K.: CAB International.

Hu, J. 2002. Simulation of crop-water relation and decision support system for field water management. Ph.D. diss. Nanjing Agric. Univ. (in Chinese with English abstract).

Hu, J., W. Cao, J. Zhang, D. Jiang, and J. Feng. 2004. Quantifying responses of winter wheat physiological processes to soil water stress for use in growth simulation modeling. *Pedosphere* 14: 509–518.

Huang, F., Z. Sun, P. Hu, and S. Tang. 1998. Present situations and prospects for the research on rice quality forming. *Chin. J. Rice Sci.* 32 (3): 372–376 (in Chinese with English abstract).

Jamieson, P. D., and M. A. Semenov. 2000. Modeling nitrogen uptake and redistribution in wheat. *Field Crops Res.* 68: 21–29.

Jiang, L., T. Dai, S. Wei, X. Gan, J. Xu, and W. Cao. 2003. Genotypic differences and evaluation in nitrogen uptake and utilization efficiency in rice. *Acta Phytoecol. Sin.* 27: 466–471 (in Chinese with English abstract).

Li, W., Y. Zhu, T. Dai, and W. Cao. 2005. An ecological model for predicting amylose concentration in rice grain. *Chin. J. Appl. Ecol.* 16: 1–5 (in Chinese with English abstract).

Lin, Z., X. Mo, and Y. Xiang. 2003. Research advances on crop growth models. *Acta Agron. Sin.* 29: 750–758 (in Chinese with English abstract).

Mae, T. 1997. Physiological nitrogen efficiency in rice: Nitrogen utilization, photosynthesis, and yield potential. *Plant Soil* 196: 201–210.

Martre, P., P. D. Jamieson, M. A. Semenov, R. F. Zyskowski, J. R. Porter, and E. Triboi. 2006. Modeling protein content and composition in relation to crop nitrogen dynamics for wheat. *Eur. J. Agron.* 25: 138–154.

Meng, Y. 2002. A process-based simulation model for rice growth. Ph.D. diss. Nanjing Agric. Univ. (in Chinese with English abstract).

Meng, Y., W. Cao, X. Liu, Z. Zhou, and Q. Jing. 2004. A preliminary study of simulation on shoot dry matter partitioning in rice. *Acta Agron. Sin.* 30: 376–381 (in Chinese with English abstract).

Meng, Y., W. W. Cao, Z. Zhou, and X. Liu. 2003. A process-based model for simulating phasic development and phenology in rice. *Sci. Agric. Sin.* 36: 1362–1367 (in Chinese with English abstract).

Ministry of Agriculture, People's Republic of China. 1988. *NY147-88 measurement methods for rice quality.* Beijing: China Standard Press (in Chinese).

Norman, R. J., D. Guindo, B. R. Wells, and C. E. Wilson. 1992. Seasonal accumulation and partitioning of nitrogen[15] in rice. *Soil Sci. Soc. Am. J.* 56: 1521–1527.

Ntanos, D. A., and S. D. Koutroubas. 2002. Dry matter and N accumulation and translocation for Indica and Japonica rice under Mediterranean conditions. *Field Crops Res.* 74: 93–101.

Pan, J., Y. Zhu, Y. D. Jiang, T. Dai, Y. Li, and W. Cao. 2006. Modeling plant nitrogen uptake and grain nitrogen accumulation in wheat. *Field Crops Res.* 97: 322–336.

Peng, S., and K. G. Cassman. 1998. Upper thresholds of nitrogen uptake rates and associated nitrogen fertilizer efficiencies in irrigated rice. *Agron. J.* 90: 178–185.

Piao, Z., L. Han, and H. J. Koh. 2003. Variations of nitrogen use efficiency by rice genotype. *Chin. J. Rice Sci.* 17: 233–238 (in Chinese with English abstract).

Ritchie, J. T., E. C. Alocilja, U. Singh, and G. Uehera. 1987. *IBSNAT and the CERES-Rice model.* Los Baños: IRRI.

Sheehy, J. E., M. J. A. Dionora, P. L. Mitchell, S. Peng, K. G. Cassman, G. Lemaire, and R. L. Williams. 1998. Critical nitrogen concentrations: Implications for high-yielding rice (*Oryza sativa* L.) cultivars in the tropics. *Field Crops Res.* 59: 31–41.

Sheehy, J. E., M. Mnzava, K. G. Cassman, P. L. Mitchell, P. Pablico, R. P. Robles, H. P. Samonte, J. S. Lales, and A. B. Ferrer. 2004. Temporal origin of nitrogen in the grain of irrigated rice in the dry season: The outcome of uptake, cycling, senescence and competition studied using a ^{15}N-point placement technique. *Field Crops Res.* 89: 337–348.

Su, Z., P. Zhou, N. Xu, and Y. Zhang. 2001. Effects of nitrogen and planting density on N absorption and yield of rice. *Chin. J. Rice Sci.* 15: 281–286 (in Chinese with English abstract).

Wang, W. 1994. Studies on rice uptake and distribution of ^{15}N fertilizer in different varieties of Indica rice and at different growth stages. *Acta Agron. Sin.* 20: 476–680 (in Chinese with English abstract).

Wei, C., S. Lan, and Z. Xu. 2002. Formation of protein bodies in the developing endosperm cell of rice. *Acta Agron. Sin.* 5: 591–594 (in Chinese with English abstract).

Yang, H., H. Zhang, L. Yang, S. Zhang, Q. Dai, and Z. Huo. 2002. Effects of nitrogen operations according to leaf age on yield and quality in good-quality rice. *J. China Agric. Univ.* 7: 19–26 (in Chinese with English abstract).

Yin, X., A. H. C. M. Schapendonk, M. J. Kropff, M. van Oijen, and P. S. Bindraban. 2000. A generic equation for nitrogen-limited leaf area index and its application in crop growth models for predicting leaf senescence. *Ann. Bot.* 85: 579–585.

Ying, J., S. Peng, G. Yang, N. Zhou, R. M. Visperas, and K. G. Cassman. 1998. Comparison of high yield rice in tropical and subtropical environments, II: Nitrogen accumulation and utilization efficiency. *Field Crops Res.* 57: 85–93.

Zhou, G., M. Xu, Z. Tan, and X. Li. 1997. Effects of ecological factors of protein and amino acids of rice. *Acta Ecol. Sin.* 17: 538–542 (in Chinese with English abstract).

Zhuang, H. 2001. Dynamic simulation and management decision-making of nitrogen, phosphorus and potassium in crop-soil system. Ph.D. diss. Nanjing Agric. Univ. (in Chinese with English abstract).

Zhuang, H., W. Cao, S. Jiang, and Z. Wang. 2004. Simulation on nitrogen uptake and partitioning in crops. *Syst. Sci. Compr. Studies Agric.* 20: 5–8 (in Chinese with English abstract).

8 Modeling Water and Nitrogen Uptake Using a Single-Root Concept: Exemplified by the Use in the Daisy Model

Søren Hansen and Per Abrahamsen

CONTENTS

8.1   Introduction 170
8.2   General Overview of the Daisy Model 170
8.3   Generic Crop Model... 173
      8.3.1   Photosynthesis... 174
      8.3.2   LAI and Canopy Structure 175
      8.3.3   Assimilate Partitioning, Respiration, and Net Production 176
              8.3.3.1   Maintenance Respiration.. 176
              8.3.3.2   Assimilate Partitioning 176
              8.3.3.3   Growth Respiration 176
      8.3.4   Senescence 177
      8.3.5   Phenological Development... 177
      8.3.6   Root Production and Development 178
      8.3.7   Water Uptake and Water Stress 178
      8.3.8   Nitrogen Uptake and Nitrogen Stress 179
8.4   Extraction of Soil Water by Roots 180
8.5   Solute Movement to Root Surfaces 185
8.6   Model Application.. 190
8.7   Conclusion... 191
References 193

8.1   INTRODUCTION

The loss of nitrogen (N) from agriculture into aquifers and surface waters is recognized as a threat to surface and groundwater quality. Therefore, agroecological
models capable of simulating the N dynamics in agricultural soils have been developed to support decisions and management related to N in agriculture. In Denmark,
the problem of managing N in agricultural systems was recognized in the 1980s,
and this prompted the development of the physically based Daisy model (Hansen
et al. 1991a). The model was successfully tested in a number of international work-
shops (Vereecken et al. 1991; Willigen 1991; Hansen et al. 1991a, 1991b; Diekkrüger et al. 1995; Svendsen et al. 1995; Smith et al. 1997; L. S. Jensen et al. 1997)
and separate studies (Djurhuus et al. 1999; Hansen et al. 1991c; Mueller et at. 1997;
Mueller et al. 1998). Later, the model was further developed into an open software
system (Abrahamsen and Hansen 2000). This development allowed for the introduction of alternative process descriptions (Van der Keur et al. 2001), distributed
large-scale application (Refsgaard et al. 1999; Thorsen et al. 2001), and applications in connection with assimilation of remote sensing data (Boegh et al. 2004).
Furthermore, the new software architecture allows for an extension of the model
into a two-dimensional version, a work that is ongoing. The Daisy model has been
used extensively for administrative and management-related purposes (e.g., Blicher-
Mathiesen et al. 1990, 1991; C. Jensen and Østergaard 1993; C. Jensen et al. 1992,
1994a, 1994b, 1996; H. E. Jensen et al. 1993; Magid and Kølster 1995; Børgesen
et al. 1997, 2001; Nielsen et al. 2004). The present chapter gives an overview of
the Daisy model, with special focus on the model for plant uptake of water and N,
and related processes. The Daisy model is freely available at the Daisy homepage,
http://www.dina.kvl.dk/~daisy/.

8.2   GENERAL OVERVIEW OF THE DAISY MODEL

The present version of Daisy is a one-dimensional agroecosystem model that, in
brief, simulates crop growth, water and heat balances, organic matter balance, the
dynamics of ammonium and nitrate in agricultural soil, and the fate of pesticides
based on information on system characteristics, management practices, and weather
data (Figure 8.1). The minimum weather data requirement is daily values of incident solar irradiance, air temperature, and precipitation. However, the model can
also make use of more detailed weather data, e.g., hourly values of solar irradiance,
net radiation, temperature, humidity (relative humidity or vapor pressure), wind
speed, and precipitation. Management data includes information on field operations such as tillage, irrigation, fertilization, spraying, sowing, and harvesting. The
time of the field operation can be defined by date, time within the year, or specified
criteria relating to the conditions in the field, e.g., soil temperature, soil moisture
content, or soil water pressure potential. This information is fed to the model by
applying a special input language, which is developed as a part of the Daisy software (Abrahamsen 2007). This makes it possible to build rotations and rather complex management schemes.

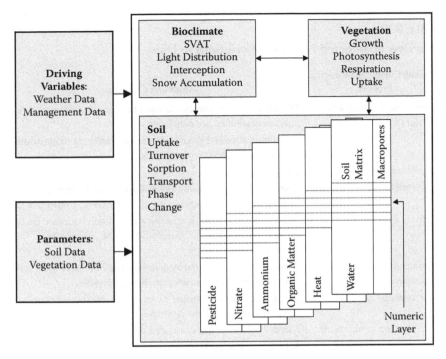

FIGURE 8.1   Schematic representation of the agroecosystem Daisy model, which comprises three main modules: a bioclimate, a vegetation, and a soil component.

As indicated in Figure 8.1, the Daisy model comprises a module for simulating the bioclimate, i.e., submodels for

1. The soil-vegetation-atmosphere transfer (SVAT)
2. Interception of water, N, and pesticides in the vegetation canopy
3. Snow accumulation and melting
4. Light distribution within the canopy

Table 8.1 gives an indication of process descriptions included in Daisy. The model's ability to make use of various process descriptions is used in the SVAT modeling. Evapotranspiration can be modeled by applying a conventional evapotranspiration model based on potential or maximum evapotranspiration, where surface-water balance and surface-energy balance are decoupled, or the evapotranspiration can be modeled by applying the two-source concept for sparse canopies, in which case surface-water balance and surface-energy balance are fully coupled (Van der Keur et al. 2001). The latter approach requires more detailed weather data than the former. The SVAT model predicts the upper limit for the transpiration, which is used by the water-uptake model. The bioclimate module requires information on soil surface conditions and vegetation, the latter being characterized by leaf area index (LAI) and vegetation height and simulated by the vegetation module.

The vegetation module includes a number of different vegetation descriptions. A simple generic crop model is described in a subsequent section.

TABLE 8.1

Overview of Selected Process Descriptions in Daisy

Physical/Chemical/Biological Process	Modeling Concept
Soil-vegetation-atmosphere transfer (SVAT)	Decoupling surface energy and water balance using reference evapotranspiration and crop factors
	Coupling surface energy and water balance using the Shuttleworth and Wallace (1985) two-source model
Interception model	Capacity approach
Snow model	Capacity approach; freezing and melting within the snow pack, depending on air temperature, solar radiation, and ground heat flux
Soil water dynamics (matrix)	Numeric solution to the Richards equation; lower boundary can be selected as free drainage, a groundwater table, or a tile drain condition
Soil water dynamics (macropore)	Bypass flow; flow initiated by potential condition; recharge to soil matrix determined by macropore depth distribution
Water uptake by plants	Single-root concept including a root surface resistance; linked to the SVAT model through the crown potential
Solute transport in soil matrix	Numeric solution to the convection-dispersion equation
Solute transport in macropores	Mass flow determined by the water flow in macropores and the solute concentration at the point of inlet
Soil heat dynamics	Extended Fourier equation taking conduction, convection, and freezing and melting into account
Organic matter turnover	Multipool model including two pools for the soil microbial biomass; the model is driven by C content; pools have fixed C:N ratios; default initialization based on equilibrium assumptions; turnover influenced by soil water pressure potential, soil temperature, and clay content; output is CO_2 evolution, net ammonification, and NO_3 immobilization
Nitrification	Michaelis-Menten type kinetics affected by soil water pressure potential and soil temperature
Denitrification	Autotrophic denitrification is a fixed part of the nitrification; heterotrophic denitrification depends on CO_2 evolution, relative water content, and the availability of NO_3
Nitrogen uptake by plants	Single-root concept taking mass flow and diffusion into account; uptake limited by crop N-demand; NH_4 is taken up in preference to NO_3
Ammonium adsorption	Double Langmuir adsorption isotherm
Crop model	Compartments: leaf, stem, root, storage organ
Photosynthesis	Single leaf photosynthesis, light distribution (Beer's law), diurnal light distribution
Respiration	Growth respiration (conversion efficiency) and temperature affected maintenance respiration
Partitioning	Function of development stage
Development stage	Function of temperature and photoperiod

TABLE 8.1 (Continued)
Overview of Selected Process Descriptions in Daisy

Physical/Chemical/Biological Process	Modeling Concept
Canopy model	LAI and LAI-distribution vs. crop height
LAI	Function of leaf weight and development stage
Plant height	Function of development stage
Root model	
Rooting depth	Function of allocation of assimilates, soil temperature, soil type, and impeding soil layers
Root density distribution	Function of rooting depth and root weight
	Function of rooting depth, root weight, and development stage

The soil module comprises models for

1. Soil water dynamics
2. Soil heat
3. Movement of solutes
4. Sorption of solutes
5. Degradation of solutes (e.g., nitrification, denitrification, decay of pesticides)
6. Production of solutes (e.g., nitrification)
7. Turnover of organic matter (Figure 8.1 and Table 8.1)

Furthermore, the module includes models for the uptake of water and solutes. Strong interaction exists between the models. For example, biologically mediated turnover processes (the organic matter turnover rates, nitrification rates, and denitrification rates) depend on soil temperature (simulated by the soil heat model) and soil moisture content expressed by either volumetric soil moisture content or soil water pressure potential (Table 8.1).

8.3   GENERIC CROP MODEL

The overall objective of the crop model is to simulate crop production and the uptake of water and N by the crop. The crop model is based on carbon flows and assimilation of carbon in carbon pools (Figure 8.2). The model comprises five carbon pools:

1. Assimilate
2. Leaf carbon
3. Stem carbon
4. Root carbon
5. Storage organ carbon (e.g., tuber in a potato crop or ear in a grain crop)

Carbon enters the system by the process of photosynthesis, which produces assimilate. The assimilate may be consumed by maintenance respiration or released

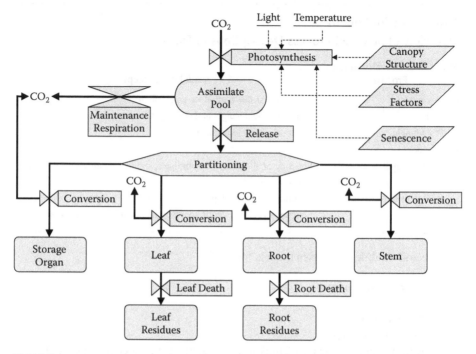

FIGURE 8.2 Overview of the structure of the default crop model included in Daisy. Solid lines represent flows of matter (e.g., carbon). Dashed lines represent flows of information. Partitioning, leaf and root death, senescence, nitrogen stress (stress factors), and canopy structure (LAI and LAI vs. height distribution) are all influenced by the development stage (not shown in the diagram).

and used for growth processes. Maintenance respiration is assumed to have priority over growth processes. The fraction of released assimilate transferred to the individual carbon pools (leaf, stem, root, and storage organ) is determined by the partitioning. During growth, a part of the assimilate is lost as growth respiration during conversion of assimilate to plant material. Finally, carbon can be lost to the senescence of leaves and roots, and other forms of rhizodeposition.

8.3.1 PHOTOSYNTHESIS

The photosynthesis model is based on the calculation of light distribution within the canopy (or composite canopy if more than one plant species is grown in the field) and single light-response curves. The light distribution within the canopy is calculated by the bioclimate module and is based on Beer's law. The extinction coefficient is assumed to be a characteristic for a given crop, and the extinction coefficient for a composite canopy is calculated as a weighted average of the individual crop extinction coefficients. The weight factors are based on the LAI of the individual crops. Reflection coefficients for light are also crop-specific, and the reflection coefficient for a composite crop is calculated in a similar manner as the extinction coefficient. The LAI distribution as a function of plant height for the composite canopy is calculated by adding the contributions of the individual plant species. In the calculation of light distribution, the (composite) canopy is divided into n distinct layers, each containing $1/n$ of the total LAI. The number of layers can be user specified; if not specified, it is by default set to $n = 30$. By applying Beer's law, the absorption of light within layer i, counted from the top of the canopy, can be calculated as

$$S_{a,i} = \left(1 - \rho_c\right) S_{v,0} \left(e^{-k_c(i-1)\Delta L_{ai}} - e^{-k_c i \Delta L_{ai}}\right) \tag{8.1}$$

where $S_{a,i}$ is the absorbed light in layer i, ρ_c is the reflection coefficient of the canopy, $S_{v,0}$ is the incident light above the canopy, k_c is the extinction coefficient and $\Delta L_{ai} = L_{ai}/n$ is the LAI within each canopy layer. When a canopy consists of more than one crop, the absorbed light allocated to each of the crops in a given canopy layer is proportional to the considered crop's contribution to the total LAI within the layer.

Gross photosynthesis is calculated by the vegetation module for each individual crop, layer by layer, by applying a light-response curve:

$$\Delta F_i = x \, \Delta L_{ai} F_m \left(1 - \exp\left(-\frac{\rho}{F_m} \frac{S_{a,i}}{\Delta L_{ai}}\right)\right) \tag{8.2}$$

where ΔF_i is the gross photosynthesis for layer i for the considered crop, x is the LAI fraction of the considered crop, F_m is a crop-specific leaf photosynthetic rate at saturated light intensity, and ε is a corresponding initial light use efficiency at low intensity. F_m is not a constant, but is assumed to be a function of temperature, as illustrated by the indicated influence of temperature on photosynthesis in Figure 8.2. The gross photosynthesis is calculated by accumulating the contribution from the individual layers. The time step in this part of the model is 1 h, and the produced assimilates are transferred hourly to the carbohydrate reserves (Figure 8.2).

The gross photosynthesis may be reduced due to senescence, water stress, and N stress. This reduction is introduced by applying stress factors.

8.3.2 LAI AND CANOPY STRUCTURE

LAI is calculated by equation (8.3):

$$L_{ai} = S_{la} W_{leaf} \tag{8.3}$$

where L_{ai} is LAI; S_{la} is specific leaf area, which is assumed to be a function of the development stage (DS); and W_{leaf} is leaf weight. Stem and storage organs may also contribute to effective LAI. Their contribution is calculated analogous to the contribution from the real leaves, using specific area and weight for stem and storage organs, respectively. In addition, a weight factor accounting for the different photosynthetic efficiencies of stem, storage organs, and leaf is used in the calculation of the effective LAI. The weight factor is calculated as the ratio between the photosynthetic rate at saturated light intensity for the stem or storage organ to the corresponding value for the leaf. If the required parameters are missing, then the contribution from stem or storage organs is neglected. If more than one plant species is grown in the field, then information on the leaf area distribution with height (LAD) of the individual plant species is needed. LAD is calculated from efficient LAI and a predefined relative LAD distribution, which is a function of the DS. The composite canopy LAD is obtained by adding the individual crop LAD distributions.

The simulated LAI influences the simulation of light interception, the transpiration process, and the interception of water and matter. Hence, it is a key element in the model.

8.3.3 Assimilate Partitioning, Respiration, and Net Production

Respiration is assumed to comprise growth and maintenance respiration (McCree 1974). Maintenance respiration is assumed to have priority over growth respiration; hence production only takes place if the available carbohydrate reserves exceed the required maintenance respiration. If a surplus of carbohydrate reserves exists, then this surplus is partitioned between the considered crop components—root, stem, leaf, and storage organs—and growth respiration is subtracted to calculate net production.

8.3.3.1 Maintenance Respiration

Maintenance respiration is assumed to be proportional to the dry weight of the plant components, and each component is assumed to be characterized by a maintenance respiration coefficient, which is temperature-dependent:

$$R_m^{component} = r_m^{component}\left(T\right) W_{component} \tag{8.4}$$

where R_m is the maintenance respiration, $r_m(T)$ is the maintenance respiration coefficient at the temperature T, and W is the dry weight of the considered crop component. The crop maintenance respiration is the accumulated maintenance respiration originating from the maintenance respiration of the individual crop components.

8.3.3.2 Assimilate Partitioning

The model only considers determinate crops. Furthermore, it is assumed that stress factors do not influence the assimilate partitioning; hence it can be assumed that partitioning is a function of the DS only. In the model, the partitioning is described by piecewise linear functions—$\gamma_r(DS)$, $\gamma_s(DS)$, $\gamma_l(DS)$, and $\gamma_o(DS)$—representing the allocation to root, stem, leaf, and storage organ, respectively. Note that first $\gamma_r(DS)$ is allocated to the root, and then $1 - \gamma_r(DS)$ is allocated to the shoot, which is assumed to comprise stem, leaf, and storage organs. Then the allocation to the shoot is distributed among stem, leaf, and storage organ, hence $\gamma_s(DS) + \gamma_l(DS) + \gamma_o(DS) = 1$.

8.3.3.3 Growth Respiration

The growth respiration rate is assumed to depend only on the end product formed; hence it can be characterized by a conversion efficiency. After subtraction of growth respiration, the net production for a specific crop component yields

$$\left(\frac{\Delta W_{\text{component}}}{\Delta t} \right) = E^{\text{component}} \gamma^{\text{component}} \left(F - \sum_{j}^{\text{components}} R_{\text{m}}^{j} \right) \qquad (8.5)$$

where $(\Delta W / \Delta t)$ is the net production rate, E is the conversion efficiency, γ is the fraction of assimilate allocated to the considered crop component (component = root, leaf, stem, and storage organ), and F is the assimilate flow from the carbohydrate reserves. F is released from the carbohydrate reserve pool by first-order kinetics, i.e., the release rate is proportional to the size of the carbohydrate reserve pool.

8.3.4 SENESCENCE

As indicated in Figure 8.2, it is assumed that root and leaf material is lost during growth due to senescence and shading. The rate at which matter is lost is assumed to be proportional to the leaf weight. The proportionality factor is divided into two components. One component is assumed to be a piecewise linear function of DS. Another component, being constant, is only brought into play when the irradiance received by the lower shaded leaves falls below a certain threshold, i.e., when transmission of light falls below a predefined value, typically around 5% (Monteith and Unsworth 1990). Furthermore, gross photosynthesis can be reduced due to senescence. This is introduced by multiplying the photosynthetic rate at saturated light intensity (F_{m}) by a reduction factor that is a function of DS.

8.3.5 PHENOLOGICAL DEVELOPMENT

The development stage of a crop, DS, influences many of the internal plant processes, as it quantifies the physiological age of the plant. It is related to the morphological appearance of the plant. In the Daisy crop model, DS has the value of 0 at emergence, 1 at flowering, and 2 at maturation. In the real world, the rate of phenological development is influenced by a number of environmental factors. However, the model only takes the effect of temperature and day length into account. The former is assumed to influence development from emergence (DS = 0) to maturation (DS = 2), while the latter may only influence the vegetative stage of the crop, i.e., from DS 0 to DS 1. It is a basic assumption that crop growth *per se* has no influence on the rate of phenological development.

Daily increments of the development stage, ΔD, are calculated from equation (8.6):

$$\Delta D = d \, f_{t} \left(T_{a} \right) f_{d} \left(D_{l} \right) \qquad (8.6)$$

where d is the development rate at reference temperature and reference day length, $f_{t}(T_{a})$ and $f_{d}(D_{l})$ are modifiers accounting for air temperature, T_{a} (daily values), and day length, D_{l}, respectively. The modifier functions are obtained by linear interpolations between tabulated values of response versus environmental factor. The adopted approach is flexible and allows for a description corresponding to a simple degree-day approach as well as much more complex responses.

Spatial root length density distribution and crop N demand are simulated by the Daisy crop model and form an important input to the uptake model for water and N.

8.3.6 ROOT PRODUCTION AND DEVELOPMENT

The root system is characterized by root weight, rooting depth, and root length density distribution. Root penetration is assumed to take place if the following conditions are fulfilled:

1. Daily net root production is positive
2. The soil temperature at the root tip is above a certain threshold temperature, which is an input parameter (default value 4°C)
3. The actual rooting depth is less than a maximum rooting depth

Maximum rooting depth is determined either by the plant species itself or by the chemical or mechanical properties of the particular soil considered. Daily root penetration rate, $(\Delta d_r/\Delta t)$, is calculated according to Jakobsen (1976):

$$\left(\frac{\Delta d_r}{\Delta t}\right) = \begin{cases} 0 & T_s \leq T_p \\ \alpha_r(T_s - T_p) & T_s > T_p \end{cases} \tag{8.7}$$

where α_r is a root penetration parameter, T_s is the soil temperature at the root tip, and T_p is the threshold temperature. The equation is used for calculation of potential as well as actual root penetration. Potential rooting depth is the rooting depth that would have occurred if the root penetration were not hampered by the soil. The total root length is assumed to be proportional to the root weight. The potential root length density distribution is described in accordance with a logarithmic root length density distribution (Gerwitz and Page 1974), assuming that the root length density at the potential rooting depth is 0.1 cm·cm^{-3}. If the actual rooting depth equals the potential one, then the actual root length density distribution equals the potential one. If this is not the case, then the actual root length density distribution is calculated again by assuming the Gerwitz and Page distribution and setting the root length density at the actual rooting depth equal to the density obtained from the potential distribution at this depth.

A version where the root distribution is made a function of rooting depth and DS is also included in the Daisy model as an option.

8.3.7 WATER UPTAKE AND WATER STRESS

The water stress model is based on the assumption that transpiration as well as CO_2 assimilation is governed by stomata responses. Furthermore, it is assumed that stomata are open when intercepted water is evaporated from the leaf surfaces. These assumptions lead to the approximation:

$$F_w = F_p \frac{E_t + E_i}{E_{t,p} + E_{i,p}} \tag{8.8}$$

where F_w is water-limited photosynthesis, F_p is potential photosynthesis, E_t and $E_{t,p}$ are actual and potential transpiration, respectively, and E_i and $E_{i,p}$ are actual and potential evaporation of intercepted water, respectively. The estimation of E_t is described in detail in section 8.4, Extraction of Soil Water by Roots.

8.3.8   NITROGEN UPTAKE AND NITROGEN STRESS

The upper limit for N uptake by the crop is determined by the difference between a potential N content, N_c^p, in the crop and the actual N content in the crop, N_c^a, i.e., $N_c^p - N_c^a$. The value for N_c^p is calculated as

$$N_c^p = \sum_{j}^{\text{components}} {}^pC_j W_j \tag{8.9}$$

where ${}^pC_j$ is crop-specific potential N concentrations in the considered crop component (j = root, leaf, stem, and storage organ), and W_j is the corresponding dry-matter weight. The potential N concentrations are functions of DS. The potential uptake determined by the crop demand, U_d, is

$$U_d = \frac{N_c^p - N_c^a}{\Delta t} \tag{8.10}$$

where Δt is the considered time step. The actual uptake by the crop is described later. As long as the actual N content of the crop, N_c^a, exceeds a critical value, N_c^c, no N stress exists. The critical value is calculated as

$$N_c^c = \sum_{j}^{\text{components}} {}^cC_j W_j \tag{8.11}$$

where ${}^cC_j$ is the crop-specific critical N concentration in each crop component (root, leaf, stem, and storage organ). The crop-specific critical N concentrations are functions of DS. However, if N_c^a falls below N_c^c, then N stress occurs. It is assumed that N stress influences the gross photosynthesis

$$F_N = F_w \frac{N_c^a - N_c^n}{N_c^c - N_c^n} \tag{8.12}$$

where F_N is the N-limited gross photosynthesis and N_c^n is a so-called nonfunction N content of the crop, which are calculated analogous to the critical N content, just replacing the critical concentrations in equation (8.11) by corresponding nonfunction concentrations. Parameters governing the critical and nonfunctioning level are obtained by calibration.

8.4   EXTRACTION OF SOIL WATER BY ROOTS

The calculation of the extraction of soil water by plant roots is based on the following assumptions:

1. The root extracts water from a cylindrical soil volume around it, and the radius of this volume corresponds to half the average distance between roots.
2. Flow toward the root is radial and can be described by the Darcy equation.
3. The pressure potential at the outer boundary of the considered soil cylinder equals the bulk pressure potential as obtained from the solution to the Richards equation.
4. The potential drop toward the root surface can be approximated by a series of steady-state profiles.
5. The plant determines the pressure potential at the root surface; however, this potential is limited by the permanent wilting point.
6. At the root surface, a contact resistance exists that can be evaluated according to Herkelrath et al. (1977), i.e., by the ratio between the soil water content at the root surface and the soil water content at saturation (θ_r/θ_s).

These assumptions lead to the following expression:

$$S_r = 4\pi L \frac{\theta_r}{\theta_s} \frac{M(h_p) - M(h_r)}{-\ln(r_r^2 \pi L)} \tag{8.13}$$

where S_r is a volumetric sink term (water uptake by roots), L is the root length density, θ_r is soil water content at h_r, θ_s is the soil water content at saturation, h_r is the soil water pressure potential at the root surface, r_r is the root radius, and M is the matrix flux potential, which is a function of the pressure potential

$$M(h_p) = \int_{-\infty}^{h_p} K dh \tag{8.14}$$

where h_p is the pressure potential at the outer boundary of the considered soil cylinder, which is approximated by the bulk pressure potential as estimated from the solution to the Richards equation.

Figures 8.3 and 8.4 illustrate the dynamics predicted by equation (8.13). The figures are based on a root length density of $L = 3$ cm·cm^{-3}, a root radius of $r_r = 0.01$ cm, and hydraulic parameters described by the Campbell/Burdine model (Campbell 1974; Burdine 1952). The hydraulic properties and the corresponding volumetric sink term are

Retention: $$\frac{\theta}{\theta_s} = \left(\frac{h_b}{h_p}\right)^{1/b}$$

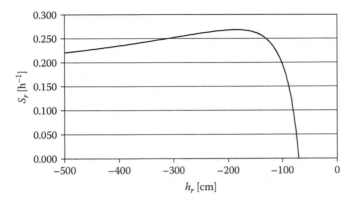

FIGURE 8.3   Volumetric sink term as a function of root surface potential h_r.

Hydraulic conductivity: $K = K_s \left(\dfrac{h_b}{h_p} \right)^{2+\frac{3}{b}}$

Matrix flux potential: $M(h) = \dfrac{-h_b K_s}{1+3/b} \left(\dfrac{h_b}{h} \right)^{1+\frac{3}{b}}$

Volumetric sink term: $S_r = \dfrac{4\pi L h_b K_s}{(1+3/b)\ln(\pi r_r^2 L)} \left(\dfrac{h_b}{h_r} \right)^{\frac{1}{b}} \left[\left(\dfrac{h_b}{h_p} \right)^{1+\frac{3}{b}} - \left(\dfrac{h_b}{h_r} \right)^{1+\frac{3}{b}} \right]$

where b is Campbell-b (3, corresponding to a loamy soil), h_b is the bubbling pressure (-10 cm), and K_s is the saturated hydraulic conductivity (3 cm·h^{-1}). Figure 8.3 shows a rapidly increasing water uptake when the potential at the root surface, h_r, decreases below the soil bulk pressure potential, h_p (the present case being -70 cm). This continues until a maximum uptake rate is reached, after which the uptake rate slowly decreases as the potential at the root surface, h_r, decreases further. If the contact resistance at the surface (θ_r/θ_s) were omitted, then equation (8.13) would be a monotonically increasing function. Figure 8.4 illustrates the maximum uptake as a function of the bulk soil pressure potential. Between field capacity and the wilting point, the uptake changes around five orders of magnitude. In the present example, the maximum uptake rate occurs at

$$h_r = (4+b)^{(b+1)/(b+3)} h_p$$

corresponding to $h_r \approx 3.66 \, h_p$.

During a simulation, L, r_r, h, and hence $M(h)$ are known, while S_r, h_r, and hence θ_r and $M(h_r)$ are unknown. Applying equation (8.13) to a soil profile, h_r must be known at any point along the roots within the soil profile. It is assumed that h_r can be estimated on the basis of the crown potential ψ_x, i.e.,

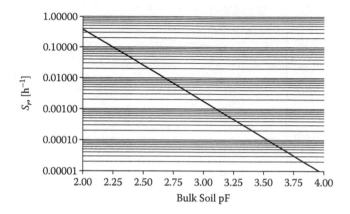

FIGURE 8.4 Maximum volumetric sink term as a function of bulk soil water pressure potential, h_p.

$$h_r = \psi_x + \left(1 + R_x\right) z \tag{8.15}$$

where R_x is a vertical transport resistance coefficient and z is the vertical coordinate. (In the present case, z assumes the value zero at the soil surface, where the crown potential is also assessed; z is positive upward.) The crown potential, ψ_x, reflects the state of the plant.

The crown potential, ψ_x, can be calculated by a comprehensive detailed SVAT model, taking the water balance of the leaves into account, or by a simpler SVAT based on the concept of potential evapotranspiration. In the latter case, ψ_x is just an internal model variable, which is assessed as indicated below.

The total or integrated uptake can be calculated by equation (8.16):

$$U_w = \int_0^{z_r} S_r dz \tag{8.16}$$

If changes of the water content of the plant are neglected, then the integrated uptake corresponds to the transpiration, i.e., $E_t = U_w$.

Two different cases exist:

1. Transpiration at a potential rate
2. Transpiration at a lower rate than the potential rate

In the first case, it is the climatic conditions that determine the water uptake by the plants, and the potential transpiration, $E_{p,t}$, can be calculated by the SVAT, and we get

$$U_w = E_{p,t} \tag{8.17}$$

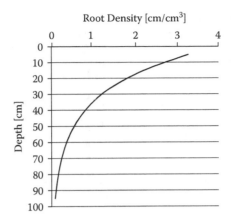

FIGURE 8.5 Root profile per the model of Gerwitz and Page (1974).

The required uptake corresponds to a specific value of the crown potential ψ_x. The problem is now to find this value by combining equations (8.13), (8.15), (8.16), and (8.17). In general, this has to be done by iteration. In Daisy, we apply this model and use a time step of 1 h.

In the second case, it is assumed that it is the transport of water from the bulk soil to the root surface that determines the water uptake. In this case, it is assumed that a common pressure potential exists along the root (h_r) and that the value of both ψ_x and h_r can be approximated by the pressure potential at wilting point.

To calculate the water uptake from a soil profile, the root length density distribution is required. Figure 8.5 shows a root distribution predicted by the model of Gerwitz and Page (1974), i.e., a logarithmic root density distribution. This distribution is used in the following predictions. Figure 8.6 illustrates the uptake pattern predicted by equation (8.13) when combined with equation (8.15). The model is parameterized as in Figures 8.3 and 8.4. The transport resistance coefficient of equation (8.15) is assumed to be 0.1 (dimensionless). Figures 8.6a, 8.6c, and 8.6e illustrate a wet, medium-dry, and dry soil profile, respectively, and Figures 8.6b, 8.6d, and 8.6f show the corresponding uptake patterns. The maximum uptake rates are shown for the wet profile in Figure 8.6b. The figure indicates that the uptake capacity of the root system by far exceeds any realistic demand (the integrated max uptake is 41 mm/h, equation [8.16]). In addition, the uptake corresponding to a crown potential of −137 cm, which corresponds to a total uptake of 0.51 mm/h or a latent heat flux of 347 W/m², is shown. It is noted that uptake is predicted only to take place in the upper part of the profile.

For the medium-dry profile in Figures 8.6c and 8.6d, the model predicts that a reasonable demand can be met. The integrated maximum uptake is 0.39 mm/h, corresponding to a latent heat flux of 265 W/m². A crown potential of −1700 cm nearly provides this uptake (0.37 mm/h, 252 W/m²). A crown potential of −1900 cm would provide an uptake very close to the maximum uptake (not shown). Even though the upper part is the drier part of the soil, most of the uptake takes place here. Lowering the crown potential to −1000 cm changes the uptake pattern, and the lower, wetter

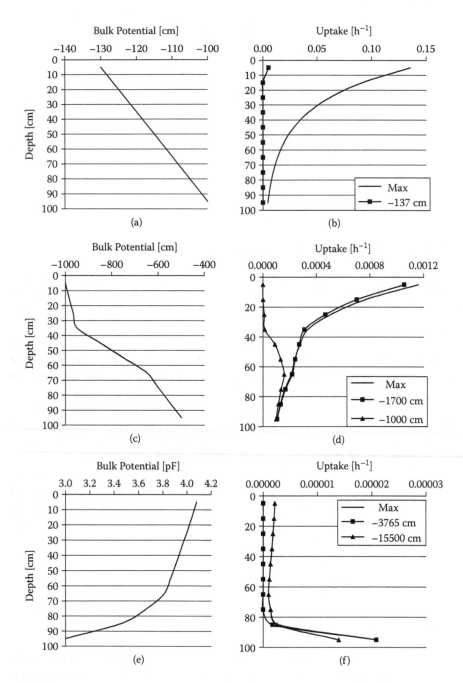

FIGURE 8.6 Soil water pressure potential profiles (a, c, and e) and corresponding volumetric soil volumetric sink terms (b, d, and f). Maximum volumetric sink terms and volumetric sink terms corresponding to selected crown potentials are shown: a and b, wet soil; c and d, medium-dry soil; e and f, dry soil.

part of the soil contributes solely to the uptake. However, in this case the integrated uptake only corresponds to 0.08 mm/h or 52 W/m².

Figures 8.6e and 8.6f show a typical prediction of the uptake in a dry soil. The crown potential of −15,500 cm corresponds to a plant at the wilting point. Most of the uptake occurs in the lower, less dry part of the profile. The integrated uptake amounts to 0.0029 mm/h (2 W/m²), which is a very low value. However, it is close to the integrated maximum uptake of 0.0036 mm/h. The crown potential of −3765 cm corresponds to an optimum h_r at the root surface at 95-cm depth, corresponding to the lowest, moistest soil layer. This crown potential yields an uptake of 0.0023 mm/h. In the present case, no crown potential will yield the maximum uptake in all soil layers within the profile.

The present model of water uptake (equation 8.13) is aimed at being a part of a soil moisture model based on a solution to the Richards equation. Information required in addition to that required by the Richards equation is a characterization of the plant, i.e., a root length density distribution and a characteristic root radius, both of these having physical meaning.

8.5 SOLUTE MOVEMENT TO ROOT SURFACES

Both ammonium and nitrate are taken up by plants by an active process. It is assumed that ammonium has priority over nitrate. Along this line, it is assumed that the plant regulates the uptake by regulating the nutrient concentration at the root surface. Furthermore, it is assumed that both mass flow and diffusion may contribute to the movement of the solute to the root surfaces.

The transport of a solute (ammonium or nitrate) from the bulk soil to the root surfaces is based on a number of assumptions similar to those adopted for water flow:

1. Each root may exploit an average effective volume of soil which is assumed to be a cylinder around the root.
2. The radius of this cylinder is assumed to correspond to the average half distance between the roots.
3. The solute is transferred to the root surface by both mass flow and diffusion.
4. The concentration–distance profile around a root develops in time in a step-wise manner, and at each time step, it approximates to a steady-state profile (Baldwin et al. 1973).

Based on these assumptions, the solute-flux toward the root surface is

$$I = \begin{cases} 4\pi D(C - C_r)\left[\dfrac{\beta^2 \ln \beta^2}{\beta^2 - 1} - 1\right]^{-1} & \alpha = 0 \\[2em] 4\pi D \dfrac{(\beta^2 - 1)C - \ln(\beta^2)C_r}{(\beta^2 - 1) - \ln(\beta^2)} & \alpha = 2 \\[2em] 2\pi D\alpha \dfrac{(\beta^2 - 1)(1 - \alpha/2)C - (\beta^{2-\alpha} - 1)C_r}{(\beta^2 - 1)(1 - \alpha/2) - (\beta^{2-\alpha} - 1)} & \text{else} \end{cases} \qquad (8.18)$$

$$\alpha = \frac{S_r}{2\pi LD} \qquad (8.19)$$

$$\beta = (r_r^2 \pi L)^{-1/2} \qquad (8.20)$$

where I is the solute uptake per unit length of the root; D is the diffusion coefficient of the nutrient in the soil; C is the bulk concentration of the nutrient in solution, which is obtained from the solution of the convection-dispersion equation; C_r is the concentration at the root surface; S_r is a volumetric sink term obtained from equation (8.13); r_r is the root radius; and L is the root length density.

It is noted that the solute uptake per unit length of the root is proportional to the diffusion coefficient. Furthermore, if water uptake occurs, then I is approximately proportional to α and hence to the water uptake. The model predictions do not depend on the total amount of N present in the soil, but only on the concentration in the soil solution. The model itself is static (based on steady-state conditions). However, applied to a transient system, the model works as a series of quasi-steady states; hence soil N content changes in time and, therefore, concentrations also change in time. As a result, soil N content is important when applying the model to a transient system, although only in an indirect way.

In the soil, diffusion is influenced by the water content of the soil both in terms of the diffusion cross-section and the tortuous pathway followed by the solute through pores. The bulk soil diffusion coefficient is calculated as

$$D = \theta D_1 f_1 \qquad (8.21)$$

where θ is the volumetric soil water content, D_1 is the diffusion coefficient in free solution, and f_1 is a so-called tortuosity factor, which can be estimated in several ways. In Daisy, the default tortuosity factor is a stepwise linear model:

$$f_1 = \begin{cases} f_1^0 & \theta \le \theta_0 \\ f_1^0 + a\left(\theta - \theta_0\right) & \theta > \theta_0 \end{cases} \qquad (8.22)$$

where f_1^0, θ_0, and a are constants. A value of f_1^0 equal to 10^{-6} is selected, while a and θ_0 are parameters characterizing the soil (default values: $a = 2$, $\theta_0 =$ soil water content at the permanent wilting point).

Calculation of the dimensionless parameters α and β does not require additional information, as compared with the model for extraction of soil water by roots, except for the diffusion coefficient of the solute in free solution; hence they can be considered known. The diffusion coefficient of nitrate in free solution is around 0.072 cm$^2 \cdot$h^{-1}. The parameter α reflects the relation between the effectiveness of mass transfer to diffusion; $\alpha = 0$ corresponds to an instance where the mass-transfer contribution is zero and diffusion is the only mechanism contributing to transfer; α typically assumes a value between 0 and 0.1. The value $\alpha = 2$ has no special physical meaning but appears due to a mathematical singularity. The β value characterizes the geometry of the system; high values of β correspond to sparse root systems and low values correspond to the high root length density. In the example of Figure 8.5, β assumes a value around 30 in the top of the soil profile, and a value of around 190 at the bottom of the root zone. From a modeling point of view, the unknown variables of equation (8.18) are I and C_r.

A linear relation exists between I and C_r, i.e.,

$$I = a_z + b_z C_r \tag{8.23}$$

where

$$
a_z = \begin{cases}
4\pi D \left[\dfrac{\beta^2 \ln \beta^2}{\beta^2 - 1} - 1 \right]^{-1} C & \alpha = 0 \\[3ex]
4\pi D \dfrac{\left(\beta^2 - 1\right)}{\left(\beta^2 - 1\right) - \ln\left(\beta^2\right)} C & \alpha = 2 \\[3ex]
2\pi D \dfrac{\alpha\left(\beta^2 - 1\right)\left(1 - \alpha/2\right)}{\left(\beta^2 - 1\right)\left(1 - \alpha/2\right) - \left(\beta^{2-\alpha} - 1\right)} C & \text{else}
\end{cases}
\tag{8.24}
$$

and

$$
b_z = \begin{cases}
-4\pi D \left[\dfrac{\beta^2 \ln \beta^2}{\beta^2 - 1} - 1 \right]^{-1} & \alpha = 0 \\[3ex]
-4\pi D \dfrac{\ln\left(\beta^2\right)}{\left(\beta^2 - 1\right) - \ln\left(\beta^2\right)} & \alpha = 2 \\[3ex]
-2\pi D \dfrac{\alpha\left(\beta^{2-\alpha} - 1\right)}{\left(\beta^2 - 1\right)\left(1 - \alpha/2\right) - \left(\beta^{2-\alpha} - 1\right)} & \text{else}
\end{cases}
\tag{8.25}
$$

From a mathematical point of view, the coefficients a_z and b_z can be considered a function of depth. The information required to calculate the coefficients comprises the information to calculate the water uptake supplemented with information on the concentration of the solute as a function of depth and of the diffusion coefficient of the solute in free solution.

Estimating solute uptake from equation (8.23) requires information on C_r. As mentioned in the beginning of this section, N uptake is assumed to be an active process. Subsequently, it is assumed that the plant can regulate the uptake by regulating C_r. Furthermore, it is assumed that the regulation will result in a common value of C_r along all root surfaces of the root system. However, in the case of very high N concentrations in the soil solution in combination with a high N demand, which may occur after a split application of fertilizer, it is assumed that the uptake I is limited by a maximum absorption rate, I_{Max}, which cannot be exceeded. The value of I_{Max} can be obtained from hydroponic experiments. The default value in the Daisy code for this parameter is 0.25 µg/cm/h. Hence, the uptake at a given location in the soil is now calculated as

$$S_s = L \, \mathrm{Min} \left\{ I; I_{Max} \right\} = \mathrm{Min} \left\{ La_z + Lb_z C_r ; L I_{Max} \right\} \qquad (8.26)$$

Finally, the uptake within a soil profile, U_s, is calculated as

$$U_s = \int_0^{z_r} S_s dz = \int_0^{z_r} L \, \mathrm{Min} \left\{ a_z + b_z C_r ; I_{Max} \right\} dz \qquad (8.27)$$

Using equation (8.27), two different instances can occur:

1. The uptake is limited by the plant's ability to take up N, i.e., the crop demand.
2. The uptake is limited by the ability of the soil to deliver N to the plant.

In the former case, the total uptake, U_s, is known, and the task is to distribute the uptake over the entire root zone. This is done by solving equation (8.27) and subsequently applying equation (8.26). Soil layers in which $C < C_r$ are assumed not to contribute to the solute uptake. In the latter case, when the uptake is limited by the availability of the solute, it is assumed that C_r is equal to a predefined minimum value, which may be zero, and in this case the root acts as a zero sink. In this case, the total uptake of the solute is calculated by integrating I over the entire root system.

The uptake pattern predicted by the uptake model, equation (8.26), is illustrated in Figures 8.7 and 8.8. The root length density distribution is shown in Figure 8.5, and the nitrate–N concentration is shown in Figure 8.7a. The N concentration profile corresponds to a N content of 35.5, 18.7, and 9.2 kg N/ha in the wet soil (Figure 8.6a), medium-dry soil (Figure 8.6c), and dry soil (Figure 8.6e), respectively.

The N uptake in Figure 8.7b corresponds to the soil water potentials and water uptake in Figures 8.6a and 8.6b, respectively (wet soil). Note that the uptake shows a strong dependency on the regulating nitrate concentration at the root surface, C_r. The values of C_r equal to 5 and 15 mg NO_3–N/L correspond to total N uptakes of 1.44 kg N/ha/h and 0.71 kg N/ha/h, respectively. Furthermore, it is noted that the patterns of water uptake and N uptake are quite different, highlighting that diffusion is both an efficient and important mechanism in the uptake process. In both cases, nearly two-thirds of the N uptake takes place in soil layers with no water uptake.

The N uptake in Figure 8.7c, which is calculated for $C_r = 5$ mg NO_3–N/L, corresponds to the soil water potentials and water uptake in Figures 8.6c and 8.6d, respectively (medium-dry soil). It is noted that even though the patterns of water

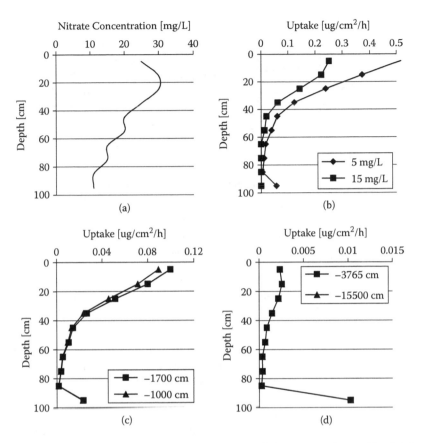

FIGURE 8.7   Soil nitrate concentration profile (a) and nitrogen-uptake profiles for a wet soil (b, corresponding to Figure 8.6b and two different values of C_r), a medium-dry soil (c, corresponding to Figure 8.6d and $C_r = 5$ mg NO_3–N/L), and a dry soil (d, corresponding to Figure 6f and $C_r = 5$ mg NO_3–N/L).

uptake differ significantly, the patterns of N uptake differ only marginally. The total N uptake is 0.32 kg N/ha/h and 0.29 kg N/ha/h for the cases with the large water uptake (0.37 mm/h) and the small water uptake (0.08 mm/h), respectively. Again, this highlights the importance of the diffusion mechanism.

The N uptake in Figure 8.7d, which is calculated for $C_r = 5$ mg NO_3–N/L, corresponds to the soil water potentials and water uptake in Figures 8.6e and 8.6f, respectively (dry soil). Again it is noted that the pattern of N uptake only differs marginally, as does the total uptake (less than 2% at a low insignificant level around 0.02 kg N/ha/h). Furthermore, it is noted that the uptake is dominated by the uptake in the bottommost part of the soil profile. This part of the profile is the moistest, and hence the diffusion coefficient in the soil is the highest in the profile, supporting transfer by diffusion as being of prime importance. (At this depth, mass flow contributes only about 2% to the total uptake.) Comparing Figures 8.7b, 8.7c, and 8.7d—the wet, medium-dry, and dry soils, respectively—shows an appreciable difference in the uptake rates.

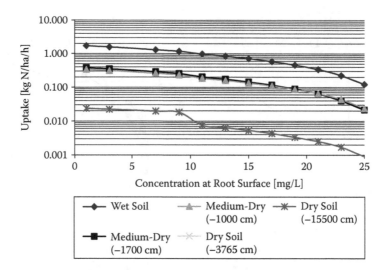

FIGURE 8.8 Nitrate uptake as a function of the nitrate–N concentration at the root surfaces for the five cases: wet soil, medium-dry soil (crown potential –1000 cm), medium-dry soil (crown potential –1700 cm), dry soil (crown potential –3765 cm), and dry soil (crown potential –15,500 cm).

Figure 8.8 show the N-uptake rate as a function of the nitrate–N concentration at the root surfaces for the five considered cases, i.e., wet soil, medium-dry soil (crown potential –1000 cm and –1700 cm), and dry soil (crown potential –3765 cm and –15,500 cm). Note that the ordinate axis is logarithmically transformed and covers the range of approximately three decades. It appears that the size of the crown potential has limited influence on the N-uptake rate. For example, in the figure the curve representing the crown potential of –15,500 cm completely covers the curve representing the crown potential of –3765 cm. However, the wetness of the soil and the nitrate–N concentration at the root surfaces are of prime importance for the uptake. Furthermore, it is noted that, in the case of dry soil, a jump in the N-uptake rate appears around the nitrate–N concentration at the root surfaces of 10 mg/L, which approximately corresponds to the threshold for diffusion-facilitated uptake in the lowest and wettest layer of the soil profile (Figure 8.7a). Again, this can be attributed to the importance of the diffusion process.

8.6 MODEL APPLICATION

The presented uptake model has been applied to highly diversified conditions since the first international presentations of the Daisy model at a workshop in Haren, The Netherlands, in 1991 (Hansen et al. 1991a). In general, the approach has proven itself robust, and the conclusion from the workshop has been supported.

At the Haren workshop, celebrating the centennial of The Institute for Soil Fertility Research, the model test was based on data sets originating from winter wheat experiments at three locations during two years (Groot and Verberne 1991). Each experiment comprised three different N treatments. The observations included measurements of soil mineral N content, soil water content, groundwater table, dry-matter production and distribution, N uptake, N distribution, and root length density.

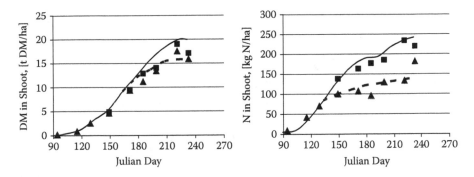

FIGURE 8.9 Comparison of measured (symbols) and simulated (lines) aboveground biomass production (top) and nitrogen content in aboveground biomass of winter wheat (bottom). Squares and solid lines represent high nitrogen fertilization (80+120+40 kg N/ha). Triangles and dashed lines represent low nitrogen fertilization (80 kg N/ha). (Source: Hansen, S., et al. 1991a. *Fert. Res.* 27: 245–259.)

Results obtained by the Daisy model at the workshop were published by Hansen et al. (1991a), and further results obtained with Daisy based on the Haren data set can be found in Hansen et al. (1995). A common assumption in these simulations was that the roots could act as a zero sink, i.e., the lowest N concentration at the root surfaces was 0 mg/L. General conclusions from the workshop were that

1. In general, simulation of the aboveground variables yielded better results than that of the belowground variables (soil water and mineral N content).
2. Dry-matter production and N uptake can be simulated satisfactorily in an independent way, i.e., without parameter fitting, as is done in models like Daisy, and hence this approach seems to be an appropriate choice (Willigen 1991).

Figure 8.9 shows the course of simulated and observed biomass production and N uptake for wheat at the Dutch experimental farm PAGV, located in a polder. The corresponding distribution of soil mineral N is shown in Figure 8.10, from which the spatial uptake pattern can be deduced. It is noted that very good simulations with respect to both the soil mineral N and the dry-matter content and N content in the aboveground part of the plant are obtained in the case of fertilizer level N1 (140 kg N/ha). Acceptable results are obtained at the fertilizer level N2 (240 kg N/ha, not shown), while not quite satisfactory results are obtained at fertilizer level N3 (300 kg N/ha). However, it is obvious that the model fails to account for the loss of soil mineral N occurring shortly after application of fertilizer in late spring and early summer. At the workshop, the reason for this loss was discussed, and although microbial immobilization was favored, no rational explanation could be given.

8.7 CONCLUSION

A model for simulating uptake of soil–water and N has been proposed in this chapter. The model is based on the single-root approach. The water-uptake submodel requires the following information:

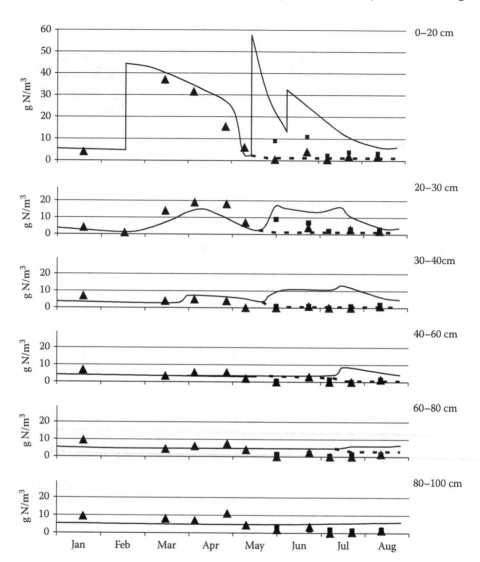

FIGURE 8.10 Comparison of measured (symbols) and simulated (lines) soil mineral nitrogen content at PAGV, 1984. Legends are as given in Figure 8.9. (Source: Hansen, S., et al. 1991a. *Fert. Res.* 27: 245–259.)

- Soil information
    - Soil hydraulic properties (retention and hydraulic conductivity functions)
    - Spatial distribution of soil water pressure potential
- Plant information
    - Root length density distribution
    - Effective root radius
- Driving force
    - Potential transpiration or crown potential

The N-uptake submodel requires the following additional information:

* Soil information
   Diffusion coefficient as a function of soil moisture content
   Spatial distribution of bulk soil solute concentration
* Plant information/driving force
   The crop N demand (potential uptake)

In respect to both uptake pattern and total uptake, water uptake is very sensitive to spatial distribution of soil water pressure potential, whereas N uptake is relatively insensitive to the simulated pattern of water uptake, indicating that diffusion can be an overall important mechanism in N uptake.

REFERENCES

Abrahamsen, P., 2007. Daisy Program Reference Manual. Royal Veterinary and Agricultural University, Copenhagen. http://www.dina.kvl.dk/~daisy/ftp/daisy-ref.pdf.

Abrahamsen, P., and S. Hansen. 2000. Daisy: An open soil-crop-atmosphere system model. *Environ. Model. Software* 15: 313–330.

Baldwin, J. P., P. H. Nye, and P. B. Tinker. 1973. Uptake of solutes by multiple root system from soil, 3: A model for calculating the solute uptake by a randomly dispersed root system developing in a finite volume of soil. *Plant Soil* 38: 621–635.

Blicher-Mathiesen, G., R. Grant, C. Jensen, and H. Nielsen. 1990. Landovervågningsoplande. *Faglig rapport fra DMU* no. 6.

Blicher-Mathiesen, G., H. Nielsen, M. Erlandsen, and P. Berg. 1991. Kvælstofudvaskning og udbytte ved ændret landbrugspraksis, Modelberegninger med rodzonemodellen DAISY. *Faglig rapport fra DMU* no. 27.

Boegh, E., M. Thorsen, M. B. Butts, S. Hansen, J. S. Christiansen, P. van der Keur, P. Abrahamsen, H. Soegaard, K. Schelde, A. Thomsen, C. B. Hasager, N. O. Jensen, and J. C. Refsgaard. 2004. Incorporating remote sensing data in physically based distributed agro-hydrological modelling. *J. Hydrol.* 287: 279–299.

Børgesen, C. D., A. Kyllingsbæk, and J. Djurhuus. 1997. Modelberegnet kvælstofudvaskning fra landbruget: Betydningen af reguleringer i gødningsanvendelsen og arealanvendelsen indført fra midten af 80érne og frem til August 1997. SP-rapport no. 19. Danmarks Jordbrugsforskning.

Børgesen, C. D., J. Djurhuus, and A. Kyllingsbæk. 2001. Estimating the effect of legislation on nitrogen leaching by upscaling field simulations. *Ecological Modelling* 136: 31–48.

Burdine, N. T. 1952. Relative permeability calculations from pore-size distribution data. *Trans. AIME* 198: 35–42.

Campbell, G. S. 1974. A simple method for determining unsaturated conductivity from moisture retention data. *Soil Sci.* 117: 311–314.

Diekkrüger, B., D. Söndgerath, K. C. Kersebaum, and C. W. McVoy, 1995. Validity of agro-ecosystem models: A comparison of results of different models applied to the same data set. *Ecological Modelling* 81: 3–29.

Djurhuus, J., S. Hansen, K. Schelde, and O. H. Jacobsen. 1999. Modelling the mean nitrate leaching from spatial variable fields using effective parameters. *Geoderma* 87: 261–279.

Gerwitz, S., and E. R. Page. 1974. An empirical mathematical model to describe plant root systems. *J. Appl. Ecol.* 11: 773–781.

Groot, J. J. R., and E. L. J. Verberne. 1991. Response of wheat to nitrogen fertilization: A data set to validate simulation models for nitrogen dynamics in crop and soil. *Fert. Res.* 27: 349–383.

Hansen, S., H. E. Jensen, N. E. Nielsen, and H. Svendsen. 1991a. Simulation of nitrogen dynamics and biomass production in winter wheat using the Danish simulation model Daisy. *Fert. Res.* 27: 245–259.

Hansen, S., H. E. Jensen, N. E. Nielsen, and H. Svendsen. 1991b. Daisy: A soil plant system model. In *Soil and groundwater research report II: Nitrate in soils.* Final report of contracts EV4V-0098-NL and EV4V-00107-C. Commission of the European Communities, 258–261.

Hansen, S., H. E. Jensen, N. E. Nielsen, and H. Svendsen. 1991c. Simulation of nitrogen dynamics in the soil plant system using the Danish simulation model Daisy. In *Hydrological interactions between atmosphere, soil and vegetation,* ed. G. Kienitz, P. C. D. Milly, M. Th. van Genucten, D. Rosbjerg, and W. J. Shuttleworth. IAHS Publication No. 204: 185–195.

Hansen, S., M. J. Shaffer, and H. E. Jensen. 1995. Developments in modelling nitrogen transformations in soil. In *Nitrogen fertilization and the environment,* ed. P. Bacon, 83–107. New York: Marcel Dekker.

Herkelrath, W. N., E. E. Miller, and W. R. Gardner. 1977. Water uptake by plants, 2: The root contact model. *Soil Sci. Soc. Am. J.* 41: 1039–1043.

Jakobsen, B. F. 1976. Jord, rodvækst og stofoptagelse. In *Simuleret planteproduktion.* Hydroteknisk Laboratorium, Den Kgl. Veterinær- og Landbohøjskole, København.

Jensen, C., and H. S. Østergård. 1993. Nitratudvaskning under forskellige dyrkningsforhold. In *Oversigt over landsforsøgene,* ed. C. Å. Pedersen. Landsudvalget for Planteavl: Årmus.

Jensen, C., B. Stougaard, and N. H. Jensen. 1992. The integration of soil classification and modelling of N-balances with the Daisy model. In *Integrated soil and sediment research: A basis for proper protection.* ed. H. J. P. Eijsackers and T. Hamers, 512–514. Dordrecht: Kluwer Academic.

Jensen, C., B. Stougaard, and P. Olsen. 1994a. Simulation of nitrogen dynamics at three Danish locations by use of the DAISY model. *Acta Agric. Scand., Sect. B, Soil Plant Sci.* 44: 75–83.

Jensen, C., B. Stougaard, and H. S. Østergaard. 1994b. Simulation of the nitrogen dynamics in farm land areas in Denmark (1989–1993). *Soil Use Manage.* 10: 111–118.

Jensen, C., B. Stougaard, and H. S. Østergaard. 1996. The performance of the Danish simulation model Daisy in prediction of Nmin at spring. *Fert. Res.* 44: 79–85.

Jensen, H. E., S. Hansen, B. Stougaard, C. Jensen, K. Holst, and H. B. Madsen. 1993. Using GIS-information to translation of soil type patterns to agro-ecosystem management: The Daisy model. In *Integrated soil and sediment research: A basis for proper protection,* ed. H. J. P. Eijsackers and T. Hamers, 401–428. Dordrecht: Kluwer Academic.

Jensen, L. S., T. Mueller, N. E. Nielsen, S. Hansen, G. J. Crocker, P. R. Grace, J. Klír, M. Körschens, and P. R. Poulton. 1997. Simulating trends in soil organic carbon in long-term experiments using the soil-plant-atmosphere model Daisy. *Geoderma* 81: 5–28.

Magid, J., and P. Kølster. 1995. Modelling nitrogen cycling in an ecological crop rotation: An explorative trial. *Nitrogen Leaching Ecol. Agric.* 77–87.

McCree, K. J. 1974. Equation for the rate of dark respiration of white clover and grain sorghum, as a function of dry weight, photosynthesis rate and temperature. *Crop Science* 14: 509–514.

Monteith, J. L., and M. H. Unsworth. 1990. *Principles of environmental physics.* New York: Routledge, Chapman and Hall.

Mueller, T., L. S. Jensen, J. Magid, and N. E. Nielsen. 1997. Temporal variation of C and N turnover in soil after oilseed rape straw incorporation in the field: Simulations with the soil plant–atmosphere model Daisy. *Ecological Modelling* 99: 247–262.

Mueller, T., J. Magid, L. S. Jensen, H. S. Svendsen, and N. E. Nielsen. 1998. Soil C and N turnover after incorporation of chopped maize, barley straw and bluegrass in the field: Evaluation of the Daisy soil–organic-matter submodel. *Ecological Modelling* 111: 1–15.

Nielsen, K., M. Styczen, H. E. Andersen, K. I. Dahl-Madsen, J. C. Refsgaard, S. E. Pedersen, J. R. Hansen, S. E. Larsen, R. N. Poulsen, B. Kronvang, C. D. Børgesen, M. Stjernholm, K. Villholth, J. Krogsgaard, V. Ernstsen, O. Jørgensen, J. Windolf, A. Friis-Christensen, T. Uhrenholdt, M. H. Jensen, I. S. Hansen, and L. Wiggers. 2004. Odense Fjord: Scenarios for reduction of nutrients. Danish National Environmental Research Institute (NERI). NERI Technical Report no. 485 (in Danish).

Refsgaard, J. C., M. Thorsen, J. Birk Jensen, S. Kleeschulte, and S. Hansen. 1999. Large scale modelling of groundwater contamination from nitrogen leaching. *J. Hydrology* 221: 117–140.

Smith, P., J. U. Smith, D. S. Powlson, J. R. M. Arah, O. G. Chertov, K. Coleman, U. Franko, S. Frolking, H. K. Gunnewiek, D. S. Jenkinson, L. S. Jensen, R. H. Kelly, C. Li, J. A. E. Molina, T. Mueller, W. J. Parton, J. H. M. Thornley, and A. P. Whitmore. 1997. A comparison of the performance of nine soil organic matter models using datasets from seven long-term experiments. *Geoderma* 81: 153–222.

Svendsen, H., S. Hansen, and H. E. Jensen. 1995. Simulation of crop production, water and nitrogen balances in two German agro-ecosystems using the DAISY model. *Ecological Modelling* 81: 197–212.

Thorsen, M., J. C. Refsgaard, S. Hansen, E. Pebesma, J. B. Jensen, and S. Kleeschulte. 2001. Assessment of uncertainty in simulation of nitrate leaching to aquifers at catchment scale. *J. Hydrology* 242: 210–227.

Van der Keur, P., S. Hansen, K. Schelde, and A. Thomsen. 2001. Modification of DAISY SVAT model for use of remotely sensed data. *Agric. For. Meteorol.* 106: 215–231.

Vereecken H., E. J. Jansen, M. J. D. Hack-ten Broeke, M. Swerts, R. Engelke, S. Fabrewitz, and S. Hansen. 1991. Comparison of simulation results of five nitrogen models using different datasets. In Soil and groundwater research report II: Nitrate in soils. Final report of contracts EV4V-0098-NL and EV4V-00107-C. Commission of the European Communities, 321–338.

Willigen, P. de. 1991. Nitrogen turnover in the soil-crop system: comparison of fourteen simulation models. *Fert. Res.* 27: 141–149.

9 Modeling Plant Nitrogen Uptake Using Three-Dimensional and One-Dimensional Root Architecture

Lianhai Wu, Ian J. Bingham, John A. Baddeley, and Christine A. Watson

CONTENTS

9.1 Introduction 198
9.2 Materials and Methods 199
    9.2.1 SPACSYS Model Description 199
        9.2.1.1 Plant Components 200
        9.2.1.2 The 3-D Root System 202
        9.2.1.3 The 1-D Root System 204
        9.2.1.4 Nitrogen Uptake 204
        9.2.1.5 Nitrogen Demand 205
    9.2.2 Field Experiment 206
        9.2.2.1 Site and Husbandry Details 206
        9.2.2.2 Soil Sampling 206
        9.2.2.3 Plant Sampling 207
    9.2.3 Model Inputs and Parameterization 207
        9.2.3.1 Meteorological Data 207
        9.2.3.2 Physical and Hydraulic Parameters of Soils 207
        9.2.3.3 Parameter Selection 208
9.3 Results 208
    9.3.1 Grain Yield and Aboveground Dry Matter 209
    9.3.2 N Offtake 210
    9.3.3 Mineral N Dynamics 210
    9.3.4 Root Penetration and Distribution 212
    9.3.5 Biomass Dynamics 213
9.4 Discussion and Conclusions 213
Acknowledgment 216
References 216

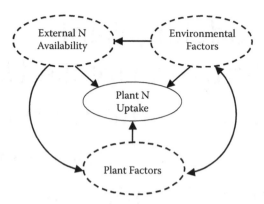

FIGURE 9.1 Interactions between plant N uptake, environmental conditions (including above- and belowground), and plant factors.

9.1 INTRODUCTION

Understanding the processes that govern plant nitrogen (N) uptake is of major importance with respect to both crop production and environmental protection. The dynamics of plant N uptake during the growing season in both arable land and grassland is a major factor influencing the fate of inorganic N (NO_3^-–N and NH_4^+–N forms) within the soil profile (Cuttle and Bourne 1993; Dowdell and Webster 1980; Recous et al. 1988; Torstensson et al. 2006). The process of plant N uptake is associated with environmental conditions (both above- and belowground) as well as plant factors (Figure 9.1). The process is often controlled by external nitrogen availability, which in turn is influenced by environmental factors such as soil temperature and soil moisture. In addition, plant factors such as plant N demand, root morphology, and root architecture are also involved in regulating N uptake. There are numerous interactions among these factors. The environmental conditions above- and belowground may determine the rate of plant growth and N demand, but in turn, plant growth has a feedback effect on the environment. For example, the shape of the root system affects the redistribution of soil moisture within and beyond the rhizosphere. One approach for dealing with the complexity of these interactions is to estimate plant N uptake dynamically within the soil–plant–atmospheric continuum, thus linking N uptake with other processes in the system.

Mathematical simulation of root nutrient uptake embedded in plant growth models has been approached in various ways, and several reviews have been published (Gregory 1996; Mmolawa and Or 2000; Rengel 1993). The process of nutrient uptake from the soil by a root depends on the concentration of nutrient at the root surface and the kinetics of uptake (Jungk 2002). In turn, the concentration depends on nutrient transport from soil to root, and on the physical root interception of the resource by a root growing toward the site where the nutrients are located in the soil. To simulate the complex interactions between carbon (C) and N cycling, and water and energy transformation between the atmosphere, plants, soils, and microbes, we developed the SPACSYS model, which includes a detailed three-dimensional (3-D) representation of plant root systems (Wu et al. 2007). Although a simulation model with 3-D root architecture may help us to better understand the process of root N uptake, the disadvantage is that more parameters to describe root architecture and growth are needed. This may introduce greater uncertainty into the simulation results.

In this chapter, we determine the merits of simulating a 3-D root system for improving estimates of plant N uptake, and the effect of replacing the 3-D root system with a one-dimensional (1-D) root system with few parameters. To do this, we ran the SPACSYS model first with its detailed 3-D root component and then with a new 1-D root-system simulation, described here for the first time. We then compared the results from the two options of the model with the same data set produced in our field experiment.

9.2 MATERIALS AND METHODS

9.2.1 SPACSYS Model Description

The SPACSYS model is a multidimensional, field-scale, weather-driven, and daily-step dynamic simulation model. It includes a plant-growth-and-development sub-model with detailed representation of the root system, in addition to components for C and N cycling in the soil with links to the plant, a soil water component, and a heat transfer component. All components are relatively independent of each other, the linkage between components (information exchange) relying on "helper" functions written within the components. The components themselves and linkages among the components are designed using object-oriented techniques and implemented in C++ with the component object model (COM). Since its original publication, the model has been modified slightly to minimize information flows among the components. The components of aboveground plant C and N have been merged so that the model now consists of six components: aboveground plant, root system, soil C, soil N, soil water, and heat (Figure 9.2).

Nitrogen cycling coupled with carbon cycling in the SPACSYS model covers the transformation processes for organic matter (OM) and inorganic N. The organic matter pool is further divided into fresh OM, dissolved OM, a litter pool, and a humus pool; inorganic N includes a nitrate pool and an ammonium pool. Fluxes between OM pools occur in a particular way according to physical and biological

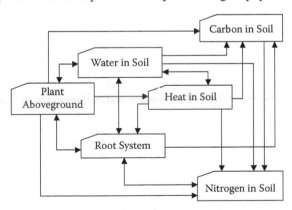

FIGURE 9.2 Components in the SPACSYS model and data flow between the components.

conditions in the source and destination pools, as described elsewhere (Wu et al. 2007). The model considers the main soil nitrogen dynamic processes, namely external N application (including chemical fertilizer, manure or slurry, and atmospheric N deposition), mineralization/immobilization, nitrification, denitrification, nitrate leaching, and N runoff loss. Nitrogen uptake by the plant (see section 9.2.1.4) and N fixation by legumes are included in the root-system component.

The structure and processes of soil water and heat transfer in the SPACSYS model follow the SOIL model (Jansson 1998) to simulate soil water and heat flows through and between layers in the soil profile. The Richards equation (Richards 1931) for water potential and Fourier's equation for heat conduction are used to simulate water and heat fluxes. Water in the soil profile is held mainly in the micro- and mesopores of the soil matrix, but if the water content in a layer rises above a specified value, a proportion is held in macropores from where rapid downward water (and solute) movement takes place due to gravitational forces alone. Water flow from the soil profile to a drainage pipe occurs when the groundwater table is above the bottom level of the pipe and the soil below the groundwater table is saturated. The Hooghoudt drainage-flow equation (Hooghoudt 1940) with modification (Jansson 1998; Youngs 1980) is adopted for the subsurface drainage flow.

The van Genuchten model (van Genuchten 1980) is used to describe the relationship between soil water content and soil water tension (soil water retention curve). When the water content is close to saturation, a linear relation is applied. When the water content is close to the wilting point, a log-linear relationship is used. The effects of load and hysteresis on the soil water retention curve are included as well. Within the soil profile, upper boundary conditions include infiltration of precipitation/irrigation water into the top soil layer, heat transfer at the soil surface, and evaporation from the soil surface, while lower boundary conditions consider water flow to deep groundwater and flows out of soil layers into field drains. Equations for soil water and heat processes are almost identical to those in the SOIL model (Jansson 1998), so they are not described in detail here.

9.2.1.1 Plant Components

9.2.1.1.1 Plant Development
Development stages are estimated based on the plant's response to heat, expressed as accumulated temperatures and threshold temperatures during different developmental phases. Three consecutive periods are defined over the whole plant life cycle: from sowing to plant emergence, from plant emergence to heading (or the end of the vegetative stage), and from heading to maturity. A development index v_{dev} (over the i range 0–3) linked to photosynthate allocation and N demand is expressed as

$$v_{dev,i} = v_{dev,i-1} + \frac{T_{add}}{AT_i} \qquad (AT_i \geq T_{add}) \qquad (9.1)$$

where AT_i is the accumulated degree-days requirement to complete period i, and T_{add} is the accumulated degree-days from the beginning of period i to the current time, expressed as

$$T_{add} = \sum T_a \times f(d) \qquad (T_a \geq T_c) \qquad (9.2)$$

where T_a is air temperature (°C), T_c is a threshold temperature for a given development stage below which no degree-days are accumulated, and $f(d)$ is the function to account for the effect of photoperiod on plant phenological development during the second period of the plant life cycle (Wu et al. 2007).

9.2.1.1.2 Plant Growth

In general, living plant tissue is considered to be divided into four components: leaves, stems, seeds, and roots. Dead material contributes to the aboveground litter and root litter pools. Live leaves immobilize atmospheric CO_2 by photosynthesis. Fixed C then flows from leaves to roots and stems. Nitrogen absorbed by roots is allocated to live leaves, stems, and roots. During the reproductive stage, a portion of the C and N from leaves and stems retranslocates to seeds.

The actual growth rate of a plant is considered as the potential growth rate regulated by the water supply–demand balance and leaf N concentration. The potential growth rate is calculated based on the Hurley Pasture Model (Thornley 1998, eq. 3.2j). The water balance effect described by a water response function is expressed as the ratio of the actual transpiration rate to potential transpiration rate. The N response function that is used to represent the influence of leaf N concentration on the plant growth rate has been described previously (Wu and McGechan 1998).

Roots receive photosynthate with the highest priority, followed by leaves and then stems. Photosynthate allocation to roots, leaves, stems, and seeds, as dependent on the plant developmental stage, has been described previously (Wu and McGechan 1998). Based on the Johnson model (Johnson 1985), the fraction of photosynthate received by roots, f_{root}, is estimated as

$$f_{root} = \frac{C_f \times W_{shoot}}{C_f \times W_{shoot} + p_f \times N_f \times W_{root}} \qquad (9.3)$$

where W_{shoot} and W_{root} are shoot biomass and root biomass at the previous time step, N_f is the ratio of plant N content to whole-plant biomass, C_f is the relative molecular mass of C in carbohydrate (approximately 0.3), and p_f is a fitted partitioning parameter that can be estimated by the best fit of the equation to the data set. Johnson (1985) described how to estimate the parameter p_f and suggested a typical value of 15.

If there is surplus photosynthate after allocation to roots, a proportion will be partitioned to leaves:

$$P_{leaf} = f_{leaf} W_{shoot} - W_{leaf} + \frac{f_{leaf}\left(1 - f_{root}\right) \cdot P_a}{C_f} \qquad (9.4)$$

where W_{leaf} is the leaf biomass at the previous time step, P_a is the canopy net photosynthesis rate, and f_{leaf} is the ratio of leaf weight to aboveground biomass.

When a plant is in the reproductive stage, dry matter in leaves and stems will translocate to seeds, in which case the actual translocation rate is the product of a potential daily translocation rate factor that is a user-specified parameter and the current biomass of the leaves or stems.

Plant respiration is divided into two components: growth respiration and maintenance respiration. The first component has been considered in the estimate of daily net photosynthesis, while the second is estimated by means of a Q_{10} expression (Eckersten et al. 1998):

$$R_{resp} = W_{biomass} Q_{10}^{\frac{T_a - T_{base}}{10}} \tag{9.5}$$

where $W_{biomass}$ is the aboveground component of plant biomass (leaves, stems, or seeds), and T_{base} is the temperature at which the Q_{10} function has a value of unity.

9.2.1.2 The 3-D Root System

As root systems are central to the acquisition of water and nutrients by a plant and a main feature of this new model, the following processes are used to describe root growth and development: branching, extension, architecture, mortality, water uptake, and nutrient uptake. We describe root development with two categories of processes: branching processes and growth processes. Branching processes include branching position and branching orientation, and growth processes include root elongation direction, elongation rate, dynamics of root diameter, root mortality, and nitrogen and water uptake. Developmental schemes categorize roots into several types, and the terms used to describe roots are defined below. A root is called an axis when it has developed from the seed (seminal axis) or the stem (nodal axis). Roots arising from the axis are designated first-order laterals; those branching from first-order laterals are designated second-order laterals, etc. For branching processes, we use three parameters to control the position: the basal nonbranch distance, the apical nonbranch distance, and the interbranch distance. The system implemented in the model has been fully described by Wu et al. (2007).

9.2.1.2.1 Root Growth and Elongation Direction

The growth of a root system is expressed as the elongation rates of various root types. The elongation rate V_a for each type is estimated as a maximum growth rate for various root categories reduced by a response function (which differs between types). Three response functions are considered: soil temperature f_{temp}, soil strength f_{SS}, and soil water solute concentration f_{sol}, where

$$V_a = V_p \cdot f_{SS} \cdot f_{temp} \cdot f_{sol} \tag{9.6}$$

and V_p is the potential (unimpeded) elongation rate for a given root age and root order. The expressions used by Clausnitzer and Hopmans (1994) to describe these effects have been implemented.

Root growth relies on a supply of carbohydrate from photosynthesis. In the model, photosynthate partitioned to a root will support both elongation and volume expansion. The model distinguishes three alternative scenarios for root growth. If the quantity of new assimilate is smaller than the potential requirement for root elongation, that allocated to all new segments is scaled down by the same factor to make the actual assimilation equal to that allocated to the root system. If the quantity is equal to the potential requirement, all assimilate is allocated to all new segments. If the allocation exceeds the requirements, the extra assimilate is allocated to established segments by the same factor as for root-volume expansion:

$$S_{b,i} = S_{b,i-1} + \frac{\Delta_a}{S_{no}} \tag{9.7}$$

where $S_{b,i}$ is the biomass of a segment at time i, $S_{b,i-1}$ is the biomass of the same segment at the previous time step, Δ_a is the extra assimilate, and S_{no} is the total number of established segments.

The growth direction of a root segment is computed using three directional components: the initial direction of the root at the previous time step, a vertical vector representing geotropism, and a vector representing mechanical constraints. The length of the first vector is set to 1, and the last two are expressed as the product of the elongation during the current time step together with weighting factors.

9.2.1.2.2 Root Branching

In the current model, we assume that lateral roots develop in an acropetal order. Lateral roots commonly originate in the pericycle of the parent root, adjacent or slightly offset to each protoxylem pole, and the distance between laterals along protoxylem-based ranks (interprimordium distance) is fairly consistent (Charlton 1975). For simplicity, we assume that interbranch distance is a constant within a particular lateral order. Near the apex and in some instances near the root base, there are two zones that lack lateral roots. We use three parameters to estimate the number of branches N_b on a root:

$$N_b = \begin{cases} 0 & L_1 < (L_{anz} + L_{bnz}) \\[3mm] \left\lfloor \dfrac{L_1 - L_{anz} - L_{bnz}}{L_{ib}} \right\rfloor + 1 & L_1 \geq (L_{anz} + L_{bnz}) \end{cases} \tag{9.8}$$

where L_1 is the length of a lateral root, L_{anz} is the length of the apical nonbranching zones of a lateral root, L_{bnz} is the length of the basal nonbranching zones of a lateral root, and L_{ib} is the interbranch distance along the lateral root. The symbol $\lfloor \rfloor$ represents a mathematical function to represent the largest integer that is less than or equal to the quotient.

Two parameters are used to determine branch orientation: insertion angle (vertical angle or branch angle) and radial angle relative to the horizontal, i.e., the azimuth (Lynch et al. 1997; Pagès et al. 1989). The insertion angle is specified as a parameter and modified from an initial value by a random variation within a predefined range. The method to calculate the radial angle ϕ from Pagès et al. (1989) is used in the current model:

$$\phi = \frac{2\pi N}{X} \tag{9.9}$$

where X is the total number of xylem poles, and N is a randomly chosen integer between 1 and X.

9.2.1.2.3   Root Mortality

The process of root mortality has a very large influence on nutrient cycling, especially in perennial plants. It causes dynamic changes in both total root volume and root-system architecture, which influence C and N cycling and water movement. At the same time, dead root material is added into the litter pool for subsequent decomposition. Interactions between root mortality and endogenous and exogenous environments are poorly understood. In the current model, root maturity is controlled by the maximum length for a root of a given order. In each time step, the whole root system is scanned. If a branch has reached its specified maximum length, it will stop growing. The matured branch is added to the litter pool according to a first-order relationship with its biomass. Before moving to the litter pool, potential litter may be withdrawn to the live roots with a specific coefficient.

9.2.1.3   The 1-D Root System

The 1-D root system is represented by rooting depth, the vertical distribution of root length density and root biomass. Downward progress of rooting depth is estimated by using a genotype-specific maximum root extension rate, modified by soil temperature. Following the Daisy model (Svendsen et al. 1995), a linear temperature response function is used when the average soil temperature within the root zone is greater than a critical temperature. The distribution of root-length density with depth is estimated by the well-cited Gerwitz and Page (1974) equation:

$$\frac{dP}{dx} = e^{-fx+C} \tag{9.10}$$

where dP is the percentage of root within a horizon of thickness dx at a depth of x cm, and f and C are constants. It is assumed that the root-length density at the potential penetration depth is 1000 m·m^{-3}.

9.2.1.4   Nitrogen Uptake

Nutrient uptake is the result of a complex system comprising different transport and uptake mechanisms. Transport of nutrients to the root surface is controlled by transpiration-driven mass flow of the soil solution and by ionic diffusion, which is dependent on nutrient concentration gradients (Barber 1962; Hopmans and Bristow

2002; Nissen 1974), but it is impossible to determine quantitatively what proportion of total nutrients taken up by a crop is supplied by mass flow and by diffusion (Tinker and Nye 2000). The diffusion process may contribute a certain quantity of nutrients for root uptake, but mass flow is the dominant process to transport nutrients from soil to the root system (Barber 1995). Since root interception provides plants with less than 1% of soil nutrients, particularly of soil N (Jungk 2002), the process can reasonably be ignored.

Uptake of nutrients by the root may be passive or active. Active uptake is commonly expressed in terms of Michaelis–Menten-type kinetics. Multiphasic and dual mechanisms for the active uptake are widely accepted (Epstein 1966; Nissen 1971, 1974). In most plant and crop growth simulation models, the dual mechanism (even simplified into one phase) is implemented to describe N uptake.

Although the process of plant N uptake is considered as the combination of passive and active uptake, whether these two uptake processes work in parallel or one process triggers another is poorly understood. Following Somma et al. (1998), we assume that both active and passive uptake are functioning simultaneously, but the contribution of each process to total uptake is controlled by a user-specified partition fraction:

$$N_u = f_p \cdot N_p + (1 - f_p) \cdot N_a \qquad (9.11)$$

where N_u is the N-uptake rate, f_p is a user-specified partition fraction for passive uptake, N_p is passive N uptake through water uptake, and N_a describes the rate of active uptake represented by a Michaelis–Menten expression:

$$N_a = \left(\frac{J_{max}}{K_m + N_{con}} \right) R_d \qquad (9.12)$$

where J_{max} is the maximum uptake rate, K_m is a Michaelis–Menten constant, N_{con} is soil available inorganic N concentration, and R_d is the root surface area. It is assumed that plants have no preference regarding uptake of N in the form of ammonium or nitrate, as plant preferential absorption for a specific N form is controversial (Bailey 1998; Kronzucker et al. 2000; Watson 1986). The partition fraction for passive uptake might depend on nutrient availability and plant nutrient demand, but little is known about its magnitude (Hopmans and Bristow 2002). In the current simulation, the fraction is set as 0.5.

9.2.1.5 Nitrogen Demand

The quantity of available N in the simulated soil profile is an input variable to the plant growth component, and in turn the N uptake estimated by the plant component is fed back to the soil N balance. The accuracy of the plant growth simulation affects the accuracy of other components, especially the flows of N and C to different pools.

Plant N demand (N_{demand}) is proportional to the daily growth, P_a, modified by the available N concentration in the root zone and the growth development index v_{dev} (Wu and McGechan 1998):

$$N_{\text{demand}} = P_{\text{a}} \cdot n_{\text{up}} \left(1 - e^{\frac{N_{\text{con}}}{N_{\text{min}}}} \right) e^{h\left(1 - v_{\text{dev}}\right)} \tag{9.13}$$

where N_{min} is the available N concentration at which N uptake ceases, n_{up} is the upper threshold of plant N concentration above which growth is unconstrained by soil N, and h is an empirical coefficient.

Atmospheric N_2 fixed by legumes, or absorbed soil N, is assumed to first satisfy the demand for mycorrhizal growth. The remainder is partitioned to roots, leaves, and stems in order. Allocation to roots, leaves, and stems depends on the N demand of the component in question. When a crop is in the reproductive stage, N in leaves and stems is retranslocated to seeds. In each case the translocation rate is the product of a user-specified rate coefficient and the N content of the leaves or stems.

9.2.2 FIELD EXPERIMENT

9.2.2.1 Site and Husbandry Details

The field experiment was conducted at Gourdie Farm, Dundee, U.K. (latitude 56.48° N, longitude 3.06° W). The field was on a south-facing slope at an altitude of 105 m. The soil was a freely drained fine sandy loam overlying loamy sand of the Buchany-hill series. The crop grown in the previous 2 years was spring barley, and the stubble was ploughed and harrowed prior to sowing the experimental plots on 6 April 2006. Sixteen spring barley genotypes, selected to cover a range of dates of commercial introduction from 1930 to 2005, were sown in 6 × 1.5-m plots at a seed rate of 360 seeds m^{-2} and a row spacing of 15 cm. Each genotype was grown with and without N fertilizer, and the experiment was laid out as a split-plot design with N as the main plots and genotypes as the subplots. There were four replicate blocks. Guard plots of cv. Optic were grown without N fertilizer between each of the main experimental ±N plots to minimize edge effects and cross-contamination when applying the fertilizer.

A maintenance dressing of P and K fertilizer (230 kg·ha^{-1} 0:20:30) N:P:K was applied to all plots on 5 May, and N was applied to designated plots as ammonium nitrate at a rate of 110 kg N·ha^{-1} on 10 May 2006. In addition, all plots received full herbicide and fungicide applications, according to commercial practice, and a foliar application of Mn to alleviate symptoms of Mn deficiency. Plots were harvested on 28 August using a small-plot combine, and seeds were dried to a moisture content of 13% before weighing. Meteorological data were collected from a weather station 2.8 km from the site.

9.2.2.2 Soil Sampling

Soil samples were taken on 4 May, shortly after crop emergence, to provide a measure of the initial soil mineral N supply prior to application of fertilizer. Samples were taken to a depth of 20 cm from guard plots in a diagonal transect across the site. Samples were pooled, stored in a cooler during transport to the laboratory, and then frozen and stored at −20°C until analysis. Postharvest soil samples were taken from each of the experimental plots to a depth of 80 cm. Samples were stored in a cool box during transport, and then held overnight in a cold room at 4°C before processing the next day. Moisture contents of the initial and postharvest samples were determined gravimetrically. Soil mineral N (NH_4^+ and NO_3^-) concentrations in 1-M KCl extracts were quantified by continuous-flow colorimetric analysis. Dissolved organic carbon was analyzed by water extraction, followed by filtration through a 0.45-μm Millipore filter, ultraviolet oxidation, and quantification by infrared gas analysis in a Rosemount Dohrmann DC 80 total organic C analyzer.

9.2.2.3 Plant Sampling

Plants were sampled at crop maturity just prior to harvesting. At four equally spaced locations along the length of each plot, a grab sample of approximately 10 shoots was taken. Plants were pulled from the ground, and belowground sections were excised and discarded. The individual grab samples from within a plot were pooled and dried in a forced-draft oven at 80°C for 48 h. After drying, ears were cut from the stem and each fraction weighed. Ears were then threshed and the chaff added to the stem fraction. The threshed grain was then weighed and the harvest index calculated as the ratio of grain weight to total (grain, chaff, and stem) weight. The total aboveground biomass per unit area was calculated from the combine-harvested grain yield and the sample harvest index. Grain and stem plus straw fractions were ball milled, and the tissue N concentration was determined by mass spectrometry. Grain and total-N offtakes were calculated as the product of the N concentration and grain yield and aboveground biomass, respectively. In this chapter, differences between genotypes are not considered, and the data presented are the mean values from all of the varieties.

9.2.3 MODEL INPUTS AND PARAMETERIZATION

9.2.3.1 Meteorological Data

Similar to other models, weather data are required to run the SPACSYS model. As the temporal iteration step of the model is daily-time-step based, daily average or cumulative values of weather elements are expected. Daily maximum and minimum air temperatures, wind speed, and precipitation are essential in the weather data set. Although daily solar radiation above the canopy and net radiation are vital for a simulation, both of them can be estimated from other relevant weather elements if a user cannot present the values. In the model, there is a toolkit to estimate missing radiation amounts based on availability of weather elements presented. If daily solar radiation is unavailable, it will be estimated by either the Ångström method (Ångström 1924)—daily sunshine duration or cloudiness, whichever is available—or the Bristow-Campbell method (Bristow and Campbell 1984). If daily net radiation on the ground is not presented, the Brunt equation (Brunt 1932) is used to estimate it, assuming that the relative humidity (or actual vapor pressure) and sunshine duration (or cloudiness) are accessible in the data set.

9.2.3.2 Physical and Hydraulic Parameters of Soils

Soil water movements depend on the soil water retention characteristic and soil hydraulic conductivity. The following parameters are required for each soil type:

TABLE 9.1

Estimated Physical and Hydraulic Properties of the Sandy Loam/Loamy Sand Soil

	Top Layer	Second Layer	Third Layer	Fourth Layer
Layer depth (m)	0.0–0.1	0.1–0.3	0.3–0.65	0.65–1.0
Residual water content (%)	4.1	4.1	4.1	4.1
Porosity (%)	54.6	52.4	55.6	35.8
Permanent wilting point (%)	10.0	11.0	16.0	10.5
Saturated hydraulic conductivity (including macropore flow) (cm h^{-1})	125	5.0	2.1	0.125
Hydraulic conductivity at break point (cm h^{-1})	48.5	17.5	7.1	0.125
Macroporosity (%)	4.0	4.0	4.0	4.0
Pore size distribution index	0.16	0.13	0.11	0.11
Soil water tension at air entry (kPa)	2.2	2.5	2.2	3.1

porosity, macroporosity, soil water tension at air entry, permanent wilting point, dry bulk density, pore size distribution index, residual water content, and saturated hydraulic conductivities with and without macropore flow.

Since data on soil physical properties were not available for the field site, the physical and hydraulic values for a comparable soil type in Edinburgh (McGechan et al. 1997) were used in the simulation, and these are shown in Table 9.1. The profile reported by McGechan et al. (1997) was subdivided into thinner layers for the simulation.

9.2.3.3 Parameter Selection

Many of the parameter values used in this study, such as plant growth rates and rate coefficients for transformation processes in the soil, are based on a previous study (Wu et al. 1998). Parameter values related to the processes of soil water movement and heat transformation are from McGechan et al. (1997). All parameters related to plant aboveground processes and nitrogen cycling are identical for both the 3-D and 1-D root-system models. Parameter values that describe root growth for these two systems are shown in Tables 9.2 and 9.3, respectively. Upon soil coring for N determination after harvest at the Gourdie site, it was observed that there was a region of greater soil strength below 0.65 m, and few roots were found below this depth. The maximum rooting depth used in the simulation was therefore set at 0.65 m.

9.3 RESULTS

For each treatment, we ran the model twice, once with the detailed description of the root system and another time with the 1-D root system. All initial values of state variables were treated identically for both model options and for both treatments. The same set of parameter values related to plant belowground growth and development was used in all model runs.

TABLE 9.2

Parameter Values Used for Root Architecture in the SPACSYS Model

	Axial Root	1st-Order Root	2nd-Order Root	3rd-Order Root
Minimum biomass per length (gDM cm^{-1})	0.003	5.0×10^{-5}	1.0×10^{-5}	1.0×10^{-5}
Potential elongation rate (cm d^{-1})	2.4	0.48	0.192	0.192
Branch orientation angle (°)	—	45	45	45
Minimum branching age (d)	6	8	10	—
Interbranch distance (mm)	1.5	2	2	—

TABLE 9.3

Parameter Values Used for Root Vertical Distribution and Penetration in the SPACSYS Model

Maximum root penetration depth (m)	0.65
Minimum root density at potential penetration depth (m m^{-3})	100
Root density at potential penetration depth (m m^{-3})	1000
Root penetration coefficient (m °C^{-1}d^{-1})	0.0025
Root penetration critical temperature (°C)	4
Root radius (cm)	0.1
Specific root length (m g^{-1} DM)	120

The model simulated the period from 1 January 2006 to 10 September 2006, starting three months ahead of the collection of experimental data to reduce the effect of errors in assuming initial soil nitrogen and water contents. Major transformation rates and pool sizes of state variables in all components from simulations were stored in a database on a daily basis. As dynamic samplings of dry-matter accumulation were not taken, we used average values of all genotypes and compared the results of dry-matter and N contents in plants at harvest in the following sections.

9.3.1 GRAIN YIELD AND ABOVEGROUND DRY MATTER

Both experimental and simulation results show that grain yield and aboveground dry matter with no external fertilizer application during the growing season are significantly lower than those grown with chemical fertilizer supply (Table 9.4). The results indicate that N status in the soil profile at this site was a major constraint to crop growth. The model with either the 3-D root system or the 1-D root system overestimated both grain yield and aboveground dry matter for the nonfertilizer treatment, but underestimated both for the fertilizer treatment, as shown in Table 9.4. The relative errors for the fertilizer treatment are smaller than those for the nonfertilizer treatment. Overall, the simulation with the 3-D root system improved the model performance to a small extent (1%–6%), especially for the fertilizer treatment.

TABLE 9.4
Comparison of Simulated and Measured Dry Matter

Treatment	Grain Yield (t/ha)	Relative Error (%)	Aboveground (t/ha)	Relative Error (%)
Control (no fertilizer)				
Simulated with a simplified root system	1.98	22.33	3.46	24.01
Simulated with a detailed root system	1.97	21.54	3.42	22.68
Sampled [a]	1.62 (0.26)	—	2.79 (0.35)	—
Fertilizer applied				
Simulated with a simplified root system	3.06	−19.60	5.45	−17.62
Simulated with a detailed root system	3.47	−9.02	5.89	−11.05
Sampled [a]	3.81 (0.53)	—	6.62 (0.66)	—

[a] Sampled values are averages of all varieties; values in parentheses are the standard deviation (N = 16).

9.3.2 N OFFTAKE

The measured offtakes of N in both grain and total aboveground biomass for the fertilizer treatment were three times greater than those for the nonfertilizer treatment (Table 9.5). The results of the simulation with the 3-D root system predicted N offtakes well, especially for the nonfertilizer treatment where the relative error was less than 1%. In the fertilizer treatment, the model with the 3-D root system underestimated grain N offtake by 4% and total offtake by 10%. In comparison, simulation results with the 1-D root system overestimated N offtakes in the fertilizer treatment and underestimated them in the nonfertilizer treatment. Furthermore, the accuracy of the predictions for the 1-D model was less than that provided by the detailed root model. Relative errors for simulations using the simplified root system ranged from 9% to 26%.

9.3.3 MINERAL N DYNAMICS

The model did not reproduce postharvest soil mineral N pool size except for the result from the run with the 3-D root system for the fertilizer treatment (Figure 9.3). Simulated values of soil mineral N content for the nonfertilizer treatment declined over the season. Mineral N status in the soil profile is an integration of various transformations that are controlled by weather conditions. For example, soil nitrate within a certain soil depth is subject to plant uptake, nitrification, and denitrification, plus potential swift movement with drainage and surface runoff.

In the absence of fertilizer, the main source of soil mineral N is from the mineralization of organic matter. The rate is affected by the quality of organic matter, soil

TABLE 9.5
Comparison of Simulated and Measured N Content in Grain and Aboveground Dry Matter

	Treatment	Grain (kg N/ha)	Relative Error (%)	Aboveground Biomass (kg N/ha)	Relative Error (%)
Control (no fertilizer)	Simulated with a simplified root system	21.37	−9.21	26.06	−11.63
	Simulated with a detailed root system	23.37	−0.70	29.29	−0.69
	Sampled [a]	23.54 (2.44)	—	29.49 (2.61)	—
Fertilizer applied	Simulated with a simplified root system	84.28	26.28	95.43	14.44
	Simulated with a detailed root system	64.22	−3.78	75.29	−9.72
	Sampled [a]	66.74 (7.04)	—	83.39 (6.98)	—

[a] Sampled values are average of all varieties; values in parentheses are the standard deviation (N = 16).

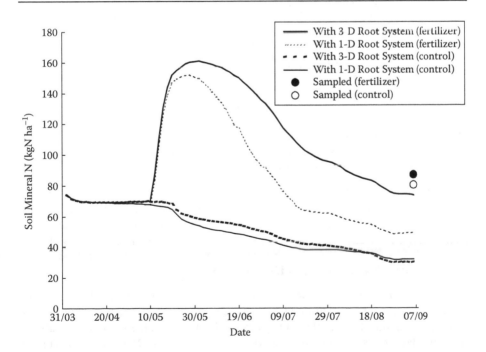

FIGURE 9.3 Dynamics of soil mineral N concentrations within the top 0.8-m depth of the soil profile from both 3-D and 1-D root-system simulations for fertilizer and nonfertilizer treatments.

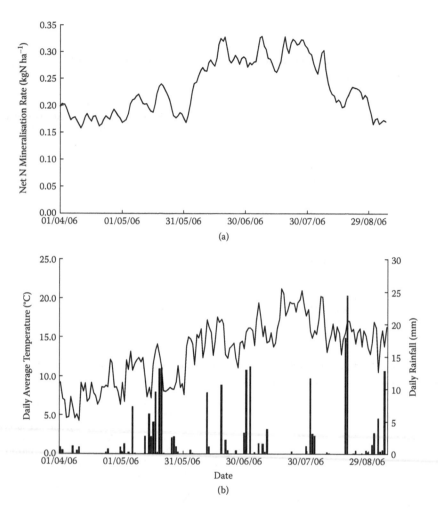

FIGURE 9.4 Dynamics of simulated net N mineralization rate for the nonfertilizer treatment (a) in relation to the prevailing weather conditions, daily rainfall (column), and daily average air temperature (line), and (b) during the growing period.

ammonium content, soil moisture, and soil temperature. Based on the soil and weather variables represented in our data set, the model suggests that weather conditions, rather than plant processes, may be the main factors controlling the rate of N mineralization, as the net mineralization rate trends with daily average temperature (Figure 9.4).

9.3.4 ROOT PENETRATION AND DISTRIBUTION

The simulation results with the 3-D and 1-D root systems (Figure 9.5) show that there is no apparent time lag for plant roots to penetrate to a certain soil level on the same growing day. Indeed, the root system arrived at the same soil depth on almost the same day for both simulation scenarios. The capability of N uptake by plant roots is determined not only by root penetration depth, but also by root distribution within the soil profile. Under the current configurations of root systems in the simulation,

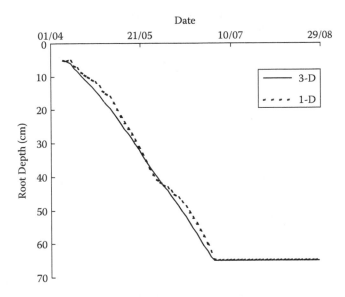

FIGURE 9.5   Dynamics of simulated root penetration depths for the fertilizer treatment run with 3-D and 1-D root systems.

many more roots occupy the upper soil layers with the 1-D root system than with the 3-D system. On the other hand, the proportion of roots located in the lower soil layers with the 3-D root system is higher (Figure 9.6).

9.3.5   Biomass Dynamics

Few samples of aboveground biomass were taken during the growing season, so the simulated dynamics of the biomass are shown in Figure 9.7. The simulation with the 3-D root systems accumulates less aboveground biomass (leaves plus stems) than that with the 1-D root system during the vegetative stage. The simulation with the 1-D root system builds up the root system in the topsoil more quickly than that with the 3-D root system in the early stage of the growing season, which results in more N uptake (Figure 9.8).

9.4   DISCUSSION AND CONCLUSIONS

When used to simulate spring barley crops, the SPACSYS model produced reasonable results with both the detailed and simplified simulations of root systems compared with sampled data. However, running the model with the detailed simulation significantly improved the accuracy of the results, especially for N offtake and mineral N in the soil profile of a high-input cropping system. Plant N uptake in most N models is based on a supply-and-demand approach, and a synthetic descriptor of root systems that describes the amount of roots in terms of biomass, length, or surface area in horizontal layers of the soil is used to represent the uptake function of root systems. This description simplifies both the heterogeneous distribution of root systems within the soil profile and the variation of the absorptive capacity among

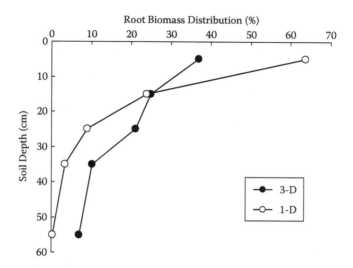

FIGURE 9.6 Simulated root biomass distribution with soil depth for the fertilizer treatment run with 3-D and 1-D root systems at the end of the vegetative stage.

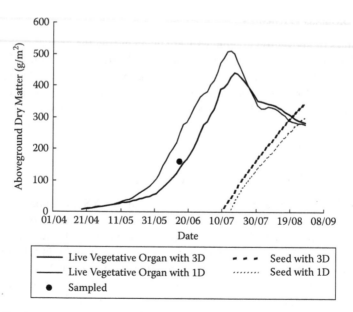

FIGURE 9.7 Dynamics of simulated aboveground dry matter for the fertilizer treatment run with 3-D and 1-D root systems. The results for the nonfertilizer treatment, which have the same trend as those for the fertilizer treatment, are not shown for simplification of the figure.

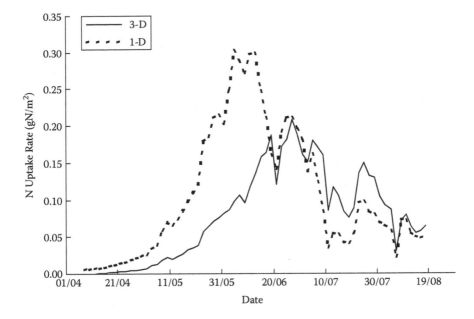

FIGURE 9.8 Simulated root N-uptake rates for the fertilizer treatment run with 3-D and 1-D root systems.

the roots of different branching order (Pagès 2002). In contrast, the 3-D root system incorporated within SPACSYS tracks the growth, morphology, and spatial position of individual roots, and also estimates N uptake by each root segment that is exposed to variable N concentrations within the soil profile. From this study it appears that the estimation of N cycling in the soil–plant–atmospheric continuum is improved by the implementation of a detailed description of plant root systems.

Although the N offtake predicted in a low-input cropping system (the nonfertilized treatment) was generally close to measured values, the simulated mineral N in the soil profile at postharvest was much lower. This may be because the current version of the model assumes that the source of N acquisition from the soil is restricted to the mineral N pool (NH_4^+ and NO_3^-), and thus may exaggerate the contribution of mineral N to plant N uptake. Plants may take up organic forms of N in addition to nitrate and ammonium (Gastal and Lemaire 2002; van Breemen 2002). Although the major source of N for plants grown in agricultural systems is mineral N (Owen and Jones 2001), plant N acquisition may be dominated by organic N forms in some circumstances, e.g., in cold natural ecosystems (Chapin III et al. 1993; Näsholm et al. 1998; Raab et al. 1999) and under conditions in which soil mineral N is limited (Bardgett et al. 2003; Jones and Darrah 1994; Michelsen et al. 1996). Thus it may be useful to incorporate the contribution of organic N to plant N uptake and N flows in soil N cycling in our simulation of a low-input agricultural system.

Both the 1-D and 3-D root-system models overestimated grain yield and aboveground biomass under nonfertilized conditions. This may reflect a weakness in the aboveground component of the model that was common to both root-system components. One possibility is that the model failed to account well enough for the effects of low N supply on tillering and its contribution to canopy size and hence light interception. Under fertilized conditions, the yield and biomass were underestimated, but with the 3-D root system the relative error was acceptable.

An inevitable consequence of implementing a detailed simulation of root architecture is that more parameters must be brought into the model. In addition, many of these parameter values, such as those that describe root elongation and branching processes, are often difficult to determine through field experiments. Thus a complex model like SPACSYS could not be used to simulate many scenarios due to a lack of parameter values. In addition, the availability of computing resources means that, at present, it is only practicable to use the model to simulate a cropping cycle of less than 10 years at the field scale.

In conclusion, models with simplified treatments of the root system as used here may be better suited to applications where predictions of crop N uptake and soil N cycling are required over many years or over a large range of soil types. Examples of such applications include predictions of crop and soil responses to climate change conducted over tens of years, or for regional comparisons where detailed soil data are not available. The greater accuracy of N-uptake predictions provided by the 3-D root-system model makes it a better choice for applications aimed to develop understanding of N cycling at the process level within crops and cropping systems.

ACKNOWLEDGMENT

We are grateful to Bill Thomas for supplying the spring barley genotypes and to the field staff of the Scottish Crop Research Institute for day-to-day management of the experiment. This work was sponsored by the Rural and Environmental Research and Analysis Directorate (RERAD) of the Scottish Government.

REFERENCES

Ångström, A. 1924. Solar and terrestrial radiation. *Q. J. R. Meteorol. Soc.* 50: 121–126.
Bailey, J. S. 1998. Varying the ratio of ^{15}N labelled ammonium and nitrate-N supplied to perennial ryegrass: Effects on nitrogen absorption and assimilation, and plant growth. *New Phytol.* 140: 505–518.
Barber, S. A. 1962. A diffusion and mass-flow concept of soil nutrient availability. *Soil Sci.* 93: 39–49.
Barber, S. A. 1995. *Soil nutrient bioavailability: A mechanistic approach.* New York: John Wiley & Sons.
Bardgett, R. D., T. C. Streeter, and R. Bol. 2003. Soil microbes compete effectively with plants for organic-nitrogen inputs to temperate grasslands. *Ecology* 84: 1277–1287.
Bristow, K. L., and G. S. Campbell. 1984. On the relationship between incoming solar radiation and daily maximum and minimum temperature. *Agric. Forest Meteorol.* 31: 159–166.
Brunt, D. 1932. Notes on radiation in the atmosphere. *Q. J. R. Meteorol. Soc.* 58:389–418.
Chapin III, F. S., L. Moilanen, and K. Kielland. 1993. Preferential use of organic nitrogen for growth by a non-mycorrhizal arctic sedge. *Nature* 361: 150–153.
Charlton, W. A. 1975. Distribution of lateral roots and pattern of lateral initiation in *Pontederia cordata* L. *Bot. Gaz.* 136: 225–235.

Clausnitzer, V., and J. W. Hopmans. 1994. Simultaneous modeling of transient three-dimensional root growth and soil water flow. *Plant Soil* 164: 299–314.

Cuttle, S. P., and P. C. Bourne. 1993. Uptake and leaching of nitrogen from artificial urine applied to grassland on different dates during the growing season. *Plant Soil* 150: 77–86.

Dowdell, R. J., and C. P. Webster. 1980. A lysimeter study using nitrogen-15 on the uptake of fertilizer nitrogen by perennial ryegrass swards and losses by leaching. *Eur. J. Soil Sci.* 31: 65–75.

Eckersten, H., P. E. Jansson, and H. Johnsson. 1998. SOILN model user's manual, version 9.2. *Communications* 98: 6. Division of Hydrotechnics, Swedish University of Agricultural Sciences, Uppsala.

Epstein, E. 1966. Dual pattern of ion absorption by plant cells and by plants. *Nature* 212: 1324–1327.

Gastal, F., and G. Lemaire. 2002. N uptake and distribution in crops: An agronomical and ecophysiological perspective. *J. Exp. Bot.* 53: 789–799.

Gerwitz, A., and E. R. Page. 1974. An empirical mathematical model to describe plant root systems. *J. Appl. Ecol.* 11: 773–781.

Gregory, P. J. 1996. Approaches to modelling the uptake of water and nutrients in agroforestry systems. *Agroforestry Syst.* 34: 51–65.

Hooghoudt, S. B. 1940. Contributions to the knowledge of some physical characteristics of soil, 7: General considerations of the problem of drainage and infiltration by means of parallel drains, furrows and canals. *Versl. Landb. Ond.* 46: 515–707 (in Dutch).

Hopmans, J. W., and K. L. Bristow. 2002. Current capabilities and future needs of root water and nutrient uptake modeling. *Adv. Agron.* 85:103–183.

Jansson, P.-E. 1998. Simulation model for soil water and heat conditions. Description of the SOIL model. *Communications* 98: 2. Division of Agricultural Hydrotechnics, Swedish University of Agricultural Sciences, Uppsala.

Johnson, J. R. 1985. A model of the partitioning of growth between the shoots and roots of vegetative plants. *Ann. Bot.* 55: 421–431.

Jones, D. L., and P. R. Darrah. 1994. Amino-acid influx at the soil-root interface of *Zea mays* L. and its implications in the rhizosphere. *Plant Soil* 163: 1–12.

Jungk, A. O. 2002. Dynamics of nutrient movement at the soil-root interface. In *Plant Roots: The Hidden Half*, ed. Y. Waisel et al., 587–616. New York: Marcel Dekker.

Kronzucker, H. J., A. D. M. Glass, M. Y. Siddiqi, and G. J. D. Kirk. 2000. Comparative kinetic analysis of ammonium and nitrate acquisition by tropical lowland rice: Implications for rice cultivation and yield potential. *New Phytol.* 145: 471–476.

Lynch, J. P., K. L. Nielsen, R. D. Davis, and A. G. Jablokow. 1997. SimRoot: Modelling and visualization of root systems. *Plant Soil* 188: 139–151.

McGechan, M. B., R. Graham, A. J. A. Vinten, J. T. Douglas, and P. S. Hooda. 1997. Parameter selection and testing the soil water model SOIL. *J. Hydrol.* 195: 312–334.

Michelsen, A., I. K. Schmidt, S. Jonasson, C. Quarmby, and D. Sleep. 1996. Leaf ^{15}N abundance of subarctic plants provides field evidence that ericoid, ectomycorrhizal and non- and arbuscular mycorrhizal species access different sources of soil nitrogen. *Oecologia* 105: 53–63.

Mmolawa, K., and D. Or. 2000. Root zone solute dynamics under drip irrigation: A review. *Plant Soil* 222: 163–190.

Näsholm, T., A. Ekblad, A. Nordin, R. Giesler, M. Högberg, and P. Högberg. 1998. Boreal forest plants take up organic nitrogen. *Nature* 392: 914–916.

Nissen, P. 1971. Uptake of sulfate by roots and leaf slices of barley: Mediated by single, multiphasic mechanisms. *Physiol. Plantarum* 24: 315–324.

Nissen, P. 1974. Uptake mechanisms: Inorganic and organic. *Annu. Rev. Plant Physiol.* 25: 53–79.

Owen, A. G., and D. L. Jones. 2001. Competition for amino acids between wheat roots and rhizosphere microorganisms and the role of amino acids in plant N acquisition. *Soil Biol. Biochem.* 33: 651–657.

Pagès, L. 2002. Modeling root system architecture. In *Plant Roots: The Hidden Half,* ed. Y. Waisel et al., 359–382. New York: Marcel Dekker.

Pagès, L., M. O. Jordan, and D. Picard. 1989. A simulation model of the three-dimensional architecture of the maize root system. *Plant Soil* 119: 147–154.

Raab, T. K., D. A. Lipson, and R. K. Monson. 1999. Soil amino acid utilization among species of the Cyperaceae: Plant and soil processes. *Ecology* 80: 2408–2419.

Recous, S., J. M. Machet, and B. Mary. 1988. The fate of labelled ^{15}N urea and ammonium nitrate applied to a winter wheat crop, 2: Plant uptake and N efficiency. *Plant Soil* 112: 215–224.

Rengel, Z. 1993. Mechanistic simulation models of nutrient uptake: A review. *Plant Soil* 152: 161–173.

Richards, L. A. 1931. Capillary conduction of liquids through porous mediums. *Physics* 1: 318–333.

Somma, F., J. W. Hopmans, and V. Clausnitzer. 1998. Transient three-dimensional modeling of soil water and solute transport with simultaneous root growth, root water and nutrient uptake. *Plant Soil* 202: 281–293.

Svendsen, H., S. Hansen, and H. E. Jensen. 1995. Simulation of crop production, water and nitrogen balances in two German agro-ecosystems using the DAISY model. *Ecol. Model.* 81: 197–212.

Thornley, J. H. M. 1998. *Grassland dynamics: An ecosystem simulation model.* Cambridge, MA: CAB International.

Tinker, P.B., and P. H. Nye. 2000. *Solute movement in the rhizosphere.* New York: Oxford University Press.

Torstensson, G., H. Aronsson, and L. Bergström. 2006. Nutrient use efficiencies and leaching of organic and conventional cropping systems in Sweden. *Agron. J.* 98: 603–615.

van Breemen, N. 2002. Natural organic tendency. *Nature* 415: 381–382.

van Genuchten, M. T. 1980. A closed form equation for predicting the hydraulic conductivity of unsaturated soils. *Soil Sci. Soc. Am. J.* 44: 892–898.

Watson, C. J. 1986. Preferential uptake of ammonium nitrogen from soil by ryegrass under simulated spring conditions. *J. Agric. Sci. Cambridge* 107: 171–177.

Wu, L., and M. B. McGechan. 1998. Simulation of biomass, carbon and nitrogen accumulation in grass to link with a soil nitrogen dynamics model. *Grass Forage Sci.* 53: 233–249.

Wu, L., M. B. McGechan, D. R. Lewis, P. S. Hooda, and A. J. A. Vinten. 1998. Parameter selection and testing the soil nitrogen dynamics model SOILN. *Soil Use Manage.* 14: 170–181.

Wu, L., M. B. McGechan, N. McRoberts, J. A. Baddeley, and C. A. Watson. 2007. SPAC-SYS: Integration of a 3-D root architecture component to carbon, nitrogen and water cycling—Model description. *Ecol. Model.* 200: 343–359.

Youngs, E. G. 1980. The analysis of groundwater seepage in heterogeneous aquifers. *Hydrol. Sci. B* 25: 155–165.

10 Simulation of Nitrogen Demand and Uptake in Potato Using a Carbon-Assimilation Approach

Dennis Timlin, Mikhail Kouznetsov,
David Fleisher, Soo-Hyung Kim,
and V. R. Reddy

CONTENTS

10.1 Introduction 220
10.2 A Model of Active and Passive N Uptake Based on Carbon
Assimilation Rate 224
    10.2.1 Objective and Approach 224
    10.2.2 Materials and Methods 224
        10.2.2.1 Growth Chambers 224
        10.2.2.2 Plant Cultivation and Environment 226
        10.2.2.3 Plant Data Collection 226
        10.2.2.4 Calculations for Passive and Active N-Uptake
                Rates 227
    10.2.3 Simulation Model Description 228
        10.2.3.1 Soil Model 228
        10.2.3.2 Plant Model 231
    10.2.4 Results 232
        10.2.4.1 Measured Data 232
        10.2.4.2 Simulations 235
    10.2.5 Discussion 237
References 240

10.1 INTRODUCTION

Nitrogen (N) is an important nutrient for plant growth but is present in soil in limited amounts for non-N-fixing agricultural crops. For this reason, soil amendment with N fertilizer is needed to obtain economic crop yields. The amendments, however, have to be carefully managed because N is highly mobile in soils and can become a pollutant when excess amounts move to groundwater.

Crop simulation models have become important tools for quantifying N dynamics, with the goal of managing N application with respect to specific soil, crop, and environmental conditions (Ahuja et al. 2002). An advantage of crop simulation models is that they can account for the relationships among environmental conditions, plant growth rate, N demand, and N availability in the soil on a dynamic basis. In order to simulate N uptake, these models, need to accurately account for carbon assimilation, N demand, and mechanisms of N uptake.

There is ample empirical evidence from experiments that the rate of growth of plants is proportional to N availability (Rodgers and Barneix 1988; Schenk 1996; Grindlay 1997). Long-term growth responses to N appear to be mainly a function of the effect of N on the increase of leaf area and resultant light interception (Grindlay 1997). Since dry-matter increase and canopy expansion (leaf area) are related via leaf growth, it is important to keep in mind that the relationships between N concentration in the plant and growth rate or biomass are largely empirical and correlative (Verkroost and Wassen 2005).

The effects of limited N on energy conversion to produce dry matter are believed to be minimal. Because leaf area controls light capture and carbon (C) assimilation, in order to utilize N, sufficient C must be available for leaf growth. In a greenhouse study, NO_3^- flux rates declined as cloud cover reduced radiances by as much as 75% (Clement et al. 1978). These results suggest that there are both feed-forward and feedback relationships between C assimilation and N uptake. This results in correlations among %N, growth rate, and dry matter (Grindlay 1997). When plant growth is limited by N supply, there is still accumulation of C compounds, resulting in reduced N percentage in the tissues (McDonald et al. 1986). This is consistent with an analysis of net primary productivity and light utilization efficiency, where Dewar (1996) concluded that N did not affect light utilization efficiency directly. This suggests that it is not photosynthesis that is ultimately limiting but, rather, the utilization of photosynthates for growth (McDonald et al. 1986; Grindlay 1997). Root sugar levels, for example, have been found to be responsible for expression of genes related to nitrate uptake (Matt et al. 2001).

There are still questions as to what determines the critical N concentration in plants (Grindlay 1997; Verkroost and Wassen 2005) and the physiological mechanisms the plant uses to regulate N uptake (Miller and Cramer 2005). This is complicated by the complex role of N in plant metabolism. Nitrogen can operate as a signaling mechanism as well as a structural component (Miller and Cramer 2005). There are numerous transporter systems in plant roots that depend on the form and concentration of N in the soil (Miller and Cramer 2005). These function under different circumstances and are subject to complex regulation. Nitrogen is also a primary component of the enzymes that control growth and development processes.

Hence, N uptake in plants involves complex processes that depend on C-assimilation rates, N demand by the plant, transpiration rate, root density and distribution, the form of N, and N concentration and water distribution in the soil.

The relationship between C assimilation and N uptake is further complicated by atmospheric CO_2 concentrations and the effects on C-assimilation rate and partitioning. Recent research with elevated CO_2 has shown that plants grown at elevated CO_2 have lower N contents than those grown at ambient levels (Kimball et al. 2002). There is still some uncertainty as to the reason for these differences. This suggests that there is not a simple relationship between C assimilation and N uptake, but within a given CO_2 level, C-assimilation rates will impact N demand.

Despite this complexity, simulation of N uptake in crop models is usually abstracted as two components: calculation of how much N the plant needs (demand), and uptake (supply) processes (see Gayler et al. 2002 for an application). According to Jeuffroy et al. (2002), there is a wide range in how crop models treat demand and uptake, and they are often simplified and combined into one process. The use of a critical N percentage is common in crop models to model N demand (Jeuffroy et al. 2002). Nitrogen limitation is calculated from the percentage of demand that can be met given maximum and minimum values of critical N. This value is expressed as a ratio from 0 to 1 and is used to adjust growth rate or radiation use efficiency. This mimics the effects of N deficiency rather than the mechanism. Nitrogen deficiency may also be used to adjust the C-assimilation rate (Verkroost and Wassen 2005) or leaf photosynthetic rate (see Jeuffroy et al. 2002).

Experimentally determined relationships between growth rate and N content have led to functional approaches to modeling N demand and uptake. One approach is to use an allometric relationship between N and biomass (Lemaire and Salette 1984; Plénet and Lemaire 1999). Another approach is the use of critical N contents in plants (Gastal and Lemaire 2002; Hirose 1988). Each plant component has a critical N content that may vary by growth stage and age. This provides a method to relate N availability to growth rate. As described in Gastal and Lemaire (2002), functional relationships between N content and growth stage are impacted by the effect of N on plant processes such as leaf photosynthesis. Nitrogen distribution in the canopy and leaf expansion rate also affects light interception, which, in turn, impacts C assimilation.

To better understand the processes of growth and N uptake, Verkroost and Wassen (2005) developed a model of N response under limited N using an N-allocation approach. Here they modeled the rate of formation of photosynthetic proteins (mainly RuBisCO, ribulose-1,5-bisphosphate carboxylase/oxygenase) as a function of N uptake. They also considered degradation of N-containing photosynthetic proteins through a first-order kinetic process. Carbon assimilation and resultant growth rate were a function of photosynthetic capacity. This model replicated linear growth rate responses to N seen in the literature, with a negative intercept (representing a growth cost for N utilization). The method of allocating N to photosynthetic proteins and allowing degradation (in which case N is recycled) was found to be a promising approach to simulation of N demand. They concluded that, when N is limited, it appeared to be allocated so as to maintain a constant photosynthetic efficiency, where N uptake and formation of photosynthetic N are balanced by degradation of photosynthetic proteins in order to maintain the highest possible C-assimilation rates.

This was a conceptual model in which light interception was not explicitly modeled. Verkroost and Wassen (2005) did show that, in N-limited situations, productivity based on N uptake is balanced between N needed to maintain photosynthesis and N left over for biomass production. However, photosynthetically active N could be an analog for leaf area. This would suggest that (under N-limiting conditions) any increase in photosynthetically active leaf area would have a cost that reduces total biomass due to the need to use C to build photosynthetically active proteins. Thus there would be an upper limit on N demand based on diminishing efficiency as N uptake increases. This concept may also be useful to model N demand as a function of $[CO_2]$, since it has been reported that the concentration of photosynthetically active proteins are reduced under elevated CO_2 (Stitt and Krapp 1999).

An important component of simulating N dynamics in plants is the quantification of N uptake. Nitrogen is present as a solute in the soil water in the forms of both ammonium (NH_4^+) and nitrate (NO_3^-), and as a result it is taken up in the transpiration stream, a process sometimes referred to as "passive uptake." This mode of uptake, however, is not completely passive, as the nutrient is required to pass through an impermeable membrane at the root surface, and an ionic balance must be maintained in the roots. It is likely that it is largely unregulated, however. The effect of transpiration rate on passive uptake may be more related to translocation from roots to the shoot than it is to the uptake process itself. In a study on the effect of transpiration rate on sodium uptake, Smith and Middleton (1980) reported that transpiration rate was more closely related to sodium translocated from the roots than total sodium content in the plant.

Plants can also actively take up N through diffusion by inducing an ionic gradient to absorb N in ionic form at a rate higher than N supplied via mass flow of water (De Willigen 1986). Active (diffusive) uptake of N is important because roots cannot always effectively intercept sufficient N when relying on mass flow alone. At limiting soil-N contents, diffusion becomes an increasingly important component of total N uptake (Plhak 2003). It has been shown that N uptake in the grass, tall fescue, does not follow transpiration, but is linked to the C-assimilation rate of the plant (Gastal and Saugier 1989). These studies suggest that the diffusive process of active N uptake is likely to be regulated by the plant.

Recently, more has been learned about the mechanisms that control uptake of NO_3^- and NH_4^+ from the soil. The active and passive components of NO_3^-–N uptake appear to be distributed over four different transport systems (Acco Lea and Azevedo, 2006):

1. Constitutive high-affinity (cHATS)
2. Nitrate-inducible high-affinity (iHATS)
3. Constitutive low-affinity (cLATS)
4. Nitrate-inducible low-affinity (iLATS)

In the case of NH_4^+, only the constitutive transport mechanisms are found. There are no ammonium-induced forms of N uptake (Acco Lea and Azevedo 2006). The high-affinity systems take up N when the concentration in the soil is low and vice versa for the low-affinity systems. Constitutive means that the mechanism for diffusive uptake is not affected by nitrate concentration in the soil. The nitrate-inducible systems, on the other hand, are affected by nitrate concentration in the soil, meaning that changes in nitrate concentration will cause changes in gene activity and production of enzymes involved in N uptake. These systems allow the rate of N movement into the roots to be regulated by the plant according to the plant's need for N. An N-uptake model for oilseed rape that explicitly accounts for these mechanisms has been developed (Malagoli et al. 2004). This model is presented in chapter 3 of this book.

Uptake of nutrients, especially N, has been an important research topic from a soil fertility management standpoint and was an early topic to be addressed in simulation models of soil processes (Bouldin 1961). Early simulation models accounted for both diffusive and mass-flow components of nutrient movement to roots (Nye and Marriott 1969). The plant component of N uptake was accounted for by two empirical parameters, the root absorbing power (demand) and water flux at the root surface (via transpiration rate). Later efforts employed a Mitscherlich-type equation to model diffusive movement of N to the roots. Much of this work has been summarized in Barber (1995).

The simulation of passive components of N uptake is straightforward, so this method is common in most crop models. The model keeps track of the amount of water taken up through transpiration and the concentration of nitrate in the soil water, and from this calculates total N uptake. Generally, when passive uptake is the main component of N uptake, there is no mechanism for the plant to regulate its N content. Since water uptake is usually dependent on soil water content, if the wettest areas of the soil have low N concentrations, the plant might not be able to take up enough N to prevent limitation of growth. On the other hand, in dry areas, lower transpiration may limit the ability of the plant to take advantage of the larger amounts of soil N that may be present.

Simulation of active uptake provides a mechanism to allow the plant to regulate N uptake. This method is more involved and requires additional parameters. The nitrate-uptake rate is typically simulated using an absorption isotherm in the form of a Michaelis–Menten (MM) equation (Claassen and Barber 1976). The use of Michaelis–Menten kinetics requires information on the maximum rate of nitrate uptake per unit of root length, root radius, root surface, and concentration of nitrate in the soil solution. Usually, average values of these parameters are used. In actuality they may vary over temperature, root age, and mechanism of nitrate uptake. For further information, see Buysse et al. (1996) and Barber (1995). The maximum rate of N uptake per unit root depends on N demand (Schenk, 1996).

There are few studies that have quantified N-uptake and C-assimilation rates under sunlit conditions using controlled temperatures and different CO_2 concentrations and soil media. Many of the studies were carried out in greenhouses or using artificial light. There is qualitative evidence, however, that the proportion of active uptake in total nitrate uptake decreases as the N in soil solution increases (Plhak 2003). Furthermore, potato plants grown under elevated CO_2 (twice ambient) have been shown to have lower N contents and transpire less water than those grown at ambient CO_2 levels (Conn and Cochran 2006; Bunce 2001). This would impact the relative rates of passive and active N uptake.

10.2   A MODEL OF ACTIVE AND PASSIVE N UPTAKE BASED ON CARBON ASSIMILATION RATE

10.2.1   Objective and Approach

Here we describe one approach to simulate active and passive N uptake based on rates of carbon assimilation, transpiration, and diffusion of N in soil using a Michaelis–Menten approach. Experimental data on relative amounts of active and passive N uptake in potatoes grown in pots in outdoor, sunlit growth chambers are used to determine patterns of passive and active N uptake and carbon assimilation rates under two levels of CO_2 and six N levels. The processes of N uptake were abstracted in a model of diffusive and active N uptake and compared qualitatively with measured data.

10.2.2   Materials and Methods

10.2.2.1   Growth Chambers

The outdoor, sunlit controlled environment chambers used for this study were constructed of clear acrylic and were 2.3 m tall and 1.5 m² in cross-sectional area with a total chamber volume of 3360 L. The space available for growing plants in the chambers is 1.0 m², excluding the internal ducting. The Soil Plant Atmosphere Research (SPAR) chambers are very similar in design to those in use at Mississippi State University (Reddy et al. 2001). The chambers and the control mechanisms were fully described by Baker et al. (2004). The air handler, mounted at the base of the chamber, contains a squirrel-cage fan that draws air and forces it past resistive heaters and a liquid-cooled heat exchanger on the return path back to the chamber. These heating and cooling elements were used to control air temperature and humidity. Constant relative humidity was maintained at 60%–70% by operating solenoid valves that inject chilled water through the cooling coils located in the air handler of each chamber. These cooling coils condensed excess water vapor from the chamber air stream in order to regulate relative humidity. The temperatures in the growth chambers were maintained to within ±0.2°C of the set points.

A feed-forward, feedback proportional–integral–differential (PID) control algorithm similar to the one described by Pickering et al. (1994) was used to control injection of CO_2. Gas flow was measured and controlled by mass-flow controllers (Omega Engineering, Stanford, CT). The amount of CO_2 injected, the air chamber volume, chamber temperature, and chamber leakage were used to calculate canopy photosynthetic and respiratory rates over 300-s intervals. Carbon dioxide was injected only during the daytime hours when there was sunlight (0600 to 2000 hours, eastern standard time). The photosynthetically active radiation (PAR, measured as photosynthetic photon flux density, PPFD, µmol photons m^{-2} s^{-1}) was integrated over the same 300-s intervals. Transpiration was calculated from measurements of water drained from the cooling coils. Some water would have come from direct evaporation from the soil, but this amount is small, since the plants were grown in pots and there was little soil surface exposed to radiation.

The facility includes a dedicated Sun SPARC 5 workstation (Sun Microsystems, Mountainview, CA) used to control chamber atmospheric [CO_2] and record C-assimilation and chamber environmental data (air and soil temperatures, humidity, CO_2 concentration, and solar radiation) every 300 s. Air temperature and relative humidity were also monitored and controlled with TC2 controllers (Environmental Growth Chambers, Chagrin Falls, OH). Incident photosynthetically active solar radiation (PAR) was monitored with a single LI 191 SB radiation sensor (LI-COR, Lincoln, NE) located outside the chambers. The acrylic plastic allows approximately 95% of the PAR to pass through (Kim et al. 2004).

Leakage rates were calculated daily (Baker et al. 2004) as

$$L = \frac{1}{r}\left(\Delta CO_2\right) \tag{10.1}$$

Here L is the leakage rate (μmol CO_2 m^{-2} s^{-1}), e.g., the CER (C exchange rate) without plant uptake, and ΔCO_2 is the gradient of internal-chamber to external-air CO_2 concentrations (μmol CO_2 mol air^{-1}). The value of the resistance, r, was obtained daily by injecting a pulse of nitrous oxide and measuring the decay (Baker et al. 2004). The gradients in equation (10.1) were calculated using ambient CO_2 measurements taken on the same temporal scale as the CO_2 measurements in the chambers. Ambient CO_2 levels varied diurnally from 350 to 550 μmol mol air^{-1} (higher values at night) and have a strong effect on leakage calculations (Baker et al. 2004).

When there were plants in the chamber, the CER represented net photosynthesis. Dark respiration, R_D (μmol CO_2 m^{-2} s^{-1}), was calculated daily as the mean CER between the hours of 0100 and 0400. Although R_D does not account for photorespiration and may be affected by [CO_2], this method has successfully been used to relate carbon assimilation to dry matter (van Iersel and Kang 2002; Timlin et al. 2006) from growth-chamber data. Therefore, R_D was used to adjust for photorespiration during daylight hours. Since this value was obtained for a temperature of 18°C, it was multiplied by 1.5 when applied to daytime C-assimilation rates (23°C). An equation (Constable and Rawson 1980; Milroy and Bange 2003) was used to model gross photosynthesis (P_G) as a function of PPFD (photosynthetic photon flux density). This equation was used to interpolate measurements of P_G for periods when gas-exchange data were missing or out of range (as when the chambers were opened for plant measurements), smooth the data, and obtain P_G values for specific light levels:

$$P_G = P_{max}\left(1 - \exp\left[1 - a \times PAR\right]\right) \tag{10.2}$$

Here P_{max} is the asymptotic rate of gross C assimilation (μmol CO_2 m^{-2} s^{-1}), PAR is photosynthetically active radiation (μmol photons m^{-2} s^{-1}), and a is a coefficient with units of μmol photons^{-1} m^2 s. P_G, gross C-assimilation rate, is calculated as net C-assimilation rate measured in the chambers with respiration added back as a positive number. Unless otherwise noted in this chapter, P_G is reported as gross C assimilation per unit surface area of the growth chamber and is not adjusted for light interception or leaf area. The two coefficients were fitted using Proc NLIN of SAS (SAS Institute 2004).

10.2.2.2 Plant Cultivation and Environment

Three certified potato (*Solanum tuberosum* cv. Kennebec) seed tubers (5 ± 10 g mean fresh weight) were planted at a depth of approximately 5 cm into 3.8-L pots. The soil type was 75% sand and 25% vermiculite by volume (Grace Construction Products, Cambridge, MA). The plants for the two CO_2 treatments were planted at different times using the same six chambers for the two experiments. The elevated CO_2 treatment was the first planting on 13 May 2005, with 50% emergence recorded on 27 May 2005. The ambient CO_2 treatment was the second planting on 22 July 2005, with 50% emergence recorded on 5 August 2005. Since total light and day length were different between the two periods, we did not compare elevated and ambient CO_2 treatments results. The tubers were germinated in pots in the outdoor chambers for the first planting, and the temperature was held constant at 23°C until full emergence. Nitrogen and CO_2 treatments were applied at planting. The pots for the second planting were initially placed in indoor lamp-lit chambers also held at 23°C to allow us to place the pots with emerged plants in the sunlit chambers immediately after harvesting the first planting. These plants were transferred to the outdoor chambers on 8 August 2005. There was about 50% emergence, and all emerged plants had four to six leaves at this time. Two days after all the plants had emerged from a pot, two of the tubers with plants were removed from the pot with as much root mass as possible, and the shoots on the remaining tuber were thinned to a single shoot.

Twelve pots were placed in each of the six chambers. The temperatures in the chambers were maintained at a constant 23/18°C ± 0.1°C day/night 16-h thermal period. Carbon dioxide concentration for the elevated treatment was 700 μmol mol air^{-1}, and 370 μmol mol air^{-1} for the ambient treatment. The value of CO_2 for the ambient treatment was close to the mean daily value measured at Beltsville, Maryland. After planting, six N levels were randomly assigned to the chambers for the elevated CO_2 treatment. The same treatment allocation was used for the ambient CO_2 treatment. This was done to minimize variation due to chamber on CO_2 comparisons within a N level. The N levels were 2, 4, 6, 8, 11, and 14 mM N. Nitrogen was manually applied as a solution three to four times a week to maintain a constant concentration of N in the soil water. Solution was applied in large enough amounts to completely saturate the pot. The N was applied as 50% NO_3 and 50% NH_4 Hoagland's solution.

Additional newly emerged shoots from the tuber were removed by pinching to maintain one main stem shoot per pot. This should not affect photosynthesis rates via source/sink effects. Oparka and Davies (1985) did not find that there was sharing of assimilates among different stems emerging from the same seed-piece.

Nighttime CO_2 was not controlled and varied around the ambient level (350–550 μmol mol air^{-1}). Graded shade cloths were adjusted around the cabinet edges to plant height to simulate shading effects found in a field crop. The plants were not staked.

10.2.2.3 Plant Data Collection

The potato plants were harvested prior to senescence to evaluate plant growth at the time of maximum canopy green leaf area. The harvest dates were 27 July 2005 for the elevated CO_2 experiment and 27 October 2005 for the ambient CO_2 experiment. The lengths of the growing seasons were 60 and 73 days, respectively. At harvest, plants were separated according to stem, green leaf, and tubers. Leaves and stems were segregated based on position as main stem, i.e., main-stem leaves, second- and third-order main-stem leaves, first- and second-order basal-stem leaves, and first- and second-order apical-stem leaves. Leaf area was measured on all components using a Li-Cor 3100 (Li-Cor, Lincoln, NB) area meter. Nitrogen and C contents were determined on main-stem stem and leaves, second-order stems, and leaves and tubers on three randomly chosen plants from each treatment. Leaves on second-order stems were younger than main-stem leaves and still actively growing at harvest, hence they provided an estimate of N in young tissue. A Perkin-Elmer 2400 CHN analyzer (PerkinElmer Life and Analytical Sciences, Wellesley, MA) was used to determine N content.

10.2.2.4 Calculations for Passive and Active N-Uptake Rates

Daily total N-uptake rates were estimated from measured C/N ratios and daily C assimilation. Here we assumed a constant C/N ratio throughout the growing period. It has been shown that the C/N ratio in potato does not vary greatly until senescence begins (Karley et al. 2002). Passive uptake of N was estimated from daily transpiration rates and the known concentration of N in the soil water. Active uptake was estimated from the difference between total and passive N-uptake rates. Note that calculation of transpiration (passive) uptake was independent of assimilation rates and C/N ratio.

There are a number of assumptions inherent in the analysis of the measured data. To correctly estimate N uptake in transpired water, we must assume that the plant does not filter N from the water and that the ammonium and nitrate forms of N were equally available. Cao and Tibbits (1998) showed that potato makes best use of N in the soil when it is partially in the form of ammonia and nitrate. Although N can be lost in nonsterile media through microbial activities (Smart et al. 1998), we did not believe this was an important mechanism of N loss in our study. Since the plant media had no organic matter, the low C substrate (only from root exudates) would have likely been insufficient to support much microbial growth.

The C/N ratio was only measured at the end of the season, and we calculated N uptake during the early growth period based on (a) current C-assimilation and transpiration rates, and (b) end-of-season N measurement. The N content of secondary-stem leaves was assumed to apply to the new leaf growth in the early season growth period. It is known that stem biomass increases over leaf biomass as N content increases, and this affects distribution of N in the canopy (Gastal and Lemaire 2002). We did scale the total N in the plant for plant component to adjust for stem and leaf differences in N content, but the exact ratio during the active growth phase may have been slightly different than that at the end of the season. We do not expect the differences to be large enough to greatly affect the results, especially the relative differences among treatments, since the ratios of stem to leaf did not change greatly over the growth period. Active uptake was calculated as the difference between two estimated values, and thus contains much of the error in the assumptions.

10.2.3   SIMULATION MODEL DESCRIPTION

10.2.3.1   Soil Model

The soil model, 2DSOIL (Timlin et al. 1996), is a two-dimensional modular finite element model of water and solute movement that was developed from SWMS-2D (Simunek et al. 1994). Additional components to simulate atmospheric processes and root and canopy growth have been taken from the model GLYCIM (Acock et al. 1985). Only the ambient CO_2 treatments were simulated, since the plant model does not account for the effects of CO_2 on stomatal conductance to water vapor.

Nitrogen may be taken up by plants by an active process or by a passive process. If the process is active, it is assumed that the plant regulates the concentration at the root surface of the considered chemical, and that both mass flow and diffusion may contribute to the movement of solutes to the root surfaces. Mathematically, however, the two cannot be differentiated. During passive uptake, the movement of solutes to the root surfaces is simulated as pure mass flow with water.

The approach to simulate advective and diffusive movement of N is to combine two equations for the N-uptake rate (I_{in}) by the plant. An absorption isotherm for N as nitrate to move to the root from the bulk soil solution can be described by a Michaelis–Menten-type equation:

$$I_{in} = \frac{I_{max}(C_R - C_{min})}{K_M + (C_R - C_{min})} \tag{10.3}$$

where

I_{in} = NO_3^--uptake rate (mg d^{-1} cm $root^{-1}$)

I_{max} = maximum (asymptotic) rate of NO_3^- uptake at the highest soil N concentration (mg d^{-1} cm $root^{-1}$)

C_R = concentration of NO_3^- at the root surface (mg cm^{-3})

K_M = concentration of NO_3^- in the soil where I_{in} is one-half I_{max} (mg cm^{-3})

C_{min} = minimum concentration of NO_3^- in the soil where plants can still take up N (mg cm^{-3})

The model assumes that solute movement takes place only in the radial soil volume surrounding the root. Furthermore, it is assumed that the concentration–distance profile around a root develops in time in a stepwise manner, and that at each time step it approximates to a steady-state profile (Baldwin et al. 1973). Each root may exploit an average effective volume of soil, which is assumed to be a cylinder corresponding to the average half-distance between the roots. Daily maximum rate of inflow of N as NO_3^- to the roots (I_{max}) depends on root age and is not considered to be regulated by the plant in this study (i.e., it is held constant throughout the simulations).

Equation (10.3) provides an estimate of N-uptake rate as a function of N concentration at the root surface (C_R) and minimum N concentration in the soil solution (C_{min}). Solute flux to the roots, however, also depends on

1. Diffusion rate of NO_3^- in the soil
2. Rate of water flow to the roots
3. Total root length and mean radius of the root
4. Concentration of NO_3^- in the soil water (C_S)

Based on these assumptions, the solute-flux toward the root surface (I_{in}) is also calculated as

$$I_{in} = \begin{cases} 4\pi D(C_S - C_R)\left|\dfrac{\beta^2 \ln \beta^2}{\beta^2 - 1} - 1\right| & a = 0 \\[3ex] q_r \dfrac{\left(\beta^2 - 1\right)C_S - \ln(\beta^2)C_R}{\left(\beta^2 - 1\right) - \ln(\beta^2)} & a = 2 \qquad (10.4) \\[3ex] q_r \dfrac{\left(\beta^2 - 1\right)\left(1 - a/2\right)C_S - (\beta^{2-a} - 1)C_R}{\left(\beta^2 - 1\right)\left(1 - a/2\right) - (\beta^{2-a} - 1)} & \text{else} \end{cases}$$

where

$a = q_r/2\pi D$

$\beta = (r_r^2 \pi L)^{-1/2}$

I_{in} = solute uptake per unit length of the root (as defined in equation 10.3)

D = diffusion coefficient in the soil (cm² d⁻¹)

C_S = bulk concentration in solution (mg cm⁻³), which is obtained from the solution of the convection dispersion equation

C_R = concentration at root surface (mg cm⁻³)

r_r = root radius (cm)

L = root density (cm root cm⁻³)

q_r = water flow toward the root surface per unit length of the root (cm cm⁻¹ d⁻¹)

The flow rate for mature, q_{r_m}, and young, q_{r_y}, roots is calculated as

$$q_{r_m} = \frac{u_m}{2\pi l_m} \quad \text{and} \quad q_{r_y} = \frac{u_y}{2\pi l_y}$$

where u_m and u_y are, respectively, water uptake from soil cells by mature and young root (cm³ cm⁻² d⁻¹), and l_m and l_y are, respectively, length of mature and young roots in soil cell (cm).

In the soil, diffusion is influenced by the water content of the soil both in terms of the diffusion cross-section and the tortuous pathway followed by the solute through pores. The bulk soil diffusion coefficient is calculated as

$$D = \lambda q_r + \theta D f$$

where

q_r = flow rate to the root (cm cm^{-1} d^{-1})
θ = volumetric soil water content (cm^3 cm^{-3})
D = diffusion coefficient in free solution (cm^2 d^{-1})
λ = soil dispersivity (cm)

f is a "tortuosity" factor that can be estimated as shown by Millington and Quirk (1960):

$$f = \frac{\theta^{7/3}}{\theta_s^2}$$

where θ_s is the soil water content at saturation.

Finally, the unknown concentration, C_R, can be found by simultaneously solving equations (10.3) and (10.4) for C_R. In this case, C_R is determined by the N concentration in soil and the parameters of the MM equation (equation 10.3). Any plant regulation would be achieved by varying the parameters of equation (10.3). If the uptake is limited by the availability of the solute ($C_S < C_R$), then C_R is assumed equal to zero and hence the root acts as a zero sink. The total uptake of the solute is calculated by integrating I_{in} over the entire root system. In the case of ample solute supply, the total solute uptake is determined by the soil supply of N.

An alternative method to simulate active uptake is to determine a value of N uptake, U_d (mg d^{-1}), for the time period based on plant need and distribute it over the entire root zone as follows:

$$\sum_{i=1}^{N} (I_{im}(C_R) \cdot l_{im} + I_{iy}(C_R) \cdot l_{iy}) = U_d \qquad (10.5)$$

I_{in} in equation (10.4) can be considered to be the sum of two components when a = 2. The first component is $I_{in}(C_S)$, and the second is $I_{in}(C_R)$ (considering that the two components are additive). Since the component for N uptake in equation (10.4) due to C_R determines plant demand, U_d (equation 10.5), we can set U_d to equal $I_{in}(C_R)$ and then solve for C_R and I_{in}. In equation (10.5), N is the number of soil cells with roots, I_{im} and I_{iy} represent N uptake by mature and young roots from cell i, and l_{im} and l_{iy} are the lengths of mature and young roots in cell i. We assume a common value of C_R to exist along the root surfaces of the entire root system. Soil cells in which $C_S < C_R$ are assumed not to contribute to the solute uptake (except in the transpiration flow). This method, however, is left for a later exercise and is not implemented in this paper.

I_{max} and K_M were determined from the measured data. I_{max} was set to the N-uptake rate for the 14-mM treatment. K_M was set to the concentration of NO$_3^-$ in the soil for the 6-mM treatment, because the inflection point in yield and uptake occurred at about this point.

10.2.3.2 Plant Model

A simple, generic, single-leaf plant model was used to calculate N demand and uptake. Carbon assimilation was a function of sunlight and leaf area. A rectangular hyperbola (Acock et al. 1971) was used to calculate C assimilation as a function of light using parameters from the experiment.

$$P_G = \frac{\alpha I \tau C}{\alpha I + \tau C} \qquad (10.6)$$

where

P_G = gross photosynthesis (g CO_2 m^{-2} h^{-1})
I = light integral (mol photons m^{-2} h^{-1})
C = external CO_2 concentration (370 μL L^{-1} [μmol mol^{-1}] air)
τ = canopy conductance to CO_2 transfer (m hr^{-1})
α = canopy light utilization efficiency (g CO_2 mol^{-1} photons)

Vegetative mass is calculated from assimilated C assuming that the vegetative dry matter is, on average, 40% C (Timlin et al. 2006). Respiration rate, R (μmol CO_2 m^{-2} s^{-1}), is a function of vegetative biomass and is calculated from assimilated C as (Timlin et al. 2006)

$$R = -0.03 + 0.008973\ V_M$$

where V_M is vegetative mass (g) and R is in units of g C m^2 h^{-1}.

The daily rates of increase of leaf area and plant height were calculated as a function of C-assimilation rate, time, and a growth coefficient. The growth coefficient was roughly calculated from a relationship between leaf area and biomass determined from the data (some of the C assimilation will produce stem as well during early growth). The purpose of the coefficient was to translate C assimilation into leaf area (units of cm^2 g^{-1}). Since constant temperatures were used in the growth chamber, no temperature dependency was used. To calculate a N deficiency index to adjust C-assimilation rates, we assumed that the plant attempts to maintain an optimum C/N ratio (derived from the experimental data). An actual C/N ratio is calculated from cumulative N uptake and cumulative C-assimilation rates. If the actual C/N ratio falls below the optimum, C assimilation is reduced by an index related to the ratio of actual C/N divided by the optimum. Carbon for root growth was partitioned from total C assimilation, as is done in the soybean model GLYCIM (Acock et al. 1985). Root growth algorithms, also taken from GLYCIM, were used to simulate root growth. These algorithms are appropriate for potato, as they can simulate a fibrous root system without a taproot. Carbon for root growth is partitioned from C assimilation. Relative root growth into a particular soil region is not affected by N content. Since we were simulating a potted experiment, the correct simulation of root dynamics was not critical.

The optimal N/C ratio was set at 0.15 (C/N = 6) at the beginning of the simulation and allowed to linearly decrease to 0.07 (C/N = 14) at the end. This was based on the experimental results for comparison of new and mature leaves.

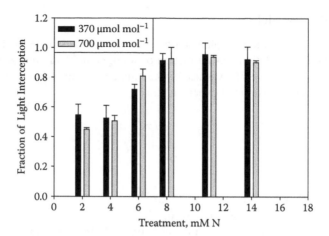

FIGURE 10.1 Maximum mean light interception at end of the season as a function of six N treatments at two levels of CO_2. The data are an average of 5 days.

10.2.4 RESULTS

10.2.4.1 Measured Data

Figure 10.1 shows maximum light interception near the end of the growing season for the six N treatments within each of the two CO_2 treatments. The trends with N are similar for the two CO_2 treatments. Note that there is a transition between 6 and 8 mM N, indicating that, below 8 mM N, a deficiency in N resulted in reduced canopy growth. There was no discernible CO_2 effect on light interception. Figure 10.2 shows mean daily C-assimilation rate as a function of N treatment during the exponential growth phase. The rates at ambient CO_2 levels reached an asymptote at about the 8-mM treatment, consistent with the light-interception data, but assimilation rates

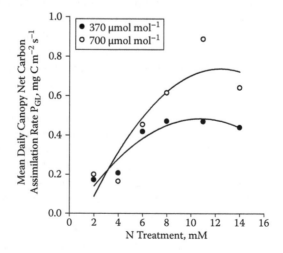

FIGURE 10.2 Mean canopy net C-assimilation rate as a function of six N treatments at two levels of CO_2.

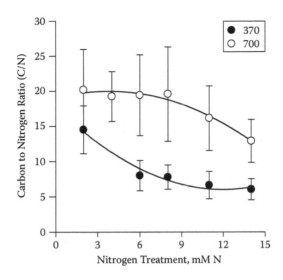

FIGURE 10.3 Mean C/N ratio of aboveground biomass as a function of six N treatments at two levels of CO_2. The error bars indicate variability among plants and plant components (stems and leaves).

were higher for the elevated CO_2 treatment and reached an asymptote at a higher level of N. This is partially due to the slightly longer day length for the first experiment (elevated), as well as CO_2 fertilization. There was little difference between C-assimilation rates for the two CO_2 treatments at the lower N levels. This suggests that there was not enough N or leaf area for the plants to take advantage of the CO_2 fertilization or extra day length.

Measured C/N ratios for the two CO_2 levels as a function of N treatment is shown in Figure 10.3. In both cases, the C/N ratio decreases with increasing N application, a result of the increasing proportion of N in the plant tissue. The C/N ratios are higher for the elevated CO_2 treatments because of relatively reduced N contents. This is consistent with other elevated CO_2 studies with potato (Conn and Cochran 2006).

The mean daily N-uptake rate during the exponential growth phase calculated from the C/N ratio and net C-assimilation rate increases with concentration of N in the soil water (Figure 10.4). The trend reaches an asymptote at higher N application rates. The asymptote appears to occur at a lower N rate for the ambient CO_2 treatment than for the elevated CO_2 treatment. Nitrogen-uptake rates estimated from C-assimilation rates and C/N ratios were higher over all N treatments for the ambient CO_2 treatment. This is primarily due to the measured low C/N ratios at the end of the season in the plants grown under ambient CO_2. Even though the C-assimilation rates (Figure 10.2) were higher for the elevated CO_2 treatment, this was not enough to offset the differences in leaf and stem N contents.

The amount of N taken up in the transpiration stream was not greatly different between the two CO_2 treatments (Figure 10.5). The amount was slightly higher in the elevated CO_2 treatment than in the ambient. This difference is likely due to longer day length and slightly higher light levels for the early summer period during which the elevated CO_2 plants were grown. Studies have reported lower water use in C_3 plants

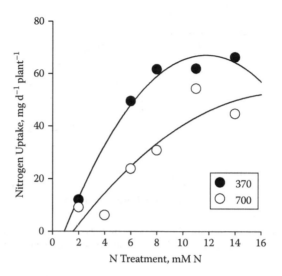

FIGURE 10.4 Total measured N uptake calculated from C/N ratio and net C-assimilation rate as a function of six N treatments at two levels of CO_2.

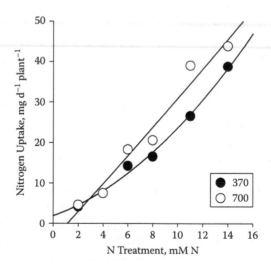

FIGURE 10.5 Mean measured passive N uptake in water (mass flow) as a function of six N treatments at two levels of CO_2.

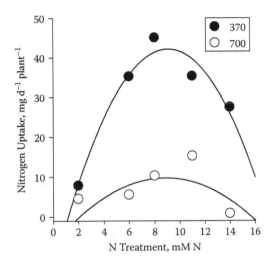

FIGURE 10.6 Mean measured active N uptake (diffusion) as a function of six N treatments at two levels of CO_2.

such as potato under elevated CO_2 (Bunce 2001). Since the plants were grown in pots, there was little exposed soil, so it is not likely there was significant contribution of soil evaporation to the condensate collected from the cooling coils. Nitrogen uptake via transpiration increased linearly with N application rate, with no asymptote at higher application rates. This is mainly a reflection of the size of the canopies and light interception as affected by N application rate. The linear increase with no asymptote results from the way passive uptake of N is calculated. Since the N content of the treatment increases more than the transpiration rate, the trend is largely a function of concentration of N in the soil water. Note that, since transpiration per unit ground area is affected by leaf cover, one would not expect a leaf area effect on transpiration at N applications greater than 8 mM when leaf area was maximal (Figure 10.1).

The rate of N uptake by diffusion calculated as the difference between N uptake in transpiration and total N uptake shows larger differences between CO_2 treatments (Figure 10.6) than for passive uptake. The trend is quadratic for both treatments, with a maximum uptake at about the medium concentration of N in soil water (6–10 mM N). The uptake decreases at the highest N levels. The ratio of N uptake in transpiration relative to total uptake increases as soil water N concentration increases (Figure 10.7). The slope is slightly steeper for the ambient CO_2 treatments than for the elevated ones, but the difference is small. In general, almost twice as much N uptake came from transpiration with elevated CO_2 than with ambient CO_2.

10.2.4.2 Simulations

Simulations were carried out using conditions and N treatments from the ambient CO_2 level. Simulated daily total N-uptake rates are similar to measured results, although the simulated total uptake continues to increase slightly with N rate, while the measured rates reach a distinct asymptote (Figure 10.8). Simulated and measured N as mass flow (passive) are similar, and both show the same increasing trend with

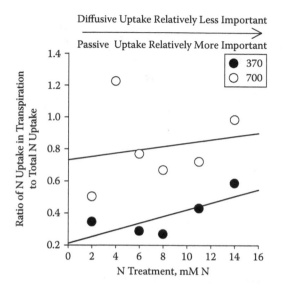

FIGURE 10.7 Ratio of measured N uptake in transpiration to total uptake for six N treatments at two levels of CO_2.

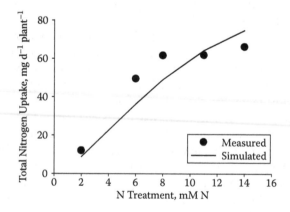

FIGURE 10.8 Simulated and measured total N uptake for the ambient CO_2 treatment.

no asymptote (Figure 10.9). The results for daily diffusive N-uptake rates show the largest differences, especially at the midrange of the treatment amount (6–10 mM). The measured diffusive uptake is about 10–15 mg d^{-1} higher than simulated. This contributes to the differences seen in total N uptake in Figure 10.8. The simulated daily diffusive N-uptake rate does begin to decrease at the higher concentrations of applied N, as seen in the measured data, but the decrease is not as marked (Figure 10.10) and begins at a higher applied N concentration. The trend in simulated total C assimilation is also similar to the measured results, where both exhibit a quadratic trend and reach a maximum at about 8–10 mM (Figure 10.11). The simulated values are higher, however, at the higher N applications.

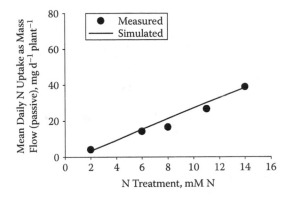

FIGURE 10.9 Simulated and measured N uptake as mass flow (passive uptake) for the ambient CO_2 treatment.

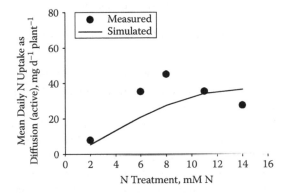

FIGURE 10.10 Simulated and measured N uptake as diffusive (active) uptake for the ambient CO_2 treatment.

10.2.5 DISCUSSION

The measured relationship between C-assimilation rate and soil water N concentration in this study is largely related to the effect of N on canopy growth and light interception (Figures 10.1 and 10.2). Previous research has demonstrated that the effect of N is to increase leaf size and area, leading to increased light interception and canopy-level C assimilation (Lawlor 2002). This is consistent with research showing that N uptake is proportional to relative growth rates (Schenk 1996) (growth rate adjusted for cumulative growth). Hence, the relationship between C assimilation and soil water N concentration is mainly related to the effects of N availability on leaf growth and resultant light interception. At lower levels of soil water N concentration (<8 mM), there was less leaf area and less light interception, resulting in lower C-assimilation rates per square meter of chamber surface. The asymptote for light interception in Figure 10.1 is due to the fact that the plants have reached maximum light interception at 8 mM N, and additional N above 8 mM increased biomass without appreciably increasing light interception. Since respiration is roughly a function

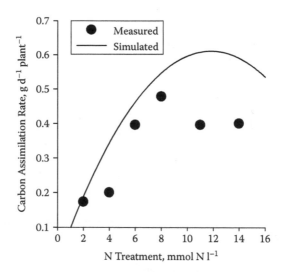

FIGURE 10.11 Simulated and measured C-assimilation rate as a function of six N treatments for the ambient CO_2 treatment.

of the amount of biomass (Amthor 2000; Timlin et al. 2006), the proportion of C lost to respiration will be higher at N levels greater than 8 mM, where N apparently increased biomass without increasing light interception.

The measured results suggest a decreasing rate of N uptake via diffusion (active uptake) as soil water N concentration increases. The differences between the ambient and elevated CO_2 results further suggest that the rate of active uptake is strongly dependent on the N demand of the plant (as defined by an optimum C/N ratio). Where N demand was low (elevated CO_2), active uptake was a smaller component of N uptake. This is in agreement with the observation that plant uptake of N is subject to down-regulation when N demand decreases (Glass et al. 2002). On a whole-plant scale, the regulation of N uptake appears to be related to the C-assimilation rate and light interception. The ability to control N uptake to match C availability would enable the plant to maximize its leaf area in the shortest time possible. Nitrogen is important in leaf growth and, in this study, was the limiting factor in obtaining full light interception for low N concentrations in soil water. Nitrogen limitation will reduce leaf growth and, in turn, reduce C assimilation. This, in fact, is the basis of modeling N in many crop models (Jeuffroy et al. 2002).

Nitrogen uptake in the transpiration stream, by calculation, was related to N concentration of the soil solution and transpiration rate. The increasing uptake rate with N level was largely related to the N concentration in solution, since transpiration rates were not greatly different, especially for treatments at full light interception. Uptake of N via the LATS system (mass-flow uptake) is more responsive to fertilization than the HATS (active uptake) system because it operates at higher concentrations of N in soil water (Malagoli et al. 2004). The response of increasing relative contributions from diffusive uptake as N level decreases, as seen in Figure 10.7 in both the measured and simulated data, has been shown to occur in the experiments as well. At low N levels or at low transpiration rates, diffusive uptake increases and can offset insufficient N uptake via transpiration (Plhak 2003). Mass-flow rates in a field study (Strebel and Duynisveld 1989) were 15%–33% of total N supply, and the ratio was highest in the surface soils (probably because that was where the water was coming from and N concentrations thus were higher). The diffusive component became more important deeper in the profile, where N contents were lower. In our simulations, the mass-flow component was about 40% at the mid N level (8 mM). Calculations given by Barber (1995, Table 4.4) suggest that the proportion of N taken up in corn as mass flow is about 80% of total uptake. In our study, the total proportion was closer to 60% at the highest N rate (Figure 10.7). Barber's (1995) analysis was based on data from regional values of N availability, water-uptake rate, and N recovery in plant tissue. It was assumed that the available N was primarily found in the upper 20 cm of soil. This would tend to bias values toward mass flow, since the analysis does not differentiate between water uptake from different parts of the soil profile and differences in N availability. There may also be crop differences as Barber's study (1995) used corn. In a study by Strebel and Duynisveld (1989), the diffusive component became more important at N rates lower than 8 mM. Mass flow and diffusive uptake calculated from the measured data, and shown in Figures 10.5 and 10.6, are approximations based on the assumptions outlined earlier. The simulation results (Figures 10.9 and 10.10), however, support these assumptions and demonstrate that the methods and approximations used here enable us to simulate total N uptake and to partition the uptake in a realistic manner to active and passive components.

There was one difference where the simulated diffusive component did not trend to lower rates at higher N levels, but instead reached an asymptote (Figure 10.10) at N levels greater than 6 mM. This may be related to the values for I_{max} and K_M and how they were derived. Only a single value of I_{max} was used to simulate diffusive movement, and the values for I_{max} or K_M may have been too high. Research suggests that they vary in potato, depending on crop growth stage and N demand (Sharifi and Zebarth 2006). The authors attributed this variance to changes in N demand per unit root length. In our model, N demand was mainly determined by how fast the crop could grow, N availability in the soil, and values of the parameters in the MM equation (equation 10.3). If the concentration of N in the soil was low, the gradient for N was high, resulting in increased diffusive uptake over mass flow. The higher gradients that controlled diffusive uptake at low N were a function of the high demand created by the I_{max} and K_M parameters and the low soil N content. Canopy growth rate was determined by the availability of N and the requirement of a specific C/N ratio to maintain C-assimilation rates. It may be worthwhile to investigate alternative measures of simulating plant demand, such as determining C_R in equations 10.3 and 10.4 as a function of plant N requirement.

Uptake of N in our model was controlled indirectly by the C-assimilation rate via the C/N ratio. If there was ample N and C to support the given C/N ratio, the canopy continued to expand and intercept light, resulting in increasing C-assimilation rates until full light interception was achieved. Sufficiency of N was determined by I_{max}, K_M, water uptake, and C-assimilation rates and soil water N concentration. Malagoli et al. (2004) found that HATS uptake (a diffusive component) of nitrate was highest at midday when C-assimilation rates were highest. Since root sugar lev-

els (a product of photosynthesis) are responsible for expression of genes related to
nitrate uptake (Matt et al. 2001), it seems reasonable to consider that photosynthetic
activity is a regulator of nitrate uptake (Malagoli et al. 2004). Nitrogen uptake via
diffusive processes, however, has been observed to drop off after flowering in oilseed rape, even though C-assimilation rates do not change greatly. Malagoli et al.
(2004) hypothesized that this may be due to a change in partitioning of C from the
roots to the shoots. We did not consider this in our model.

As shown by Malagoli et al. (2004), this research provides additional evidence
that N uptake can be modeled using factors such as light interception and C-assimilation and transpiration rates to control N uptake, in addition to an N dilution curve
(critical N content) as is used in many current models. There are other factors that
we did not consider that may also be important in N uptake; these include root distribution, temperature, senescence, and N recycling, among others. We parameterized
values for calculations of N uptake using an N response curve and assumed these
parameters were valid over the entire growth period. This may need to be investigated further.

REFERENCES

Acco Lea, P. J., and R. A. Azevedo. 2006. Nitrogen use efficiency, 1: Uptake of nitrogen from
the soil. *Ann. Appl. Biol.* 149 (3): 243–247.
Acock, B., V. R. Reddy, F. D. Whisler, D. N. Baker, H. F. Hodges, and K. J. Boote. 1985. The
soybean crop simulator GLYCIM. Model documentation. USDA, Washington, DC.
Acock, B. A., J. H. M. Thornley, and J. W. Wilson. 1971. Photosynthesis and energy conversion. In *Potential crop production*, ed. P. F. Wareing and J. P. Cooper, 43–75. London:
Heinemann.
Ahuja, L. R., L. Ma, and T. A. Howell. 2002. Whole system integration and modelling—
essential to agriculture and technology in the 21st century. In *Agricultural systems
models in field research and technology transfer*, ed. L. R. Ahuja, L. Ma, and T. A.
Howell. Boca Raton, FL: Lewis Publishers.
Amthor, J. S. 2000. The McCree–de Wit–Penning-de Vries–Thornley respiration paradigms:
30 years later. *Ann. Bot.* 86: 1–20.
Baker, J. T., S.-H. Kim, D. C. Gitz, D. Timlin, and V. R. Reddy. 2004. A method for estimating carbon dioxide leakage rates in controlled-environment chambers using nitrous
oxide. *Environ. Exp. Botany* 51: 103–110.
Baldwin, J. P., P. H. Nye, and P. B. Tinker. 1973. Uptake of solutes by multiple root systems
from soil, 3: A model for calculating the solute uptake by a randomly dispersed root
system developing in a finite volume of soil. *Plant Soil* 38: 621–635.
Barber, S. A. 1995. *Soil nutrient bioavailability.* 2nd ed. New York: Wiley.
Bouldin, D. R. 1961. Mathematical description of diffusion processes in the soil-plant system.
Soil Sci. Soc. Am. Proc. 25: 476–480.
Bunce, J. A. 2001. Direct and acclimatory responses of stomatal conductance to elevated
carbon dioxide in four herbaceous crop species in the field. *Global Change Biology* 7
(3): 323–331.
Buysse, J., E. Smolders, and R. Merckx. 1996. Modeling the uptake of nitrate by a growing
plant with an adjustable root nitrate uptake capacity. *Plant Soil* 181 (1): 19–23.
Cao, W., and T. W. Tibbitts. 1998. Response of potatoes to nitrogen concentrations differ with
nitrogen forms. *J. Plant Nutr.* 21 (4): 615–623.
Claassen, N., and S. A. Barber. 1976. Simulation model for nutrient uptake from soil by a
growing plant root system. *Agron. J.* 68: 961–964.

Clement C. R., M. J. Hopper, and L. H. P. Jones. 1978. The uptake of nitrate by *Lolium perenne* from flowing nutrient solution, 2: Effect of light, defoliation and relationship to CO_2 flux. *J. Exp. Botany* 29: 1173–1183.

Conn, J. S., and V. L. Cochran. 2006. Response of potato (*Solanum tuberosum* L.) to elevated atmospheric CO_2 in the North American subarctic. *Agric. Ecosys. Environ.* 112 (1): 49–57.

Constable, G. A., and H. M. Rawson. 1980. Effect of leaf position, expansion and age on photosynthesis, transpiration, and water use efficiency of cotton. *Aust. J. Plant Physiol.* 7: 89–100.

Dewar, R. C. 1996. The correlation between plant growth and intercepted radiation: An interpretation in terms of optimal plant nitrogen content. *Ann. Bot.* 78 (1): 125–136.

De Willigen, P. 1986. Supply of soil nitrogen to the plant during the growing season. In *Fundamental, ecological and agricultural aspects of nitrogen metabolism in higher plants*, ed. H. Lambers, J. J. Neeteson, and I. Stulen, 417–432. Dordrecht: Martinus Nijhoff.

Gastal, F., and G. Lemaire. 2002. N uptake and distribution in crops: An agronomical and ecophysiological perspective. *J. Exp. Bot.* 53: 789–799.

Gastal, F., and B. Saugier. 1989. Relationships between nitrogen uptake and carbon assimilation in whole plants of tall fescue. *Plant Cell Environ.* 12: 407–418.

Gayler, S., E. Wang, E. Priesack, T. Schaaf, and F.-X. Maidl. 2002. Modeling biomass growth, N-uptake and phenological development of potato crop. *Geoderma* 105 (3–4): 367–383.

Glass, A. D. M., D. T. Britto, B. N. Kaiser, J. R. Kinghorn, H. J. Kronzucker, A. Kumar, M. Okamoto, S. Rawat, M. Y. Siddiqi, S. E. Unkles, and J. J. Vidmar. 2002. The regulation of nitrate and ammonium transport systems in plants. *J. Exp. Bot.* 53 (370): 855–864.

Grindlay, D. J. C. 1997. Towards an explanation of crop nitrogen demand based on the optimization of leaf nitrogen per unit leaf area. *J. Agric. Sci.* 128: 377–396.

Hirose, T. 1988. Modelling the relative growth rate as a function of plant nitrogen concentration. *Physiologia Plantarum* 72 (1): 185–189.

Jeuffroy, M. H., B. Neyand, and A. Ourry. 2002. Integrated physiological and agronomic modelling of N capture and use within the plant, *J. Exp. Bot.* 53: 809–823.

Karley, A. J., A. E. Douglas, and W. E. Parker. 2002. Amino acid composition and nutritional quality of potato leaf phloem sap for aphids. *J. Exp. Biol.* 205: 3009–3018.

Kim, S.-H., V. R. Reddy, J. T. Baker, G. C. Gitz, and D. J. Timlin. 2004. Quantification of photosynthetically active radiation inside sunlit growth chambers. *Agric. Forest Meteorol.* 126: 117–127.

Kimball, B. A., K. Kobayashi, and M. Bindi. 2002. Responses of agricultural crops to free-air CO_2 enrichment. *Adv. Agron.* 77: 293–386.

Lawlor, D. W. 2002. Carbon and nitrogen assimilation in relation to yield: Mechanisms are the key to understanding production systems. *J. Exp. Bot.* 53 (370): 773–787.

Lemaire, G., and J. Salette, 1984. Relation entre dynamique de croissance et dynamique de prélèvement d'azote pour un peuplement de graminées fourragères, 1: Etude de l'effet du milieu. *Agronomie* 4: 423–430.

Malagoli, P., P. Lainé, E. Le Deunff, L. Rossato, B. Ney, and A. Ourry. 2004. Modeling nitrogen uptake in *Brassica napus* L. cv. Capitol during a growth cycle using influx kinetics of root nitrate transport systems and field experimental data. *Plant Physiol.* 134 (1): 388–400.

Matt, P., M. Geiger, P. Walch-Liu, C. Engels, A. Krapp, and M. Stitt. 2001. The immediate cause of the diurnal changes of nitrogen metabolism in leaves of nitrate-replete tobacco: A major imbalance between the rate of nitrate reduction and the rates of nitrate uptake and ammonium metabolism during the first part of the light period. *Plant Cell Environ.* 24: 177–90.

McDonald, A. J. S., T. Lohammar, and A. Ericsson. 1986. Growth response to step-decrease in nutrient availability in small birch (*Betula pendula* Roth). *Plant Cell. Environ.* 9: 427–432.

Miller, A. J., and M. D. Cramer. 2005. Root nitrogen acquisition and assimilation. *Plant Soil* 274 (1–2): 1–36.

Millington, R. J., and J. P. Quirk. 1960. Transport in porous media. In *7th Int. Congr. Soil Sci. Trans.*, 3: 97–106, Madison, WI.

Milroy, S. P., and M. P. Bange. 2003. Nitrogen and light responses of cotton photosynthesis and implications for crop growth. *Crop Sci.* 43: 904–913.

Nye, P. H., and F. H. C. Marriott. 1969. A theoretical study of the distribution of substances around roots resulting from simultaneous diffusion and mass flow. *Plant Soil* 30 (3): 459–472.

Oparka, K. J., and H. V. Davies. 1985. Translocation of assimilates within and between potato stems. *Ann. Bot.* 56: 45–54.

Pickering, N. B., L. H. Allen Jr., S. L. Albrecht, P. Jones, J. W. Jones, and J. T. Baker. 1994. Environmental plant chambers: Control and measurement using CR-10 data loggers. In *Computers in agriculture*, ed. D. G. Watson, F. S. Zazueta, and T. V. Harrison, 29–35. Proceedings of the Fifth International Conference. St. Joseph, MI: American Society of Agricultural Engineers.

Plénet, D., and G. Lemaire. 1999. Relationships between dynamics of nitrogen uptake and dry matter accumulation in maize crops: Determination of critical N concentration. *Plant Soil* 216 (1–2): 65–82.

Plhak, F. 2003. Nitrogen supply through transpiration mass flow can limit nitrogen nutrition of plants. *Plant Soil Environ.* 49 (10): 473–479.

Reddy, K. R., J. T. Baker, V. R. Reddy, J. McKinion, L. Tarpley, and J. J. Read. 2001. Soil-Plant-Atmosphere-Research (SPAR) facility: A tool for plant research and modeling. *Biotronics* 30: 27–50.

Rodgers, C. O., and A. J. Barneix. 1988. Cultivar differences in the rate of nitrate uptake by intact wheat plants as related to growth rate. *Physiologia Plantarum* 72: 121–126.

SAS Institute. 2004. *SAS 9.1.3 help and documentation.* Cary, NC: SAS Institute.

Schenk, M. K. 1996. Regulation of nitrogen uptake on the whole plant level. *Plant Soil* 181 (1): 131–137.

Sharifi, M., and B. Zebarth. 2006. Nitrate influx kinetic parameters of five potato cultivars during vegetative growth. *Plant Soil* 288 (1): 91–99.

Simunek. J., T. Vogel, and M. Th. van Genuchten. 1994. The SWMS_2D code for simulating water flow and solute transport in two-dimensional variably saturated media, Version 1.21. Research report no. 132. U.S. Salinity Laboratory, USDA, ARS, Riverside, CA.

Smart D. R., K. Ritchie, A. F. Bloom, and B. B. Bugbee. 1998. Nitrogen balance for wheat canopies (*Triticum aestivum* cv. Veery 10) grown under elevated and ambient CO_2. *Plant Cell Environ.* 22: 753–763.

Smith, G. S., and K. R. Middleton. 1980. Sodium nutrition of pasture plants, 1: Translocation of sodium and potassium in relation to transpiration rates. *New Phytologist* 84 (4): 603–612.

Stitt, M., and A. Krapp. 1999. The interaction between elevated carbon dioxide and nitrogen nutrition: The physiological and molecular background. *Plant Cell Environ.* 22: 583–621.

Strebel, O., and W. H. M. Duynisveld. 1989. Nitrogen supply to cereals and sugar beet by mass flow and diffusion on a silty loam soil. *Zeitschrift für Pflanzenernährung und Bodenkunde* 152 (2): 135–141.

Timlin, D. J., Ya. Pachepsky, and B. Acock. 1996. A design for a modular, generic soil simulator to interface with plant models. *Agronomy J.* 88: 162–169.

Timlin, D. J., S. M. L. Rahman, J. Baker, V. R. Reddy, D. Fleisher, and B. Quebedeaux. 2006. Whole plant photosynthesis, development, and carbon partitioning in potato as a function of temperature. *Agron. J.* 98: 1195–1203.

van Iersel, M. W., and J. G. Kang. 2002. Nutrient solution concentration affects whole-plant CO_2 exchange and growth of sub-irrigated pansy. *J. Am. Soc. Hort. Sci.* 127 (3): 423–429.

Verkroost, A. W. M., and M. J. Wassen. 2005. Simple model for nitrogen-limited plant growth and nitrogen allocation. *Ann. Bot.* 96 (5): 871–876.

11 Roots below One-Meter Depth Are Important for Uptake of Nitrate by Annual Crops

Hanne L. Kristensen and Kristian Thorup-Kristensen

CONTENTS

11.1 Introduction..245
    11.1.1 Root Depths of Annual Crops...245
    11.1.2 N Uptake at Deep Depths 246
11.2 Experimental Details and Data Analysis...247
11.3 Potential Nitrate Uptake at Depths 249
11.4 The EU-rotate_N Model 252
11.5 Modeling the Effect of Deep Roots on Soil N 254
11.6 Factors of Importance for Deep-Depth N Uptake 256
11.7 Conclusions 256
Acknowledgments..257
References...257

11.1 INTRODUCTION

11.1.1 ROOT DEPTHS OF ANNUAL CROPS

Recent research has shown that deep-rooted crops and cover crops (catch crops) can have high rates of nitrate uptake from deep soil layers (Kristensen and Thorup-Kristensen 2004a; Kristensen and Thorup-Kristensen 2004b). A deep-rooted crop was found to take up more than 100 kg ha^{-1} of mineral nitrogen (N) that had leached below 1-m depth (Thorup-Kristensen 2006a). Thus deep-rooted crops can recirculate nitrate that was otherwise leached and lost from the agricultural system to the water environment. However, it is often assumed that roots below 1 m are not important compared with those within 1 m of the soil depth when modeling plant N uptake and N losses from agroecosystems. Also, experimental work on root distribution and leaching losses below annual crops is often based on this assumption, and it is common practice to limit root and N-leaching studies to the soil surface layers. The rationale is that deep roots make up a relatively small percentage of the total root biomass and therefore have minor importance for N uptake of the crop. Further,

there are methodological difficulties in working at depths greater than 1 m. Therefore, data on roots below 0.6–1-m depth is often missing, and root studies often show
some roots at their deepest measuring depth. In fact, an increasing number of studies
show that some annual crops extend roots well below 1-m depth.

In the temperate regions, root depths in the range of 1- to at least 2.5-m depth
have been found for a large variety of horticultural and agricultural crops and cover
crops (Kuhlmann et al. 1989; Barraclough 1989; Barraclough et al. 1989; Wiesler
and Horst 1994; Kristensen and Thorup-Kristensen 2004a; Kristensen and Thorup-
Kristensen 2004b; Kristensen and Thorup-Kristensen 2007; Christiansen et al. 2006;
Thorup-Kristensen 2006a). Development of deep roots is an intrinsic property of the
species. For example, root depth was found to be only slightly affected by differences
in the vertical distribution of soil mineral N (Christiansen et al., 2006; Kristensen and
Thorup-Kristensen 2007), whereas extreme situations like compacted subsoils may
or may not limit root growth (Barraclough and Weir 1988; Munkholm et al. 2005).
Thus, we can expect potentially deep-rooted crops to actually grow deep roots. This
is illustrated by results obtained in a number of field studies in Denmark in Table 11.1,
showing that many annual crops extended the root system below 1-m depth.

11.1.2 N UPTAKE AT DEEP DEPTHS

The fact that the deep roots of many annual crops are ignored is critical for the
understanding and modeling of N cycling only if the deep roots actually take up
significant amounts of N. This will make the estimated leaching loss from the root
zone deviate from the actual leaching loss. Nitrogen uptake by deep roots was shown
in recent work, where mineral N uptake from below 1-m depth amounted to more
than 100 kg N ha^{-1} for white cabbage, when grown where much nitrate had been
leached to deep soil layers (Thorup-Kristensen 2006a). In this study, white cabbage
roots reached more than 2.4-m depth at harvest and left only 20 kg N ha^{-1} in the
1–2.5-m soil layer, compared with a crop of leek that reached a root depth of 0.5 m
and left 133 kg N ha^{-1} in that soil layer. Other studies have also shown significant N
uptake at depth by annual crops such as corn, sugar beet, and winter wheat (Gass et
al. 1971; Peterson et al. 1979; Kuhlmann et al. 1989). However, the conclusions were
based on studies of changes in plant and soil mineral N content over the growing
season, where other factors such as leaf loss, mineralization of soil organic matter,
and leaching may have influenced results. Further, crop N uptake may be limited by
the availability of nitrate at depth.

One approach to investigate the potential for deep N uptake of crops is to measure the N uptake of deep roots over a few days under conditions with excess N
availability. Such results may further be used to investigate the relationship between
root distribution and the potential for N uptake at depth, i.e., the N inflow at the scale
of single roots. This is needed in the effort to understand and model N uptake on the
basis of the function of single roots (e.g., Raynaud and Leadley 2004). The potential
N uptake of roots in the deep part of the root zone was measured in a range of short-
term experiments with field-grown annual crops (Kristensen and Thorup-Kristensen
2004a; Kristensen and Thorup-Kristensen 2004b; Kristensen, unpublished).

In this chapter, these results are used to investigate the importance of deep N
uptake of annual crops. We discuss the methodological basis for estimation of the

TABLE 11.1

Average Root Depth for Vegetables and Cereals at Harvest and for Cover Crops in Late Fall as Measured by Minirhizotrons

	Species	Root Depth (m)	Reference
Vegetables	Onion[a]	0.3	Thorup-Kristensen (2006b)
	Celeriac[a]	0.5	Christiansen et al. (2006)
	Leek[a]	0.5	Kristensen and Thorup-Kristensen (2007)
	Lettuce[a]	0.6	Thorup-Kristensen (2006b)
	Pea[a]	0.6–0.8	Thorup-Kristensen (1998)
	Potato[a]	0.7	Kristensen and Thorup-Kristensen (2007)
	Sweet corn[a]	0.9	Christiansen et al. (2006)
	Broccoli[a]	0.6–1.2	Thorup-Kristensen (1993)
	Cauliflower[a]	0.9–1.2	Thorup-Kristensen and Van den Boogaard (1998)
	Chinese cabbage[a]	1.3	Kristensen and Thorup-Kristensen (2007)
	Carrot[a]	1.3	Kristensen and Thorup-Kristensen (2004b)
	Rucola 1st harvest[b]	0.8	Kristensen (unpublished)
	Rucola 2nd harvest[b]	1.6	Kristensen (unpublished)
	Beetroot[a]	1.6–1.9	Christiansen et al. (2006), Kristensen and Thorup-Kristensen (2007)
	Summer squash[a]	1.9	Kristensen and Thorup-Kristensen (2007)
	Curly kale[a]	> 2.5	Kristensen (unpublished)
	White cabbage[a]	> 2.5	Kristensen and Thorup-Kristensen (2004b)
Cover crops	Ryegrass[a]	0.6–1	Kristensen and Thorup-Kristensen (2004a) Thorup-Kristensen (2006a)
	Winter rye[a]	1.2	Kristensen and Thorup-Kristensen (2004a)
	Chicory[a]	> 2.5	Thorup-Kristensen (2006a)
	Fodder radish[a]	> 2.5	Kristensen and Thorup-Kristensen (2004a)
Cereals	Spring barley[a]	1.0–1.2	Thorup-Kristensen (2006a)
	Winter wheat[a]	1.7–2.4	Thorup-Kristensen (unpublished)

[a] Grown on a sandy loam soil at the Aarslev Research Centre.
[b] Grown on coarse sand at Yding.

potential for nitrate uptake, as well as the implications for N uptake by deep roots. Further, it is our purpose to show the importance of root depths for N uptake in modeling by simulating soil nitrate left by crops at harvest by use of the EU-rotate_N model.

11.2 EXPERIMENTAL DETAILS AND DATA ANALYSIS

Results presented here were obtained in four field studies in Denmark. Three studies were performed at the Aarslev Research Centre (10°27'E, 55°18'N). The soil was a sandy loam (Typic Agrudalf), with the 0–0.5-m soil layer containing 0.9% OC (organic carbon), 5% clay, 4% silt, and 90% sand; the 0.5–1-m layer containing 0.2%

OC, 3% clay, 3% silt, and 93% sand; and the 1–2.5-m layer containing 0.1% OC, 3% clay, 3% silt, and 94% sand. The pH_{CaCl2} was 6.7, 6.5, and 6.4 in the 0–0.5, 0.5–1, and 1–2.5-m layers, respectively. Average annual precipitation (624 mm) and air temperature (7.8°C) were recorded at a meteorological station at the research center. One field study, with rucola, was performed at the Yding location (9°44'E, 55°59'N) on coarse sand, with the 0–0.5-m soil layer containing 1.1% OC, 13% clay, 15% silt, and 70% sand; the 0.5–1-m layer containing 0.3% OC, 19% clay, 13% silt, and 68% sand; and the 1–2.5-m layer containing 0.2% OC, 18% clay, 14% silt, and 68% sand. The pH_{CaCl2} was 6.3, 5.4, and 6.8 in the 0–0.5, 0.5–1, and 1–2.5-m layers, respectively.

Six deep-rooted crops, four vegetables and two cover crops, were grown in the four field studies. Carrots were sown and white cabbages planted at the end of April and May, respectively, and harvested in October 2000. The curly kale was planted in May and harvested in October 2002. These three vegetables were grown with organic farming practice using organic fertilization and irrigation at the Aarslev Research Centre. The cover crops winter rye and fodder radish were sown on 8 August 2000 at the research center and ploughed into the soil at the end of October 2000 without further management. The rucola was grown at Yding as a baby-leaf vegetable after conventional farming practice with inorganic fertilization and irrigation. The rucola was sown on 29 June 2005, harvested the first time on 19 August 2005, and the second time at the end of September 2005.

Root depths and distributions were obtained by use of minirhizotrons. These were glass tubes of 3-m length and 70-mm outer diameter. In replicate field plots in a completely randomized block design, the minirhizotrons were inserted along crop rows at an angle of 30° from the vertical and reached a depth of approximately 2.4 m into the soil. The number of replicate tubes ranged from four to eight in the different studies. Further details about installation of the minirhizotrons are given in Kristensen and Thorup-Kristensen (2004a). Each minirhizotron had two replicate counting grids (40 × 40-mm square counting grids) painted along the minirhizotron surface. A minivideo camera was used to record the roots at the minirhizotron surface. From those recordings, root depth and root intensity were obtained. Root depth was registered as the deepest root observed in each of the two 40 × 40-mm square counting grids on each minirhizotron. For calculation of root intensity, the total number of roots crossing the lines in each 40 × 40-mm square was recorded and used to calculate root intersections per meter grid line (intersections m⁻¹) per 34.6-mm soil layer (=cos(30°) × 40 mm).

The root intensity in different soil depths for the six crops was measured in the ^{15}N experiments, which were performed to measure the potential N uptake of roots at depth. For estimation of N inflow (rate of N uptake per unit of root length), the results of root intensity were used to calculate root length density following the modified Newman-line-intersect method (Tennant 1975), and by assuming a depth of view of 2 mm into the soil surrounding the minirhizotron (Kristensen and Thorup-Kristensen 2004a).

The potential N uptake of roots at depth was measured by deep-point placement of excess ^{15}N-nitrate, with subsequent analysis of ^{15}N content in aboveground and taproot biomass. The ^{15}N field experiments were performed by placing $^{15}NO_3$ at four to six different depths under the crops, followed by sampling of plant biomass and analysis of ^{15}N enrichment in the plant N pool. The ^{15}NO$_3$ was applied on 27–29 September 2000 for carrot and white cabbage, on 25–26 October 2000 for winter rye and fodder radish, and on 16 September 2005 for rucola in four replicate subplots (0.9 × 0.8 m) for each depth and species. For curly kale, the ^{15}NO$_3$ was applied on 13 August 2002 in five replicate subplots (1.5 × 2 m) for each depth.

The ^{15}NO$_3$ was injected into each subplot through four holes. The holes were placed at an angle 30° from the vertical using a piston steel rod with a diameter of 20 mm. This was done to minimize damage to the plants within the subplots while drilling. A plastic tube was inserted into each hole, and ^{15}NO$_3$ was injected into the soil as a solution of Na^{15}NO$_3$ (5 or 10 mg N mL^{-1}, 99% ^{15}N enriched). Each hole had 5 or 10 mL of ^{15}NO$_3$ solution applied, i.e., a total injection of 99 or 198 mg ^{15}N to each subplot, followed by application of demineralized water to each tube for rinsing. The amount of 198 mg ^{15}N per subplot was applied to curly kale, and 98 mg ^{15}N per subplot was applied to the other five crops. The tubes were removed, and either sand or a wooden rod was used to fill each hole to prevent roots from growing down the hole. The aboveground and taproot plant biomass in each subplot were harvested either 6 days (carrot, white cabbage, winter rye, fodder radish, rucola) or 3–30 days (curly kale) after ^{15}N injection. The samples were then rinsed, dried, and milled, and subsamples were finely ground (<0.5 mm) and analyzed for ^{15}N, total N, and C analysis using a continuous flow isotope ratio mass spectrometer consisting of an automatic nitrogen and carbon analyzer coupled to a 20–20 mass spectrometer (Europa Scientific Ltd., Crewe, U.K.).

The ^{15}N plant uptake was calculated from the ^{15}N results, as the experimental design was equivalent to the "negative discard method" (Powlson and Barraclough 1993; Kristensen and Thorup Kristensen 2004a), in which ^{13}N plant uptake was calculated as excess plant ^{15}N by subtraction of background ^{15}N abundance determined for each species. The results of ^{15}N plant uptake were used for calculation of the potential N-uptake rate per unit soil or root, by assuming that the ^{15}N solution at each injection point was distributed in 1 L soil and by adjusting the ^{15}N enrichment for the dilution effect of the soil NO$_3^-$ pool (assumed to be 0.366% ^{15}N). The calculation of the N-uptake rate per unit root, the N inflow, was based on the results of root density obtained with the minirhizotron method, assuming a depth of view of 2 mm into the soil surrounding the minirhizotron, as mentioned above. The potential N-uptake rate per unit soil was further used to calculate the potential N uptake from below 1-m depth, using the estimated number of growth days with roots below 1-m depth.

11.3 POTENTIAL NITRATE UPTAKE AT DEPTHS

The potential for N uptake below 1-m depth was found to be high for all crops studied except for winter rye. The potential N-uptake rate varied greatly between species. This variation was largest in the potential uptake calculated for the season. This was due to the variation in N-uptake rates and to the difference in the number of growth days below 1-m depth. The latter depended on the rate of root penetration and on the length of the growing season. All measurements were done at one point in time for each crop, and therefore do not show the variability that can be expected over time. In fact, some of the variability between species could be due to the differences in developmental stage and physiological state. Thus, the potential plant N uptake at deep depth ranged from 0.03 mg N·kg soil⁻¹ for winter rye grown as a cover crop, which had roots below 1-m depth only for a few days, to 112 mg N·kg soil⁻¹ for curly kale, which had roots below 1-m depth for a long period of time (Table 11.2). Part of the explanation for the large difference in N-uptake rate between these two crops is that the measurements were done at a time when winter rye was losing leaves and stopping growth due to the approaching winter, whereas the curly kale was still in vigorous growth in late summer.

The potential N-uptake rates varied greatly not only between species, but also *within* the root system of each species (Table 11.2). The N-uptake rate at a given place in a root system can be affected by the amount of roots present at that place. Therefore, the N-uptake rate per unit of root is often calculated as the N inflow. Root distributions for each species at the time of the ¹⁵N experiments (Figure 11.1) were used to calculate the N inflow. However, the N inflows show almost as large a variation as the potential N-uptake rate per unit soil for each crop (Table 11.2). The finding that the N inflow varies within a root system at a given point in time confirms laboratory experiments that N inflows vary with local soil N availability (Robinson et al. 1994; Hodge 2004). However, in the present experiment, the local mineral N availability was approximately the same after excess ¹⁵N-nitrate placement. Thus the N inflows were potential N inflows, and the differences between the potential N inflows within the root system must have been caused by other factors. These could, for example, be the physiological state of the roots due to age or the availability of nitrate prior to ¹⁵N-nitrate addition, which may stimulate the pathway for nitrate uptake (Reidenbach and Horst 1997; Lainé et al. 1998).

Another reason for the variation in potential N inflow within the root system at a given point in time may be the method used for estimation of root distribution. It has been shown by several studies that the two conventional methods—the minirhizotron method, which was used in the present study, and the root extraction method—give quite different results regarding the relative distribution of roots with depth (e.g., Heeraman and Juma 1993). Typically, the minirhizotron method gives lower estimates in the surface soil and higher estimates at depth compared with the root-extraction method, where roots are extracted from soil samples by rinsing with water. The differences in root methodology are even more important when comparing N inflows from different studies, as both the relative distribution and the magnitude of the results can differ. The minirhizotron method gives a two-dimensional measure of root density. Therefore, an assumption is made about the depth of view from the surface of the minirhizotron tube when estimating root length density from minirhizotron data. In Table 11.2, a depth of view of 2 mm was chosen in accordance with previous studies (e.g., Heeraman and Juma 1993; Kristensen and Thorup-Kristensen 2004a). In our study, an N inflow of 30 pmol m root⁻¹ s⁻¹ was calculated for curly kale at 1.6-m depth based on minirhizotron results (Table 11.2). However, the N inflow based on the root-extraction method was calculated to be 120 pmol m root⁻¹ s⁻¹ (Kristensen, unpublished), which was four times the N inflow based on the minirhizotron method. New root techniques like X-ray imaging suggest that the root density may be much larger than that found by both the minirhizotron and the root-extraction methods due to the underestimation of fine roots (Pierret et

TABLE 11.2

Potential N Uptake and N Inflow Estimated on the Basis of Field Studies of ^{15}N–Nitrate Uptake in Deep-Rooted Cover Crops and Vegetables Grown in Denmark

Crop	Time of Measurements	Soil Type	Root Depth (m)	Measuring Depth (m)	Potential N Uptake Rate[a] (mg N kg soil [dry wt]$^{-1}$ d^{-1})	Root Grow below 1 m (d)	Potential N Uptake below 1 m for the Season (mg N kg soil [dry wt]$^{-1}$)	N Inflow[b] (pmol m root^{-1} s^{-1})	Reference
					Cover Crops				
Winter rye	Late fall	Sandy loam	1.2	1.1	0.002	15	0.03	1	Kristensen and Thorup-Kristensen (2004a)
Fodder radish	Late fall	Sandy loam	>2.5	1–2.5	0.03–0.22	41	1.2–9.2	3–28	Kristensen and Thorup-Kristensen (2004a)
					Vegetables				
Rucola 2nd harvest	Summer	Coarse sand	1.5	1.1	0.01–0.07	30	0.2–2.2	0.5–5	Kristensen, unpublished
Carrot	Fall	Sandy loam	1.4	1–1.4	0.02–0.05	65	1.2–3.4	13–15	Kristensen and Thorup-Kristensen (2004b)
White cabbage	Fall	Sandy loam	>2.5	1–2.5	0.02–0.12	123	2.0–15.1	3–6	Kristensen and Thorup-Kristensen (2004b)
Curly kale	Late summer	Sandy loam	>2.4	1.6	0.91	124	112	30	Kristensen, unpublished

Note: The root distribution at the time of the ^{15}N experiments is shown in Figure 11.1.

[a] Calculated on the basis of ^{15}N uptake in the plant biomass after deep point placement of ^{15}N-nitrate. Assumptions are given in the text.

[b] Calculated by the use of minirhizotron results. Assumptions are given in the text.

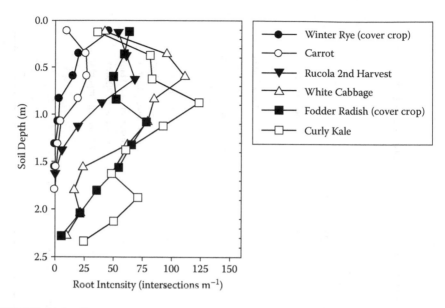

FIGURE 11.1 The root distribution of the species presented in Table 11.2 as measured by use of minirhizotrons at the time of the ^{15}N experiments.

al. 2005). Thus there is a need for development of root methodology to improve our understanding of the relationship between root distribution and N uptake.

11.4 THE EU-ROTATE_N MODEL

The EU-rotate_N is a new model-based decision-support system aimed at optimizing N use in horticultural crop rotations across Europe. The main purpose of the model is to simulate effects of different crop sequences, fertilizer rates, and other cultural practices on the cycling of N within crop rotations. The model is written in FORTRAN and runs through Microsoft's DOS emulator. The model organization is presented in Figure 11.2, and the model is available at www.warwick.ac.uk/go/euro-taten (Rahn et al. 2007). The Web site also provides details about the model.

In short, the model consists of a number of subroutines to simulate the growth below- and aboveground, N mineralization from the soil and crop residues, N uptake, and balance between N supply and demand. These are all regulated in daily time steps by weather and soil conditions relative to a user-supplied target yield for each crop. The model subroutines are based on existing models and published algorithms: For the crop-growth submodule, the N-ABLE model is used (Greenwood 2001), and for the mineralization module, the Daisy model is used (Hansen et al. 1990; Abra-hamsen and Hansen 2000). New subroutines based on new knowledge have been added, such as the root module that models root distribution in two dimensions. Horizontal root extension is modeled to simulate the effects of growing row crops, and vertical root extension is modeled to 2-m depth (Rahn et al. 2007). In the model, the soil is divided into 0.05 × 0.05-m units. The actual N uptake is calculated as a

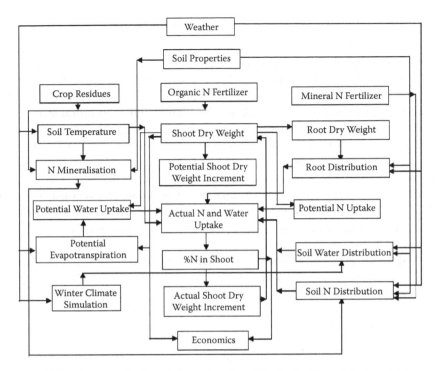

FIGURE 11.2   The organization of the main submodules in the EU-rotate_N model.

function of crop N demand on a specific day and the potential root N uptake that day. The latter is calculated as a function of the root length in each soil unit, the content of available soil N, and the root N-uptake efficiency.

The root distribution is calculated in three parts:

1. The depth of the root system over the season calculated as the root penetration rate, which is the slope of the regression line of the average root depth versus accumulated daily average temperature from sowing, with a crop-specific base temperature and a lag period from sowing until a root depth of 0.1 m is reached
2. The total root length of the crop based on crop biomass and crop-specific root shoot ratios during development, where a fixed specific root length is used for all crops
3. The vertical distribution of the root system based on a logarithmic function, $Lz = e^{-az}$, to 2-m depth (adapted from Gerwitz and Page 1974). The function is parameterized for each species by the factor a, which controls the steepness of the logarithmic decline ($a = 3$ is default). The root length density is allowed to vary at maximal root depth, which allows high root densities in the subsoil close to the maximal root depth. This differs from the way this equation has been adapted in other models such as the Daisy model, where root density is always low close to the maximal root depth.

11.5   MODELING THE EFFECT OF DEEP ROOTS ON SOIL N

When modeling plant N uptake in ecosystems, the potential N uptake is affected by crop properties, growth conditions, and crop N demand. However, the potential rate of N uptake is seldom differentiated *within* a root system in such models. This is, as discussed above, probably due to the limited knowledge about how N inflow differs within and between crops. This includes the limited ability to differentiate roots based on physiologically meaningful properties—"a root is a root" (Pregitzer 2002; Zobel 2003). Therefore, it is still a common practice to model plant N uptake by use of more-or-less-fixed root properties, e.g., the fixed potential N inflow or resource-capture coefficient (e.g., King et al. 2003). The EU-rotate_N model is an example of a crop N-uptake computer model that is parameterized with a fixed potential root N inflow represented by the N-uptake efficiency. Here the actual N uptake per root is calculated as a function of crop N demand and the root length, available soil N, and the N-uptake efficiency.

To show the effect of root depth on the modeling of mineral N left by the crop at harvest, the EU-rotate_N model was used to simulate the mineral N left by spring and winter wheat by assuming a maximal root depth at harvest of 0.9, or no (unlimited) maximal root depth. The model was parameterized at the Aarslev Research Centre with crop properties for spring and winter wheat, with target yields of 8 and 13 Mg ha^{-1} of total plant dry weight, respectively. For both crops, the root-depth penetration rate was 0.001 m d^{-1} °C^{-1}, with 0°C as base temperature and a lag period of 100 d °C. The *a* factor was 3. The maximal root depth was set either at 0.9 m or with no maximal root depth at harvest. The depth of 0.9 m was chosen because the model works in 0.3-m soil increments. The simulation was performed for the two-year period 2000–2001, with one harvest in 2001. The winter wheat was planted on the 290th day in 2000, and the spring wheat on the 100th day in 2001. On the 110th day in 2001, the amounts of 185 and 137 kg mineral-N ha^{-1} were simulated to be available in the 0–2-m profile for spring and winter wheat, respectively. Simulated soil N was distributed mainly below 0.9-m soil depth. Harvest was simulated for both spring and winter wheat on the 218th day in 2001, with amounts of 4.3 and 10.4 Mg ha^{-1} of total plant dry weight, respectively.

The effect of a maximal root depth on the simulations of root distributions of spring and winter wheat at harvest is seen in Figure 11.3. Residual soil N at winter wheat harvest was 135 kg N ha^{-1} in the 0–2-m soil profile when a maximal root depth of 0.9 m was assumed, as compared with a 92-kg N ha^{-1} residual soil N without a preassumed root depth at harvest, which showed a 43-kg N ha^{-1} overestimation of residual soil N at harvest with a maximal root depth of 0.9 m. Similarly, residual soil N was overestimated by 18 kg N ha^{-1} for spring wheat with a 0.9-m maximal root depth at harvest in the 0–2-m soil profile (166 vs. 148 kg N ha^{-1}). As expected, the decrease in soil residual N at harvest without a preset maximal root depth was smaller (10%) for spring wheat compared with winter wheat (30%) due to less root development below 0.9-m soil depth (Figure 11.4). Thus, the results from the simulations with the EU-rotate_N model show that the potential nitrate leaching during the following winter season was clearly affected by restricting root depth to 0.9 m for spring and winter wheat. The simulations also illustrate that the effect of root depth

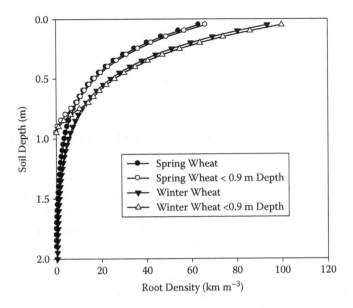

FIGURE 11.3   Simulation with the EU-rotate_N model of the root distribution of spring and winter wheat at harvest, with a maximal root depth of 0.9 m or no maximal root depth.

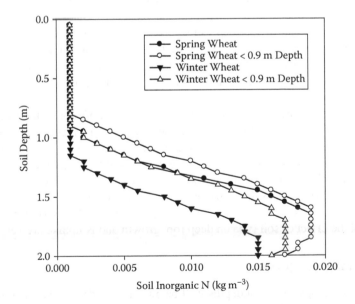

FIGURE 11.4   Simulation with the EU-rotate_N model of the effect of maximal root depth of spring and winter wheat on the distribution of soil mineral N at harvest. The difference for spring wheat of a maximal root depth of 0.9 m or no maximal root depth amounted to 18 kg N ha^{-1}. The difference for winter wheat amounted to 43 kg N ha^{-1}.

on the potential for nitrate leaching was greater for the deep-rooted crops. Assuming a shallower-than-actual root depth may lead to overestimation of the potential nitrate leaching after the deep-rooted crops. Results from experiments with winter wheat showed an estimated N uptake of 25 kg N ha^{-1} from the 0.9–1.5-m soil layer (Kuhlmann et al. 1989), which is in agreement with the simulation showing significant deep N uptake by winter wheat.

11.6   FACTORS OF IMPORTANCE FOR DEEP-DEPTH N UPTAKE

The significance of N uptake of annual crops from deep soil layers depends on a number of factors. The main factor is, of course, the root distributions. Many annual crops have roots well below 1-m depth and may, therefore, have the ability to take soil N at deep depth. The actual uptake of N depends then on the availability of soil N at deep depth. The soil N distribution, especially nitrate due to its mobility in soil water, depends strongly on the cropping and farming history, such as the use and timing of incorporation of cover crops and green manures prior to growing of deep-rooted crops, as well as other fertilization practices (Thorup-Kristensen 2006a; Kristensen and Thorup-Kristensen 2007). Only a few studies have reported the distribution of mineral N below 1-m depth, and therefore knowledge is lacking about the extent to which mineral N is available at deep depth under different conditions related to management, soil, and weather conditions.

Another factor of importance for deep N uptake is the crop N demand when the roots reach the deep soil layers. The amount of N fertilization and soil N availability in the surface soil layers during the early part of the growth season have a major influence on crop N demand during the later part of the growth season, when roots are present in the deep soil layers and may, therefore, also influence the actual N uptake from these layers. Many studies have shown that high N fertilization decreased the uptake of soil N from deep soil layers in general, but knowledge about relationships between the timing and amount of N fertilization, and on the uptake of N by deep roots, is lacking.

All factors that affect root growth and the extension of root systems are important for the ability of crops to take up N at deep depth. This includes the factors for growth in general, such as light, temperature, and the availability of water and nutrients to enable plant growth and development. The soil type and the occurrence of physical barriers for root growth may affect the development of deep roots (Munkholm et al. 2005), whereas the distribution of soil N may only have limited effect on deep-root development (Kristensen and Thorup-Kristensen 2007). However, investigations of the effect of soil type on deep root growth and N uptake are very few.

11.7   CONCLUSIONS

Many annual crops have root depths well below 1-m soil depth, and the results presented in this chapter show that the potential for nitrate uptake from below 1-m depth was significant for five of the six crops studied. This indicates that deep-rooted crops can take up deep-soil N when it is available and recycle the N within the cropping system. Therefore, deep roots and deep-soil N uptake need to be understood to model

N cycling in agroecosystems. The N uptake of crops at deep depth depends on the matching in time and space of deep roots, deep-soil mineral N, and crop N demand. An important implication to increase deep-soil N uptake is to improve the efficiency of soil N use in cropping systems. This would involve the growing of deep-rooted crops and cover crops when much N is available at deep depth, so that the amount of nitrate that is prone to leaching to the water environment can be reduced. The development of crop N models like the EU-rotate_N can help in understanding deep-soil N uptake and in improving the N-use efficiency of cropping systems.

ACKNOWLEDGMENTS

This work was supported by the Danish Councils for Independent Research; the Directorate for Food, Fisheries and Agri Business; and The Danish Research Center for Organic Food and Farming. The EU-rotate_N model was developed in the EU-rotate_N project funded by the European Commission (5th Framework Programme).

REFERENCES

Abrahamsen, P., and S. Hansen. 2000. Daisy: An open soil-crop-atmosphere system model. *Environ. Model Software* 15: 313–330.
Barraclough, P. B. 1989. Root growth, macro-nutrient uptake dynamics and soil fertility requirements of a high-yielding winter oilseed rape crop. *Plant Soil* 119: 59–70.
Barraclough, P. B., H. Kuhlmann, and A. H. Weir. 1989. The effects of prolonged drought and nitrogen fertilizer on root and shoot growth and water uptake by winter wheat. *J. Agron. Crop Sci.* 163: 352–360.
Barraclough, P. B., and A. H. Weir. 1988. Effects of a compacted subsoil layer on root and shoot growth, water use and nutrient uptake of winter wheat. *J. Agric. Sci.* 110: 207–216.
Christiansen, J. S., K. Thorup-Kristensen, and H. L. Kristensen. 2006. Root development of beetroot, sweet corn and celeriac, and soil N content after incorporation of green manure. *J. Hortic. Sci. Biotech.* 81: 831–838.
Gass, W. B., G. A. Peterson, R. D. Hauck, and R. A. Olson. 1971. Recovery of residual nitrogen by corn (*Zea mays* L.) from various soil depths as measured by ^{15}N tracer techniques. *Proc. Soil Sci. Soc. Am.* 35: 290–294.
Gerwitz, A., and E. R. Page. 1974. An empirical mathematical model to describe plant root systems. *J. Appl. Ecol.* 11: 773–781.
Greenwood, D. J. 2001. Modeling N-response of field vegetable crops grown under diverse conditions with N-ABLE: A review. *J. Plant Nutr.* 24: 1799–1815.
Hansen, S., H. E. Jensen, N. E. Nielsen, and H. Svendsen. 1990. Daisy—a soil plant atmosphere system model. NPO Research from the National Agency of Environmental Protection, No. A 10. Denmark.
Heeraman, D. A., and N. G. Juma. 1993. A comparison of minirhizotron, core and monolith methods for quantifying barley (*Hordeum vulgare* L.) and fababean (*Vicia faba* L.) root distribution. *Plant Soil* 148: 29–41.
Hodge, A. 2004. The plastic plant: Root responses to heterogeneous supplies of nutrients. *New Phytol.* 162: 9–24.
King, J., A. Gay, R. Sylvester-Bradley, I. Bingham, J. Foulkes, P. Gregory, and D. Robinson. 2003. Modelling cereal root systems for water and nitrogen capture: Towards an economic optimum. *Ann. Bot.* 91: 383–390.
Kristensen, H. L., and K. Thorup-Kristensen. 2004a. Root growth and nitrate uptake of three different catch crops in deep soil layers. *Soil Sci. Soc. Am. J.* 68: 529–537.

Kristensen, H. L., and K. Thorup-Kristensen. 2004b. Uptake of ^{15}N labeled nitrate by root systems of sweet corn, carrot and white cabbage from 0.2–2.5 meters depth. *Plant Soil* 265: 93–100.

Kristensen, H. L., and K. Thorup-Kristensen. 2007. Effects of vertical distribution of soil inorganic nitrogen on root growth and subsequent nitrogen uptake by field vegetable crops. *Soil Use Manage.* 23: 338–347.

Kuhlmann, H., P. B. Barraclough, and A. H. Weir. 1989. Utilization of mineral nitrogen in the subsoil by winter wheat. *Z. Pflanzenernahr Bodenkd.* 152: 291–295.

Lainé, P., A. Ourry, J. Boucaud, and J. Salette. 1998. Effects of a localized supply of nitrate on NO_3^- uptake rate and growth of roots in *Lolium multiflorum* Lam. *Plant Soil* 202: 61–67.

Munkholm, L. J., P. Schjønning, M. H. Jørgensen, and K. Thorup-Kristensen. 2005. Mitigation of subsoil recompaction by light traffic and on-land ploughing, 2: Root and yield response. *Soil Till Res.* 80: 159–170.

Peterson, G. A., F. N. Anderson, G. E. Varvel, and R. A. Olson. 1979. Uptake of ^{15}N–labeled nitrate by sugar beets from depths greater than 180 cm. *Agron. J.* 71: 371–372.

Pierret, A., C. J. Moran, and C. Doussan. 2005. Conventional detection methodology is limiting our ability to understand the roles and functions of fine roots. *New Phytol.* 166: 967–980.

Powlson, D. S., and D. Barraclough. 1993. Mineralization and assimilation in soil-plant systems. In *Nitrogen isotope techniques*, ed. R. Knowles and T. H. Blackburn, 209–242. San Diego: Academic Press.

Pregitzer, K. S. 2002. Fine roots of trees: A new perspective. *New Phytol.* 154: 267–273.

Rahn, C. R., K. Zhang, R. Lillywhite, C. Ramos, J. Doltra, J. M. de Paz, H. Riley, M. Fink, C. Nendel, K. Thorup-Kristensen, F. Piro, A. Venezia, C. Firth, U. Schmutz, F. Raynes, and K. Strohmeyer. 2007. Brief description of the EU-rotate_N model. http://www.warwick.ac.uk/go/eurotaten.

Raynaud, X., and P. W. Leadley. 2004. Soil characteristics play a key role in modelling nutrient competition in plant communities. *Ecology* 85: 2200–2214.

Reidenbach, G., and W. J. Horst. 1997. Nitrate-uptake capacity of different root zones of *Zea mays* (L.) in vitro and in situ. *Plant Soil* 196: 295–300.

Robinson, D., D. J. Linehan, and D. C. Gordon. 1994. Capture of nitrate from soil by wheat in relation to root length, nitrogen inflow and availability. *New Phytol.* 128: 297–305.

Tennant, D. 1975. A test of a modified line intersect method of estimating root length. *J. Ecol.* 63: 995–1001.

Thorup-Kristensen, K. 1993. Root development of nitrogen catch crops and of a succeeding crop of broccoli. *Acta Agric. Scand. B* 43: 58–64.

Thorup-Kristensen, K. 1998. Root growth of green pea (*Pisum sativum* L.) genotypes. *Crop Sci.* 38: 1445–1451.

Thorup-Kristensen, K. 2006a. Effect of deep and shallow root systems on the dynamics of soil inorganic N during 3-year crop rotations. *Plant Soil* 288: 233–248.

Thorup-Kristensen, K. 2006b. Root growth and nitrogen uptake of carrot, early cabbage, onion and lettuce following a range of green manures. *Soil Use Manage.* 22: 29–38.

Thorup-Kristensen, K., and R. Van den Boogaard 1998. Temporal and spatial root development of cauliflower (*Brassica oleracea* L. var. *botrytis* L.). *Plant Soil* 201: 37–47.

Wiesler, F., and W. J. Horst. 1994. Root growth of maize cultivars under field conditions as studied by the core and minirhizotron method and relationships to shoot growth. *Z. Pflanzenernahr Bodenkd.* 157: 351–358.

Zobel, R. 2003. Fine roots: Discarding flawed assumptions. *New Phytol.* 160: 273–280.

12 Nitrogen-Uptake Effects on Nitrogen Loss in Tile Drainage as Estimated by RZWQM

Robert W. Malone and Liwang Ma

CONTENTS

12.1 Background 259
  12.1.1 Grain Demand and N-Fertilizer Use 260
  12.1.2 Reactive N (Nr) in the Environment and the Consequences 261
  12.1.3 Balancing Crop Production and N Losses....................................262
  12.1.4 Quantifying N Budget of Agricultural Systems under
          Different Conditions 262
  12.1.5 N Fate Estimation 263
12.2 Materials and Methods 264
  12.2.1 Site Description and Management..264
  12.2.2 RZWQM Parameterization, Calibration, and Performance 264
  12.2.3 Regression Analysis...265
12.3 Results and Discussion..266
  12.3.1 RZWQM-Estimated N Uptake under Different Management
          and Years..266
  12.3.2 RZWQM-Estimated N Loss in Tile Drainage as a Function of
          Precipitation and N Uptake...268
  12.3.3 RZWQM-Estimated N Budget under Different Active N
          Uptake..270
12.4 Summary and Conclusions 272
References..273

12.1 BACKGROUND

It is estimated that about 40% of the current global population would not be able to survive (let alone thrive) without fertilizer use, and our dependence on nitrogen (N) fertilizer is likely to grow in the next few generations (Smil 2001; Fixen and West 2002). Therefore, the development of commercial N fertilizer may be the most important invention of the 20th century toward sustaining a growing population (Smil 2001). However, because excess N entering the environment has negative consequences, the use of N fertilizer for crop production should be as efficient as possible. To optimize N use and crop production under the vast number of combinations of soils, climate, and management, system models are needed. For these models to be useful, however, they need to accurately estimate the N budget under different conditions. The crop available N budget includes crop N uptake, denitrification, N leaching below the root zone, annual change in available N storage in the soil profile, net mineralization, and fixation. The largest annual component of the N budget for corn–soybean systems is generally crop N uptake; therefore, this component must be estimated correctly for accurate model simulations. Another important component to the N budget in the U.S. corn belt is N in tile drainage because of its contribution to N loading to the Mississippi River Basin, which has been implicated as a cause of hypoxia in the Gulf of Mexico (Goolsby et al. 2001; Rabalais et al. 2001).

Despite the apparent importance of crop N uptake to N in tile drainage, little information is available quantifying this effect. Possibly, our understanding may be enhanced through application of a thoroughly tested and calibrated agricultural system model. Recently, the Root Zone Water Quality Model (RZWQM) was used to better understand, quantify, and extend field experimental results for the effects of a variety of management practices on crop production and N in drainage water at Nashua, Iowa. The results of this study were published in a special issue of *Geoderma*, with a preface by Ahuja and Hatfield (2007). Our objective is to use this calibrated and tested model to quantify the effects of N uptake by crops on nitrate loss in subsurface drains (and other N budget components).

12.1.1 GRAIN DEMAND AND N-FERTILIZER USE

World rice, maize, and wheat production was 1700 Mt in 2000, 1900 Mt in 2004 (FAOSTAT 2004), and demand is projected to increase to 2400 Mt in 2025 (Cassman et al. 2003). World grain demand to satisfy human diets is projected to increase faster than world population increases, largely because of greater per capita meat consumption in developing countries (Cassman et al. 2003). Additional grain production for biomass fuels may also be necessary to displace petroleum consumption. To displace at least 30% of the U.S. petroleum consumption with biomass assumes a 50% increase in yield of corn, wheat, and other small grains (Perlack et al. 2005).

Increased global grain demand will require additional N-fertilizer input under just about all possible scenarios, and additional fertilizer application will be required for adequate grain production if no additional land is devoted to agriculture (Cassman et al. 2003). New land area devoted to agriculture may be limited to reduce encroachment on sensitive areas (e.g., rain forests) and to avoid marginal agricultural land. The U.S. population is projected to increase by about 80 million between 2000 and 2030, from about 280 to 360 million (U.S. Census Bureau 2004). Because of the increasing U.S. population, the U.S. fertilizer use is expected to increase from about 11 Mt in 2000 to nearly 20 Mt in 2030, assuming increased grain exports, U.S. census bureau population projections, standard U.S. diet, and constant fertilizer rate per grain production (Howarth et al. 2002). This projection, however, is very high because increasing nitrogen use efficiency through improved N management and improved hybrids could substantially reduce future N increases. In U.S. maize systems, nitrogen use efficiency (NUE) increased from 42 kg kg^{-1} in 1980 to 57 kg kg^{-1} in 2000, which represents a 36% increase (1.6% y^{-1}) (Cassman et al. 2003).

12.1.2   REACTIVE N (NR) IN THE ENVIRONMENT AND THE CONSEQUENCES

Nitrogen compounds in nature can be divided into reactive and nonreactive. About 99% of nitrogen on the Earth is in the form of nearly chemically inert dinitrogen gas (N_2) and is not available to most plants. Approximately 1% of nitrogen is reactive (Nr), which includes organic N and inorganic N such as nitrate (NO_3), nitrogen oxide (NO_x), ammonia (NH_3), and ammonium (NH_4). Prior to human activity, Nr did not accumulate in the environment, but recently it has been accumulating, and this accumulation is accelerating (Galloway et al. 2003). The increase in reactive N in circulation has doubled nitrate concentrations in Greenland glacial ice cores between 1955 and 1980 (Mayewski et al. 1986). In the northeastern United States, reactive N deposition has increased 5- to 10-fold over preindustrial conditions (Aber et al. 2003; Galloway et al. 1984). U.S. nitrogen deposition, however, has not been detected to increase in the 1990s (Lynch et al. 2000; Baumgardener et al. 2002), possibly due to the relatively stable U.S. consumption of N fertilizer (IFA, 2005). Although reactive nitrogen deposition may not be increasing in the United States, the cumulative effect of deposition above preindustrial levels may lead to localized N saturation, which can result in increased nitrate leaching, soil acidification, and soil-nutrient imbalances (Galloway et al. 2003).

The negative consequences of excess N use are an emerging worldwide concern (Giles 2005). As a comparison, another environmental concern is our rising releases of CO_2 from fossil fuel combustion, which has been described as carrying out an unprecedented large-scale geophysical experiment (Smil 2001; Revelle and Suess 1957). Anthropogenic carbon is clearly important because of its role in the Earth's thermal balance; however, anthropogenic emissions are only a small fraction of the natural fluxes of the element. Human activity, on the other hand, now accounts for 2/5 to 1/2 of all terrestrial reactive nitrogen, and the atmospheric deposition of NOy and NHx has roughly trebled (Smil 2001). Therefore, fertilizer-induced perturbations in nitrogen cycling "constitute an unprecedented large-scale biogeochemical experiment whose eventual impacts we can predict only poorly" (Smil 2001).

Atmospheric deposition of ammonia and nitrates is between about 20 and 60 kg N/ha per year in parts of eastern North America, northwestern Europe, and East Asia (Smil 2001). "Because N supply often limits primary production and other ecosystem processes over much of the natural world, human alteration of the N cycle has the capacity to change Earth's ecosystems substantially" (Vitousek et al. 2002). Reactive nitrogen enrichment contributes to: greenhouse gas production and depletion of stratospheric ozone; loss of soil nutrients such as calcium and potassium; the acidification of soils, streams, and lakes in several regions; increased organic carbon stored within terrestrial ecosystems; accelerated losses of biological diversity, especially plants adapted to efficient use of nitrogen, and losses of the animals and microorganisms that depend on them; and changes in the composition and functioning of estuarine and nearshore ecosystems, and long-term declines in coastal marine fisheries (Vitousek et al. 1997).

One major contributor to declines in coastal fisheries is hypoxia, which occurs when dissolved oxygen concentrations are below those necessary to sustain most animal life. Since 1993, midsummer bottom-water hypoxia in the northern Gulf of Mexico has generally been larger than 15,000 km^2, and in 1999 it was nearly 20,000 km^2, which is about the size of the state of New Jersey (CENR 2000). Hypoxia in the northern Gulf of Mexico is caused primarily by excess nitrogen delivered from the Mississippi–Atchafalaya River Basin in combination with stratification of Gulf waters. The annual flux of nitrate–N into the Gulf of Mexico from the Mississippi River has nearly tripled between the periods 1950–1970 and 1980–1996 to often higher than 1.0 Mt N/year (CENR 2000). As a comparison, North America has never consumed more than 13 Mt N-fertilizer/y prior to 2005 (IFA 2005). Sediment cores from the Gulf of Mexico hypoxic zone show that algal production and deposition, as well as oxygen stress, were much lower earlier in the 1900s and that significant increases occurred in the latter half of the 20th century (CENR 2000).

12.1.3 BALANCING CROP PRODUCTION AND N LOSSES

Agriculture is a major contributor to anthropogenic Nr in the environment. For example, the global fertilizer N consumption was about 85 Mt/y in 1999/2000 (IFA 2005), and the Nr from energy production was about 25 Mt in 2000 (Galloway et al. 2003).

But how does society simultaneously increase grain production, minimize (and possibly reduce) reactive N losses to the environment, and minimize new land devoted to crop production? Loss of N fertilizer results from gaseous plant emissions, soil denitrification, surface runoff, volatilization, and leaching (Raun and Johnson 1999). Raun and Johnson (1999) estimated the world cereal grain nitrogen use efficiency (NUE) to be 33% using the formula

$$NUE = [(total\ cereal\ N\ removed) - (N\ from\ soil) - (N\ rain)]/N\_applied$$

This is similar to the estimate of north-central U.S. maize on-farm N-recovery efficiency of 37% (Cassman et al. 2002). Therefore, increasing the nitrogen use efficiency will reduce losses while maintaining crop yield. Improving the world cereal NUE will require a systems approach that incorporates selected rotations (e.g., corn–soybean), improved hybrids, forage (e.g., preflowering winter wheat, silage), split N application timing, and N application amounts that coincide with demand and in-field variability (Raun and Johnson 1999).

12.1.4 QUANTIFYING N BUDGET OF AGRICULTURAL
SYSTEMS UNDER DIFFERENT CONDITIONS

To address the impact of agriculture on water quality, the United States Department of Agriculture (USDA) organized the Management Systems Evaluation Areas (MSEA) project "to evaluate the impact of farming systems and nitrogen inputs to crop production on groundwater quality beneath the crop" (Power et al. 1998). MSEA activities were designed to document the effectiveness of existing technologies and initiate the development of new technologies capable of increasing crop yields while improving surface and groundwater quality. "While the MSEA project stood as the largest research and demonstration activity on the effects of agriculture on water quality ever conducted in the United States, it remained too limited to provide comparisons of more than just a few complete farming systems" (Power et al. 2001). The multitude of soil, climate, and management systems result in almost limitless interactions of factors affecting N transformation, water movement, and crop production (Power et al. 1998, 2000, 2001).

The complexity of investigating management alternatives to optimize yield and NUE under different conditions will require the use of agricultural system models. Ahuja et al. (2002) reported that "integrating system modeling with field research ... [may] promote quick and accurate transfer of results to different soil and weather conditions, and to different cropping and management systems outside the experimental plots." For the MSEA research, however, models provided grossly erroneous results in some instances and were not recommended for management or regulatory decisions (Power et al. 2000). To optimize NUE using models will require reasonable estimation of the N budget (plant uptake and removal, net mineralization, denitrification, fixation, leaching).

12.1.5 N FATE ESTIMATION

The difficulties in parameterizing and estimating soil organic-matter (SOM) dynamics are well known (e.g., Bruun et al. 2003; Diekkruger et al. 1995). Bruun et al. (2003) concluded that "the fraction of SOM involved in medium-term turnover is substantially larger than previously thought and inferred by most SOM turnover models." Calibrated model parameter differences from default values resulted in estimated soil carbon differences after 100 years of agricultural management of more than 20 Mg C/ha in some instances (Bruun et al. 2003). To illustrate the sensitivity of seemingly small soil carbon changes to N available for leaching and crop uptake, consider that a net decrease in soil carbon in the top 15 cm of 1.5% to 1.45% increases available N (e.g., nitrate–N) to the system by 100 kg N/ha (loss of 1 Mg C/ha, assuming a soil C/N ratio of 10 and a bulk density of 1.4 g/cm^3).

In another modeling study, Sogbedji et al. (2006, 2001) investigated N transformation using short-term field data (≤ 3 years). The LEACHMN model accurately estimated short-term N dynamics when N transformation rates were calibrated for each treatment and transformation (denitrification, nitrification, volatilization, mineralization) (Sogbedji et al. 2006). Under other conditions, N rate of application (i.e., treatment) did not appear to affect N rate constants (Sogbedji et al. 2001).

Difficulties in quantifying soil C/N dynamics are not limited to modeling studies. Even in a tightly controlled laboratory experiment where measurement of the N budget was a goal (including measurement of N_2O and NH_3 emissions), Parkin et al. (2006) speculated that the N not accounted for was due to immobilization. Parkin et al. (2006) reported that "increased N immobilization ... of the magnitude necessary to account for our observed discrepancies in N balance (4–11 g N/m^2) is within the error ranges we measured for the soil organic N pool."

Quantification of long-term N budgets under different N uptake was beyond the scope of Sogbedji et al. (2001, 2006), Parkin et al. (2006), and Bruun et al. (2003).

In fact, the literature is sparse concerning quantification of N budgets with different N-uptake scenarios in long-term corn systems. Although sparse, some related research has been reported for other plants. For example, Pare et al. (2006) reported that increased total N uptake by grasses resulted in reduced N leaching.

12.2 MATERIALS AND METHODS

To achieve the objective, the long-term simulations of Ma et al. (2007a, 2007b) and Malone et al. (2007a) were used to:

1. Estimate N uptake under different conditions such as crop rotations, tillage, fertilizer management, and years
2. Develop multivariate regression equations that quantify the effects of precipitation and RZWQM-estimated N uptake on N loss in tile drainage
3. Investigate the RZWQM-estimated N budget under different N-uptake and N-application rates

12.2.1 SITE DESCRIPTION AND MANAGEMENT

A data set collected from 36 0.4-ha plots located at the Iowa State University Northeast Research Station near Nashua, Iowa, (43.0° N, 92.5° W) that included 14 years (1990–2003) of weather records, crop yield, tile drainage volume, and N concentration in drainage water was used for model testing by Malone et al. (2007a) and Ma et al. (2007a, 2007b). For this research, the parameters from plot 25 were used because this plot had been used by Ma et al. (2007b) for model calibration. From 1977 to 1992 the field management included no-till and 202 kg N/ha anhydrous ammonia (AA) spring application to continuous corn. From 1993 through 1998 the management included no-till and spring application of urea-ammonium nitrate (UAN-N) at 110 kg/ha to corn–soybean rotations. From 1999 through 2003 the management included no-till and spring swine manure application to corn–soybean rotations.

The soils at plot 25 are Kenyon loam (fine-loamy, mixed, superactive, mesic Typic Hapludolls) (USDA-NRCS 2006) and Readlyn loam (fine-loamy, mixed, superactive, mesic Aquic Hapludolls) (USDA-NRCS 2006). This plot has a seasonally high water table and thus benefits from subsurface tile drainage.

12.2.2 RZWQM PARAMETERIZATION, CALIBRATION, AND PERFORMANCE

Calibration, parameterization, and overall performance of RZWQM for this field site have been described in detail (Ma et al. 2007b). RZWQM adequately quantified the relative effects of corn production and N loss under several alternative management practices after calibration and thorough testing (Malone et al. 2007a; Ma et al. 2007a). We use the calibrated and tested model for plot 25 with chisel-tillage after corn harvest in a corn–soybean rotation to investigate the N budget under five spring preplant N application rates between 100 and 200 kg N/ha and high and low active N uptake. Nitrogen enters the plant through transpiration, and if plant demand is greater than N in the transpiration stream, active N uptake occurs. RZWQM simu-

lates the rate of active N uptake by crops (Ntact, g/plant/day) in a manner similar to
the Michaelis–Menten model (Ahuja et al. 2000):

$$Ntact = (u_1 \times Ntsoil)/(u_2 + Ntsoil) \qquad (12.1)$$

where Ntsoil is the amount of N available in the soil (g/cm^3 soil), u_1 is the maximum
rate of N that can be removed from the soil (g/plant/d), and u_2 is the soil concentration that can produce an active uptake rate ½-maximum N-uptake rate for the
species being simulated (g/cm^3 soil). The total amount of nitrogen available to the
plant through uptake is the sum of active and passive uptake. High and low N-uptake
scenarios were achieved by inputting "maximum N-uptake rate" in RZWQM (u_1) as,
respectively, 2.0 or 1.0 g/plant/d for corn, and as 0.6 and 0.2 g/plant/d for soybean.
The "maximum N-uptake rate" in the simulations of Ma et al. (2007a, 2007b) and
Malone et al. (2007a) was 1.5 g/plant/d for corn and 0.4 g/plant/d for soybean. Simulations were run with northeastern Iowa weather data from 1951 through 2003. The
first 10 years of model runs (1951–1960) were eliminated from analysis and used
only for initiation of soil organic carbon pools and microbial populations.

Plant N demand was simulated by using predefined maximum N concentration
(N_X) in plant, and N concentration at maturity ($N_{GS=1}$):

$$N_T = N_X \, e^{(-b*GS)} \qquad (12.2)$$

$$b = -\ln(N_{GS=1}/N_X) \qquad (12.3)$$

where GS is growth stage from 0 to 1, and N_T is targeted N concentration each day (g
N/g plant). A targeted plant N demand is the product of N_T and total plant biomass.

12.2.3 Regression Analysis

Malone et al. (2007b) developed regression equations to quantify the relationship
between nitrate leaching and corn yield and field observations, climatic variables,
and management variables. An advantage of applying an agricultural system model
(e.g., RZWQM) to field data compared with using only field results is that a complete
N budget can be obtained for a long-term period that includes N fixation and total N
uptake by both corn and soybean. A regression equation that relates nitrate loss in
tile flow over the 2-year corn–soybean rotation ($Nloss_t$) with the variable N uptake
by crops would help quantify the effect of N uptake on $Nloss_t$. Using the field data set
from Nashua, Iowa, Ma et al. (2007a) concluded that RZWQM required improved
simulation of N uptake and mineralization for improved simulation of N fate under
different agricultural management, but RZWQM generally responded accurately to
yearly corn yield variation (Ma et al. 2007b). Therefore, we assume that RZWQM-
estimated N uptake and N loss in tile drains responds reasonably to annual climate
variations under constant management. Thus, a regression equation was developed
to describe 1962–2003 RZWQM-estimated $Nloss_t$ under an application rate of 150
kg N/ha to corn as a function of precipitation during the soybean year (p_s), precipitation during the corn year (p_c), precipitation during the year prior to corn (previ-

ous soybean year, p_{yl}), soil-derived N uptake by soybean ($Nupt_s$), soil-derived N uptake by corn ($Nupt_c$), and soil-derived N uptake during the year prior to corn (prior soybean N uptake, $Nupt_{yl}$). In this chapter, soil-derived N uptake is defined as RZWQM-estimated total N uptake minus fixation minus N application rate, and it is assumed to have more direct effect on N loss in tile drains than total N uptake. For example, $Nupt_c$ is the total (RZWQM-estimated) N uptake for the year minus 150 kg N/ha, and $Nupt_{yl}$ is the total (RZWQM-estimated) N uptake for the year minus fixation because 150 kg N/ha is applied every corn year. A stepwise procedure was used for variable selection ($p < .1$). We selected a second-order multivariate polynomial to describe RZWQM-estimated $Nloss_t$:

$$\begin{aligned}
Nloss_t^{0.5} = {} & a_0 + a_1(p_s) + a_2(p_c) + a_2(p_{yl}) + a_2(Nupt_s) + a_3(Nupt_c) + a_4(Nupt_{yl}) \\
& + a_5(p_s{}^*p_c) + a_6(p_s{}^*p_{yl}) + a_7(p_s{}^*Nupt_s) + a_8(p_s{}^*Nupt_c) + a_{10}(p_c{}^*p_{yl}) \\
& + a_{11}(p_c{}^*Nupt_s) + a_{12}(p_c{}^*Nupt_c) + a_{13}(p_c{}^*Nupt_{yl}) + a_{14}(p_{yl}{}^*Nupt_s) \\
& + a_{16}(p_{yl}{}^*Nupt_{yl}) + a_{17}(Nupt_s{}^*Nupt_c) + a_{18}(Nupt_s{}^*Nupt_{yl}) \qquad (12.4) \\
& + a_{19}(Nupt_c{}^*Nupt_{yl}) + a_1(p_s)^2 + a_2(p_c)^2 + a_2(p_{yl})^2 + a_2(Nupt_s)^2 \\
& + a_3(Nupt_c)^2 + a_4(Nupt_{yl})^2
\end{aligned}$$

To achieve the best fit and achieve a relatively simple final model (e.g., to reduce the number of interaction terms in the final model), the terms $a_{15}(p_{yl}{}^*Nupt_c)$ and $a_9(p_s{}^*Nupt_{yl})$ were not included in the analysis.

12.3 RESULTS AND DISCUSSION

12.3.1 RZWQM-Estimated N Uptake under Different Management and Years

Ma et al. (2007a) and Malone et al. (2007a) investigated RZWQM-estimated effects of agricultural management on N fate over long-term simulations. The average annual RZWQM-estimated soil-derived N uptake (total N uptake − fixation − application) increased 28 kg N/ha between application rates of 100 and 200 kg/ha, and increased nearly 50 kg N/ha between continuous corn and corn–soybean rotations (Table 12.1). Also, with a 150-kg N/ha application, RZWQM-estimated total annual N uptake by corn varied from year to year by more than 50 kg N/ha (Figure 12.1). Little difference in soil-derived N uptake was estimated between some management systems such as different tillage (Table 12.1), different spring application timings (data not shown; see Malone et al. 2007a), or different drainage management (Table 12.1). Clearly, soil-derived N uptake and total annual N uptake can vary substantially between some treatments and years, which should affect N-loss pathways (e.g., leaching, denitrification). Ma et al. (2007a), however, reported that observed corn yield varied by as much as 2000 kg/ha for different tillage, but that RZWQM did not respond to these differences. The annual RZWQM-estimated corn yield across application rates and years is correlated with total N uptake (Figure 12.1, $r^2 = 0.60$). Ma et al. (2007a) concluded that RZWQM required improved N uptake and mineralization for improved estimation of N fate under different agricultural management.

TABLE 12.1

Total RZWQM-Estimated Crop N Uptake, N Fixation, and Soil-Derived N Uptake under Different Management

Study[a]	Rotation[b]	Tillage[c]	Drainage[d]	N Appli[e] (kg N/ha)	Soil-Derived N Uptake[f] (kg N/ha)	Corn N Uptake (kg N/ha)	Soybean N Uptake (kg N/ha)	Fixation (kg N/ha)
Ma	cc	nt	fd	202	22	224	—	—
Ma	cc	mp	fd	202	22	224	—	—
Ma	cc	cp	fd	202	22	224	—	—
Ma	cc	nt	cd	202	22	224	—	—
Ma	cc	mp	cd	202	22	224	—	—
Ma	cc	cp	cd	202	23	225	—	—
Ma	cs+sc	nt	fd	168	67	225	319	243
Ma	cs+sc	mp	fd	168	70	226	333	252
Ma	cs+sc	cp	fd	168	67	226	322	247
Ma	cs+sc	nt	cd	168	67	226	322	246
Ma	cs+sc	mp	cd	168	70	227	337	255
Ma	cs+sc	cp	cd	168	67	227	325	250
Malone	cs	cp	fd	100	96	211	302	219
Malone	cs	cp	fd	125	91	222	301	216
Malone	cs	cp	fd	150	85	232	301	214
Malone	cs	cp	fd	175	77	239	301	212
Malone	cs	cp	fd	200	68	244	301	211

[a] Studies are Ma et al. (2007a), where simulations were dated 1979–2002, or Malone et al. (2007b), where simulations were dated 1961–2003.

[b] Rotations are: cc = continuous corn, cs + sc = average of corn–soybean and soybean–corn, cs = corn in even years and soybean in odd years.

[c] Tillage treatments are: nt = no till, mp = moldboard plow, cp = chisel plow.

[d] Drainage treatments are: fd = free drainage, cd = controlled drainage (see Ma et al. 2007b for description).

[e] N appli is fertilizer-N application rate.

[f] Soil-derived N uptake = (total N uptake) – (biological fixation during soybean years) – (UAN-N application during corn years).

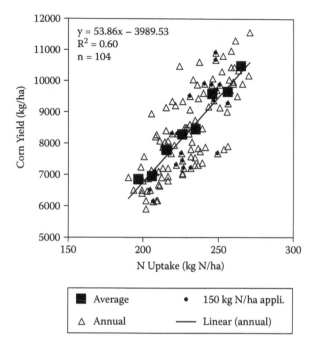

FIGURE 12.1 Annual RZWQM-estimated corn yield and N uptake from 1961 to 2003 (Malone et al. 2007a). The treatments are corn–soybean rotations with spring N application of 100, 125, 150, 175, and 200 kg N/ha. The large square symbols indicate average corn yield and average N uptake for 10-kg/ha N-uptake increments. The small dot symbols indicate annual corn yield and N uptake for applications of 150 kg N/ha.

12.3.2 RZWQM-Estimated N Loss in Tile Drainage as a Function of Precipitation and N Uptake

Applying regression to the results of Malone et al. (2007a) with an application rate of 150 kg N/ha during corn years resulted in an equation that described 96% of variation in the RZWQM-estimated N loss in tile drainage using the variables precipitation and RZWQM-estimated soil-derived N uptake (Figure 12.2). The 2-year RZWQM-estimated N loss in tile drainage ($Nloss_t$) generally decreased with increased soil-derived N uptake during the corn year and the previous soybean year (Figure 12.3) according to the equation (variable inclusion of $p < .1$)

$$Nloss_t^{0.5} = -13.60 + (0.2586 * p_s) + (0.1746 * p_{yl}) + (0.01848 * Nupt_{yl})$$
$$- (0.0003576 * Nupt_{yl} * Nupt_c) - (0.002340 * p_s * p_{yl}) \qquad (12.5)$$

With approximately average p_{yl} and p_s, $Nloss_t$ decreased from 38 to 19 kg N/ha with an $Nupt_c$ increase of 50 to 110 kg N/ha (Figure 12.3). Inclusion of N uptake in equation (12.5) is evidence that crop N uptake significantly affects N loss in tile drainage. These results also suggest that conservation practices that maximize crop production (N uptake) will contribute to reduced N loss in tile drains.

Furthermore, these results suggest that if the conditions significantly contributing to N loss in tile drainage were known (p_{yl} and $Nupt_{yl}$) or could be estimated

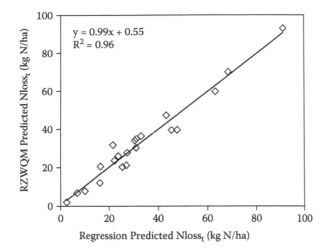

FIGURE 12.2   Regression-estimated vs. RZWQM-estimated $Nloss_t$. $Nloss_t$ is the nitrate loss in tile drainage for the 2-year corn–soybean rotation from 1962 to 2003.

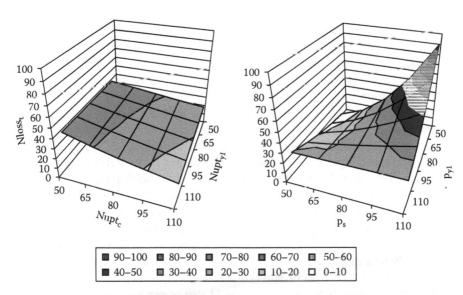

FIGURE 12.3   Three-dimensional representation of regression-estimated $Nloss_t$ as a function of $Nupt_c$, $Nupt_{y1}$, p_s, and p_{y1}. $Nloss_t$ is the nitrate loss in tile drainage for the 2-year corn–soybean rotation from 1962 to 2003, $Nupt_c$ is the soil-derived N uptake during the corn year (total N uptake minus application rate), $Nupt_{y1}$ is the soil-derived N uptake during the year prior to corn (total N uptake minus N fixation), p_s is the precipitation during the soybean year, and p_{y1} is the precipitation during the year prior to corn.

(p_s and $Nupt_c$), management could be adjusted to reduce N loss. Reducing fertilizer application from 150 to 125 kg N/ha for the years that produced the 10 greatest RZWQM-estimated $Nloss_t$ results, in average reduced $Nloss_t$ from 32.1 to 29.1 kg N/ha (9.1%), with estimated corn yield reduced by only 1.1%. Malone et al. (2007a) reported that reducing fertilizer application from 150 to 125 kg N/ha every corn year resulted in RZWQM-estimated $Nloss_t$ reduction of 16.2% and corn yield reduction of 3.6%.

12.3.3 RZWQM-Estimated N Budget under Different Active N Uptake

Clearly RZWQM-estimated N loss in tile drainage is affected by N uptake (Figures 12.2 and 12.3; equation 12.5). One of the main RZWQM parameters driving N uptake is "maximum N uptake rate," which is often calibrated (Bakhsh et al. 2001, 2004). With an application rate of 150 kg N/ha, active N uptake increases from 127 kg N/ha to 203 kg N/ha, and total N uptake increases from 520 to 542 kg N/ha with low and high active N uptake (Table 12.2). With increasing N application rates and constant "maximum N uptake rate," active N uptake decreases, but total N uptake increases (Table 12.2).

A maximum N-uptake rate of 2.0 compared with 1.0 (i.e., more active N uptake) results in greater corn yield and lower N loss in tile drains over the long-term simulations (Table 12.2). At an application rate of 150 kg N/ha with high active N uptake by corn and soybean, total N uptake (including roots) is 4% higher, corn yield is 8% higher, and nitrate loss in drains is 30% less than with low active N uptake.

Because the average 2-year N loss in tile drains and N uptake at high active N uptake were, respectively, 30 and 542 kg N/ha at an application rate of 150 kg N/ha (Table 12.2), a more substantial difference in tile drainage nitrate–N than 30% (9 kg N/ha) could be expected between high and low N uptake. That is, low N uptake has 40 kg/ha more N available for leaching over the 2-year rotation, given that it has 18 kg N/ha more fixation and 22 kg N/ha less uptake (Table 12.2). However, examining the N budget, when N loss is defined as the sum of denitrification, N in tile drains, and N seepage, reveals that the average 2-year N loss difference between low and high N uptake was 32 kg N/ha at 150-kg N/ha application rate (Table 12.2). The remaining difference between high and low N uptake is accounted for in higher mineralization from the high-N-uptake corn that is not removed from the field (Table 12.2). It should be noted that the long-term N budget balances to within 1 kg N/ha (Table 12.2, N source [fixation, application, net mineralization] – N loss [denitrification, tile, deep seepage] – crop N uptake [grain, stover, roots] <1 kg N/ha).

As N application rate increases, the corn yield difference decreases, but the difference in N loss in tile drains remains relatively constant between high and low active N uptake (Table 12.2). Because corn yield difference decreases with increasing N application rate, N loss difference could also be expected to decrease. The N source difference between high and low active N uptake, however, becomes more negative (decreases) with increasing N application (Table 12.2; Figure 12.4). In other words, with increasing N application, net mineralization increases more rapidly than fixation decreases for low N uptake, while the sum of fixation and net mineralization remains relatively constant for high N uptake (Table 12.2). RZWQM-estimated

TABLE 12.2

Average 2-Year N Budget for 42-Year Simulation (1961–2002)

N Application (kg N/ha)	Corn Yield (kg/ha)	Net Mineralization (kg N/ha)	Tile Drainage (kg N/ha)	Deep Seepage (kg N/ha)	Fixation (kg N/ha)	Denitrification (kg N/ha)	Crop N Uptake (kg N/ha)	Active N Uptake (kg N/ha)
				High Active N Uptake				
100	8183	275	20	23	215	24	523	206
125	8478	279	24	30	212	28	533	205
150	8705	282	30	38	210	32	542	203
175	8795	284	36	47	207	36	547	199
200	8857	285	43	57	205	40	551	195
				Low Active N Uptake				
100	7197	264	29	38	231	31	497	131
125	7657	269	33	44	229	35	510	130
150	7975	273	38	52	227	39	520	127
175	8309	277	44	61	226	44	529	125
200	8501	280	50	70	224	48	536	122
			Difference between High and Low Active N Uptake					
100	987	11	−9	−15	−16	−7	26	75
125	821	10	−9	−15	−17	−8	24	75
150	729	9	−9	−15	−18	−8	22	76
175	486	7	−8	−14	−18	−8	18	75
200	356	6	−7	−13	−19	−9	15	73

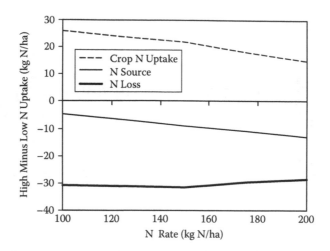

FIGURE 12.4 The N budget differences between high and low N-uptake scenarios (high–low) over the 2-year corn–soybean rotation. Crop N uptake is the sum of total N uptake (above- and belowground, active and passive N uptake) for corn and soybean. Nitrogen source is the sum of N rate, net mineralization, and fixation. Nitrogen loss is the sum of denitrification, tile flow, and deep seepage.

fixation is reduced when N is applied prior to soybean and soil N is higher (Malone et al. 2007a).

12.4   SUMMARY AND CONCLUSIONS

Hirel et al. (2001) investigated the genetic basis of nitrogen use efficiency in maize partly because research suggests that "increases in grain yield observed during the two last decades were … due to a better NUE as a result of a more efficient nitrogen remobilization." Improved understanding of the basic processes behind NUE can help improve future agricultural production and profit. Much research has occurred over the last several decades and continues today to improve our understanding of crop N uptake and model-estimated crop production (e.g., Jeuffroy et al. 2002; Plenet and Lemaire 1999; Gastal and Lemaire 2002; Devienne-Barret et al. 2000). As discussed in the background section 12.1, research concerning N loss to the environment from agriculture should continue to be an important topic for the foreseeable future. In fact, it can be argued that the dual goals of meeting food demand while protecting the environment from excess reactive nitrogen may be one of our greatest ecological challenges (Cassman et al. 2002). The complexity of this challenge will require use of agricultural systems models, and one of the biggest challenges to accurately simulate crop production and nitrogen loss is to accurately simulate nitrogen uptake by plants. One difficulty associated with accurately simulating N uptake by crops is simulating "active N uptake," which is often calibrated. Active N uptake is simulated when plant demand is higher than N entering plants through the transpiration stream. Annual N demand and uptake by plants is generally much larger than N loss to subsurface drains; therefore, small errors in estimating uptake can lead to unacceptably large errors in estimating N loss to the environment under different management, weather, soils, etc. Research literature, however, is sparse concerning estimating N budgets with different N-uptake scenarios in agricultural systems. Therefore, our objective was to quantify the effect of N uptake by a corn–soybean rotation on N loss in tile drainage (and other N-budget components).

A regression equation was developed that described 96% of variation in RZWQM-estimated N loss in tile drainage for corn–soybean rotations using the variables precipitation and RZWQM-estimated soil-derived N uptake (total N uptake − fixation − application). The regression equation estimated that the average 2-year N loss in tile drainage decreased from 38 to 19 kg N/ha, with soil-derived N uptake during corn years (total N − uptake − application) of 50 to 110 kg N/ha. For high compared with low active N uptake, the average 2-year N loss in tile drains at an application rate of 150 kg N/ha was 30% less (−9 kg N/ha) for 4% higher (+22 kg N/ha) N uptake by crops. Additional N loss to the environment for the low active N-uptake scenario included +8 kg N/ha more denitrification and +15 kg N/ha more deep seepage. The results presented clearly indicate that N uptake significantly affects N loss in tile drainage, and that seemingly small errors in estimating crop N uptake can lead to more substantial errors in estimating N loss to the environment. Although not explicit, the results imply that conservation practices that maintain high crop production (high N uptake) contribute to reduced N loss in tile drains.

REFERENCES

Aber, J. D., C. L. Goodale, S. V. Ollinger, M. L. Smith, A. H. Magill, M. E. Martin, R. A. Hallett, and J. L. Stoddard. 2003. Is nitrogen deposition altering the nitrogen status of northeastern forests? *BioScience* 53: 375–389.

Ahuja, L. R., and J. L. Hatfield. 2007. Integrating soil and crop research with system models in the Midwest USA: Purpose and overview of the special issue. *Geoderma* 140: 217–222.

Ahuja, L. R., L. Ma, and T. A. Howell. 2002. Whole system integration and modeling: Essential to agricultural science and technology in the 21st century. In *Agricultural system models in field research and technology transfer*, ed. L. R. Ahuja, L. Ma, and T. A. Howell. Boca Raton: CRC Press.

Ahuja, L. R., K. W. Rojas, J. D. Hanson, M. J. Shaffer, and L. Ma, eds. 2000. *The root zone water quality model.* Highlands Ranch, CO: Water Resources Publications.

Bakhsh, A., J. L. Hatfield, R. S. Kanwar, L. Ma, and L. R. Ahuja. 2004. Simulating nitrate drainage losses from a Walnut Creek watershed field. *J. Environ. Qual.* 33: 114–123.

Bakhsh, A., R. S. Kanwar, D. B. Jaynes, T. S. Colvin, and L. R. Ahuja. 2001. Simulating effects of variable nitrogen application rates on corn yields and NO_3–N losses in subsurface drain water. *Trans. ASAE* 44 (2): 269–276.

Baumgardener, R. E., T. F. Lavery, C. M. Rogers, and S. S. Ilil. 2002. Estimates of the atmospheric deposition of sulfur and nitrogen species: Clean air status and trends network, 1990–2000. *Environ. Sci. Technol.* 36: 2614–2629.

Bruun, S., B. T. Christensen, E. M. Hansen, J. Magid, and L. S. Jensen. 2003. Calibration and validation of the soil organic matter dynamics of the Daisy model with data from the Askov long-term experiments. *Soil Biol. Biochem.* 35: 67–76.

Cassman, K. G., A. Dobermann, and D. T. Walters. 2002. Agroecosystems, nitrogen-use efficiency, and nitrogen management. *Ambio* 31: 132–140.

Cassman, K. G., A. Dobermann, D. T. Walters, and H. Yang. 2003. Meeting cereal demand while protecting natural resources and improving environmental quality. *Annu. Rev. Environ. Resour.* 28: 315–58.

CENR. 2000. Integrated assessment of hypoxia in the Northern Gulf of Mexico. Washington, D.C.: National Science and Technology Council Committee on Environment and Natural Resources. http://www.nos.noaa.gov/products/hypox_final.pdf.

Devienne-Barret, F., E. Justes, J. M. Machet, and B. Mary. 2000. Integrated control of nitrate uptake by crop growth rate and soil nitrate availability under field conditions. *Ann. Bot.* 86: 995–1005

Diekkruger, B., D. Sondgerath, K. C. Kersebaum, and C. W. McVoy. 1995. Validity of agroecosystem models: A comparison of results of different models applied to the same data set. *Ecol. Model.* 81: 3–29

FAOSTAT. 2004. Data from FAO. http://faostat.fao.org/.

Fixen, P. E., and F. B. West. 2002. Nitrogen fertilizers: Meeting contemporary challenges. *Ambio* 31: 169–176.

Galloway, J. N., J. D. Aber, J. W. Erisman, S. P. Seitzinger, R. W. Howarth, E. B. Cowling, and B. J. Cosby. 2003. The nitrogen cascade. *BioScience* 53: 341–356.

Galloway, J. N., G. E. Likens, and M. E. Hawley. 1984. Acid precipitation: Natural versus anthropogenic components. *Science* 226: 829–831.

Gastal, F., and G. Lemaire. 2002. N uptake and distribution in crops: An agronomical and ecophysiological perspective. *J. Exper. Bot.* 53: 789–799.

Giles, J. 2005. Nitrogen study fertilizes fears of pollution. *Nature* 433: 791.

Goolsby, D. A., W. A. Battaglin, B. T. Aulenbach, and R. P. Hooper. 2001. Nitrogen input to the Gulf of Mexico. *J. Environ. Qual.* 30: 329–336.

Hirel, B., P. Bertin, I. Quilleré, W. Bourdoncle, C. Attagnant, C. Dellay, A. Gouy, S. Cadiou, C. Retailliau, M. Falque, and A. Gallais. 2001. Towards a better understanding of the genetic and physiological basis for nitrogen use efficiency in maize. *Plant Physiol.* 125: 1258–1270.

Howarth, R. W., E. W. Boyer, W. J. Pabich, and J. N. Galloway. 2002. Nitrogen use in the United States from 1961–2000 and potential future trends. *Ambio* 31: 88–96.

International Fertilizer Industry Association (IFA). 2005. Nitrogen fertilizer nutrient consumption. http://www.fertilizer.org/ifa/statistics.asp.

Jeuffroy, M. H., B. Ney, and A. Ourry. 2002. Integrated physiological and agronomic modelling of N capture and use within the plant. *J. Exper. Bot.* 53: 809–823.

Lynch, J. A., V. C. Bowersox, and J. W. Grimm. 2000. Acid rain reduced in eastern United States. *Environ. Sci. Technol.* 34: 940–949.

Ma, L., R. W. Malone, P. Heilman, D. B. Jaynes, L. R. Ahuja, S. A. Saseendran, R. S. Kanwar, and J. C. Ascough II. 2007a. RZWQM-simulated effects of crop rotation, tillage, and controlled drainage on crop yield and nitrate-N loss in drain flow. *Geoderma* 140: 260–271.

Ma, L., R. W. Malone, P. Heilman, D. L. Karlen, R. S. Kanwar, C. A. Cambardella, S. A. Saseendran, and L. R. Ahuja. 2007b. RZWQM simulation of long-term crop production, water and nitrogen balances in Northeast Iowa. *Geoderma* 140: 247–259.

Malone, R. W., L. Ma, P. Heilman, D. L. Karlen, R. S. Kanwar, and J. L. Hatfield. 2007a. Simulated N management effects on corn yield and tile-drainage nitrate loss. *Geoderma* 140: 272–283.

Malone, R. W., L. Ma, D. L. Karlen, T. Meade, D. Meek, P. Heilman, R. S. Kanwar, and J. L. Hatfield. 2007b. Empirical analysis and prediction of nitrate loading and crop yield for corn–soybean rotations. *Geoderma* 140: 223–234.

Mayewski, P. A., W. B. Lyons, M. J. Spencer, M. S. Twickler, W. Dansgaard, B. Koci, C. I. Davidson, and R. E. Honrath. 1986. Sulfate and nitrate concentrations from a South Greenland ice core. *Science* 232: 975–977.

Pare, K., M. H. Chantigny, K. Carey, W. J. Johnston, and J. Dionne. 2006. Nitrogen uptake and leaching under annual bluegrass ecotypes and bentgrass species: A lysimeter experiment. *Crop Sci.* 46: 847–853.

Parkin, T. B., T. C. Kaspar, and J. W. Singer. 2006. Cover crop effects on the fate of N following soil application of swine manure. *Plant Soil* 289: 141–152.

Perlack, R. D., L. L. Wright, A. Turhollow, R. L. Grahm, B. Stokes, and D. C. Erbach. 2005. Biomass as feedstock for a bioenergy and bioproducts industry: The technical feasibility of a billion-ton annual supply. ORNL/TM-2005/66. http://feedstockreview.ornl.gov/pdf/billion_ton_vision.pdf.

Plenet, D., and G. Lemaire. 1999. Relationships between dynamics of nitrogen uptake and dry matter accumulation in maize crops: Determination of critical N concentration. *Plant Soil* 216: 65–82.

Power, J. F., D. Flowerday, R. A. Wiese, and D. G. Watts. 1998. Agricultural nitrogen management to protect water quality. IDEA No. 4. http://idea.exnet.iastate.edu/idea/marketplace/mseancr/publications.asp.

Power, J. F., R. Wiese, and D. Flowerday. 2000. Managing nitrogen for water quality-lessons from management systems evaluation areas. *J. Environ. Qual.* 29: 355–366.

Power, J. F., R. Wiese, and D. Flowerday. 2001. Managing farming systems for nitrate control: A research review from management systems evaluation areas. *J. Environ. Qual.* 30: 1866–1880.

Rabalais, N. N., R. E. Turner, and W. J. Wiseman. 2001. Hypoxia in the Gulf of Mexico. *J. Environ. Qual.* 30: 320–329.

Raun, W. R., and G. V. Johnson, 1999. Improving nitrogen use efficiency for cereal production. *Agron. J.* 91: 357–363.

Revelle, R., and H. E. Suess. 1957. Carbon dioxide exchange between atmosphere and ocean and the question of an increase of atmospheric CO_2 during the past decades. *Tellus* 9: 18–27.

Smil, V. 2001. *Enriching the Earth.* Cambridge, MA: MIT Press.

Sogbedji, J. M., H. M. van Es, and K. L. Agbeko. 2006. Modeling nitrogen dynamics under maize on Ferralsols in western Africa. *Nutrient Cycling Agroecosystems* 74: 99–113.

Sogbedji, J. M., H. M. van Es, and J. L. Hutson. 2001. N fate and transport under variable cropping history and fertilizer rate on loamy sand and clay loam soils, 1: Calibration of the LEACHMN model. *Plant Soil* 229: 57–70.

U.S. Census Bureau. 2004. U.S. interim projections by age, sex, race, and Hispanic origin. http://www.census.gov/ipc/www/usinterimproj/.

U.S. Department of Agriculture, National Resource Conservation Service (USDA-NRCS). 2008. Soil Survey Division. Official soil series descriptions. Washington, D.C.: USDA-NRCS. http://ortho.ftw.nrcs.usda.gov/cgi-bin/osd/osdname.cgi (verified May 2, 2008).

Vitousek, P. M., J. D. Aber, R. W. Howarth, G. E. Likens, P. A. Matson, D. W. Schindler, W. H. Schlesinger, and D. G. Tilman. 1997. Human alteration of the global nitrogen cycle: Sources and consequences. *Ecological Appl.* 7: 737–750.

Vitousek, P. M., S. Hattenschwiler, L. Olander, and S. Allison. 2002. Nitrogen and nature. *Ambio.* 37: 97–101.

13 Simulated Soil Water Content Effect on Plant Nitrogen Uptake and Export for Watershed Management

Ping Wang, Ali Sadeghi, Lewis Linker,
Jeff Arnold, Gary Shenk, and Jing Wu

CONTENTS

13.1 Introduction 278
    13.1.1 Overview of N Cycle and Plant N Uptake Simulations in
        Watershed Models 278
    13.1.2 Importance of Soil Moisture in Plant N Uptake and DIN
        Export 280
    13.1.3 Plant N-Uptake Simulations in HSPF and SWAT 282
        13.1.3.1 Target-Based Plant N Uptake in HSPF 282
        13.1.3.2 Demand-Based Plant N Uptake in SWAT..................... 283
        13.1.3.3 First-Order Kinetics of Plant N Uptake in HSPF........... 284
        13.1.3.4 Michaelis–Menten Saturation Kinetics of Plant N
                Uptake 284
13.2 Method and Approach 286
    13.2.1 Brief Description of the Approach 286
    13.2.2 Michaelis–Menten Plant N-Uptake Simulation for Forest in
        HSPF and Our Modification 286
        13.2.2.1 Using Mg/L as the Unit of Maximum Uptake Rate 286
        13.2.2.2 Using Mass as the Unit of Maximum Uptake Rate 287
        13.2.2.3 Michaelis–Menten Plant N-Uptake Simulation Used
                in This Chapter 288
    13.2.3 Demand-Based Plant N-Uptake Simulation for Crop in SWAT
        and Our Modification 289
    13.2.4 Target-Based Plant N-Uptake Simulation for Crop in HSPF
        and Our Modification 289

13.3   Results 291
        13.3.1   Comparing Michaelis–Menten Plant N-Uptake Simulations
                 for Forest by Initial and Modified HSPF 291
        13.3.2   Comparing Demand-Based Plant N-Uptake Simulations for
                 Crop by Initial and Modified SWAT 293
        13.3.3   Comparing Target-Based Plant N-Uptake Simulations for
                 Crop by Initial and Modified HSPF 296
                 13.3.3.1   Compare Plant N Uptakes Simulated by Three
                            Methods of Moisture Effect 296
                 13.3.3.2   Compare N Uptake and Export between
                            Precipitation Factors 1.0 and 1.1 298
13.4   Discussion 300
        13.4.1   Unit of the Maximum Uptake Rate in Michaelis–Menten
                 Equation: Mass or Concentration? 300
        13.4.2   Applicability of Michaelis–Menten Equation in Forest
                 Simulations in HSPF 301
        13.4.3   Improvement of N Uptake for HSPF Target-based Module 301
        13.4.4   Additional Control of Uptake by Michaelis–Menten Equation
                 in SWAT Model 302
13.5   Conclusion 302
Acknowledgment 303
References 303

13.1   INTRODUCTION

13.1.1   Overview of N Cycle and Plant N Uptake Simulations in Watershed Models

In watershed nutrient management for water quality assessment, it is important to understand the critical pathways of nutrient cycles and nutrient transport processes from lands to the receiving water bodies.

Most mechanistic watershed models, such as the Soil and Water Assessment Tool (SWAT) (Arnold et al. 1998; Arnold and Sammon 1988) and the Hydrological Simulation Program—FORTRAN (HSPF) (Bicknell et al. 1997), simulate important processes of the nitrogen (N) cycle, including N uptakes. The HSPF model, for example (Figure 13.1), simulates various forms of nitrogen, including nitrate, ammonia, and labile and refractory organics (in soluble and particulate forms). The HSPF model simulates adsorption/desorption of soluble forms of nitrogen between soil and water. Plants uptake dissolved inorganic nitrogen (DIN), e.g., ammonia and nitrate, from soil and transform it to organic nitrogen and ultimately to plant biomass. The organic form of plant biomass then returns to the soil through leaf fall or plant decay, or it is removed out of the land segment by harvest. In the soil, the organic form of N transforms to ammonia and/or nitrate through mineralization by chemical reactions and microorganism activities; conversely, other microorganism activity transforms nitrate and ammonia to organic forms through immobilization. Ammonia can transform to nitrate through nitrification or it can volatilize to the atmosphere. Nitrate can

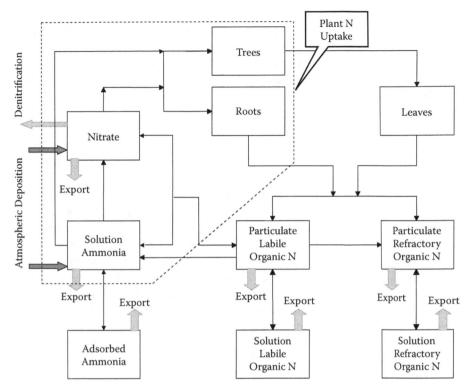

FIGURE 13.1 Flow chart of N-cycling simulations in HSPF (after HPPF User's Manual, v.12).

transform to nitrogen gas through denitrification in anaerobic conditions. Provisions can be made in the HSPF model for certain plants called legumes (e.g., soybeans), which can utilize nitrogen gas from the atmosphere and transform it to the organic form through a process called N fixation. Both the HSPF and SWAT models simulate all of the above processes. Many of the above reactions are simulated using simple equations, e.g., the first-order rate with temperature correction. The parameters of the reactions used in the models are normally obtained from the literature and through model calibration. Some of the reactions, such as plant nitrogen uptake, are simulated using more complex processes. However, the HSPF and SWAT models use different approaches to simulate these processes. Detailed explanations of each model approach are discussed in section 13.1.3.

The HSPF and SWAT models track nutrient storages within the soil and plant biomass and simulate exchanges between these storages on a relatively short time scale, typically one hour or one day. Both models consider inputs of nitrogen by human application and from atmospheric deposition. All of the N-component forms in soil have the potential for export, i.e., leaching and runoff from lands to receiving water bodies, which is driven by hydrology. Water is both a solvent for dissolving inorganic nitrogen for plant uptake and a medium of hydrologic force for the transport of nitrogen. Both models simulate critical processes such as precipitation, interception, evapotranspiration (ET), infiltration, percolation, surface run off, interflow/lateral flow, groundwater flow, and runoff. The ET is simulated by the considerations of potential evaporation and transpiration, which are determined by weather and plant conditions and by the availability of water in the system. The excess water then infiltrates, percolates, and transports in the simulated soil layers and groundwater. The DIN export is calculated by outflow times its DIN concentration. Both models simulate multiple layers of the soil profile. However, the depths and other physical properties of the soils are specified by users.

Plant N uptake significantly affects the proportion and distribution of nitrogen-cycle components (Likens and Bormann 1995; Fenn and Poth 1999). An accurate simulation of plant N uptake is important in computer models of the nitrogen cycle in watershed management.

Plant N uptake is dependent on many soil and environmental parameters, including soil DIN storage, DIN concentration, soil moisture, plant productivity and biomass, stage of plant growth, root development, temperature, the amount and temporal distribution of precipitation, and soil hydraulic conductivity. These factors, as individually or lumped factors, are considered in various ways in different computer models (Willigen 1991), including HSPF and SWAT. Most of the models consider N uptake as driven principally by plant demand, such as SWAT and the target-based module in HSPF. Some models use a water-budget approach for plant demand and growth (Timlin et al. 2001), and others consider N uptake to be controlled by DIN concentration in soil solution, such as in the first-order rate (Wagenet and Hutson 1989; Ramos and Carbonell 1991) or Michaelis–Menten type models (Grant 1991; Tinker and Nye 2000).

13.1.2 Importance of Soil Moisture in Plant N Uptake and DIN Export

Soil water is a solvent in the soil matrix for dissolved inorganic nitrogen (DIN), allowing plant uptake, and is also a carrier for transporting DIN. Therefore, soil water content plays important roles in both DIN uptake and export.

Figure 13.2 illustrates the relationships among soil moisture and plant water uptake, DIN uptake, and DIN export. In the model approach, plant uptake of water and nitrogen begins when soil moisture is above the wilting point. Plant productivity and N uptake increase with increasing soil water content until the field capacity is approached (Law et al. 2000; Aber et al. 1995). Below the field capacity, the assumption is that there is no water or nitrogen export, as all is being held by the soil matrix or is consumed by plant demand. Due to the depletion of DIN storage by uptake, the subsequent DIN export potential is reduced with the increases in soil moisture.

The gentle slopes of the graph near and above the field capacity (Figure 13.2B) indicate that the change of N uptake is rather insignificant with the change in soil moisture level. This also agrees with observations by Aber and Federer (1992) in Coweeta, an area without water stress (comparable to high-moisture conditions). In their assessment, increasing precipitation did not change the net biomass production much. Above the field capacity, root damage due to anaerobiosis can also reduce productivity (Kimmins 1987; Tinker and Nye 2000). As Tinker (personal communication) pointed out, the highest uptake usually occurs around the field capacity. Below this moisture level, lower moisture reduces DIN uptake. Above the field

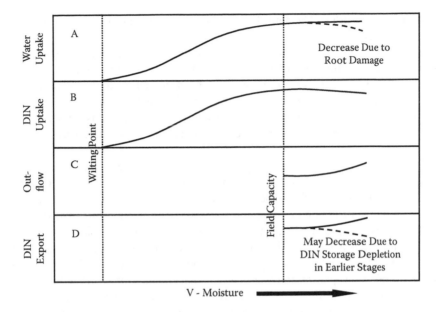

FIGURE 13.2 Schematic relationships of plant water, DIN uptakes, outflow, and DIN export with soil moisture.

capacity, however, additional inputs of water may flow into the next soil layer or be discharged as subsurface flow and as dilute DIN solution, causing a reduction in N uptake (Wang and Linker 2006).

An increase in soil moisture beyond the field capacity is most likely to decrease DIN concentration, resulting in less DIN uptake by plants and more DIN available in the soil matrix for export, thereby increasing DIN export due to the increase in outflows. Observations generally show that more water outflow is accompanied by higher DIN export, as shown in Figure 13.3. The data in Figure 13.3 are from the observations obtained at a forested catchment at Young Woman Creek, Pennsylva-

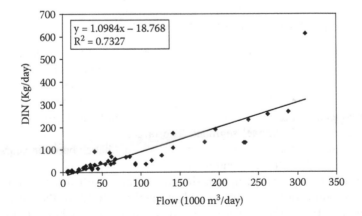

FIGURE 13.3 An example of relationship between DIN export and water discharge from a forested catchment.

nia, from 1984 to 2000. The data show that DIN export is positively correlated with water discharge.

We have discussed two notable but different trends in the relationship between soil moisture and DIN uptake/export. Below the field capacity, the increase in soil moisture generally increases DIN uptake and reduces DIN export potential; above the field capacity, the increase in soil moisture does not increase DIN uptake but does increase DIN export (Figure 13.2). However, as will be discussed below, many watershed models fail to simulate these relationships between soil moisture and nitrogen uptake.

In the model simulation for watershed nutrient management, one of the key tasks is to correctly estimate plant nutrient uptake and export as well as their relationships with soil water and flow discharge (Bhuyan et al. 2003). The intention of this chapter is not to discuss how the models simulate the mechanisms involved in the uptake processes on solute/nutrient transport through the soil–plant system (e.g., roots, stems, leaves). Rather, this chapter focuses on the importance of soil moisture on plant N uptake and the consequent effect on DIN export as evaluated from the viewpoint of water quality and watershed management. Our effort is focused primarily on the relationships between the available soil water and plant N uptake and DIN export. This chapter provides examples of improvements in N-uptake/export simulation by incorporating the effect of soil moisture into the existing models, such as the target-based N uptake in HSPF, the demand-based N uptake in SWAT, and the Michaelis–Menton N uptake in HSPF.

13.1.3   PLANT N-UPTAKE SIMULATIONS IN HSPF AND SWAT

Various approaches to plant N-uptake simulation are used by different watershed models. HSPF, for example, supplies three different modules of plant N uptake for user selection. They are: first-order kinetics, target-based uptake, and the Michaelis–Menten uptake kinetics. The SWAT, however, uses an approach based on plant growth and demand, which is similar to the target-based uptake of HSPF, but with more detailed simulations of plant growth and nitrogen demand.

13.1.3.1   Target-Based Plant N Uptake in HSPF

The target-based plant N uptake in HSPF is generally used for cropland simulations. The user typically assigns a monthly target of N uptake (in mass per unit land area) as an input to the model. The monthly target is then interpolated into daily or finer time intervals. The monthly targets for a specific model segment are differentiated based on collective information of crop productivity in typical growing seasons in a region at or near the model segment. The target, however, is not linked to the simulated plant biomass in the model, and it does not change between years. The N uptake is unrelated to DIN concentration.

In the HSPF target-based uptake simulation, there are two options associated with soil moisture. In the first option, plant uptake is 100% of the daily (or finer time interval) target without consideration of soil moisture condition. In the second option, plant N uptake is a step function of soil moisture at the wilting point. That is, when moisture is below the wilting point, there is no uptake; when moisture is above the wilting point, 100% of the uptake target is obtained if sufficient soil nitrogen is available. The zero N uptake at soil moisture content below the wilting point would cause a deficit in N uptake. In the following time step, if the soil moisture content is above the wilting point, the plant will achieve the uptake target plus the deficit in the previous time step, assuming that sufficient soil DIN is available.

Variations in temporal distribution of precipitation can cause variations in soil moisture conditions. Some days are below the wilting point, and some days are above the wilting point. This may cause different DIN uptakes and exports in different times.

There are two limitations to the HSPF target-based N uptake. One limitation is the lack of a function to differentiate the amount of N uptake with changes in soil moisture. Although HSPF differentiates nitrogen uptake in soil moisture contents above and below the wilting point, it does not differentiate the amount of uptake in the range of soil moisture contents above the wilting point, including conditions above the field capacity. The second limitation is that the target is not linked with simulated plant conditions such as the plant biomass. The targets are user-defined, fixed values in the model input, largely independent of other conditions of plant growth.

13.1.3.2 Demand-Based Plant N Uptake in SWAT

In the SWAT model, plant N uptake considers plant growth stages within a planting–harvest cycle that are dependent on a plant's growing stages, the biomass, and the amount of nitrogen in the plant biomass (Neitsch et al. 2002).

The plant nitrogen uptake is controlled by the plant nitrogen equation. The plant nitrogen equation calculates the fraction of nitrogen in the plant biomass as a function of growth stage given optimal growing conditions:

$$frn = (frn,a - frn,c)[1 - \frac{frph}{frph + \exp(na - nb * frph)}] + frn,c$$

where

frn = fraction of nitrogen in the plant biomass on a given day of optimal growth condition,

frn,a = normal fraction of nitrogen in the plant biomass at emergence,

frn,c = normal fraction of nitrogen in the plant biomass at maturity,

$frph$ = fraction of potential heat units accumulated for the plant on a given day in the growing season, and

na and nb are shape coefficients.

The optimal mass of nitrogen stored in plant material for the current growth stage (BIOn,opt) is acquired by

$$BIOn,opt = frn * Biomass$$

The plant nitrogen demand (Nup,opt) for a given day is determined by taking the difference between the optimal nitrogen of the plant biomass for the plant's growth stage (BIOn,opt) and the actual nitrogen content (BIOn):

$$Nup,opt = BIOn,opt - BIOn$$

Generally, in SWAT, the plant growth is dependent on the availability of water and nutrient supply, solar radiation and temperature, canopy and root development, growing stages (maturity), etc. They are also determined by daily precipitation, potential evapotranspiration, plant water demand, canopy development, etc. Therefore, the SWAT's plant N-uptake simulation has certain connections to water supply, but only indirectly. It does not directly simulate the effect of soil moisture and DIN concentration in soil water solution on plant N uptake. The SWAT demand-based N uptake is different from the HSPF target-based N uptake, as it is correlated with plant growth that is simulated in SWAT.

The target-based N uptake in HSPF does not differentiate passive or active plant N uptake, while the SWAT demand-based N uptake simulates mainly the active uptake by plants. However, through model calibration of N-cycle simulation and mass balance, the simulated N uptake by SWAT can be regarded as representing the sum of passive and active uptakes.

13.1.3.3   First-Order Kinetics of Plant N Uptake in HSPF

Plant roots uptake nitrogen primarily from soil water solution through the xylem of roots, with subsequent transport through stems and leaves. Nitrogen concentration in the soil solution therefore affects the amount of N uptake (Tinker and Nye 2000). The first-order N-uptake kinetics (i.e., uptake to be proportional to concentration) is utilized in some models (Wagenet and Hutson 1989). The first-order kinetics assumes that uptake is proportional to DIN concentration. Although the first-order kinetics of plant N uptake considers the role of DIN concentration, it is loosely related to plant demands. Another problem is that the first-order relationship is approximately valid at low concentrations, but not at higher concentrations, as will be addressed in the next subsection. HSPF adopts this approach in its first-order kinetics plant-uptake module. However, since such a module is seldom used, we will not discuss it here further.

13.1.3.4   Michaelis–Menten Saturation Kinetics of Plant N Uptake

People have recognized that N uptake is not a pure first-order rate with DIN concentration, but acts more like a Michaelis–Menten type (Grant 1991; Bicknell et al. 1997), with rates increasing nearly linearly with concentration at low N concentrations, although the increase in uptake slows down at higher concentrations, as demonstrated by the experimental isotherm method for plant uptake (Glass and Siddiqui 1984). Figure 13.4 presents three hypothetical Michaelis–Menten curves of plant-uptake rate (u) in response to DIN concentration (C) at a defined maximum uptake rate (U) for three different half-saturation constants (C_s). The rate–concentration relation is similar to the growth in a limiting substrate or nutrient in a container, such as bacterial (Metcalf and Eddy 1979) or algal growth (Bowie et al. 1985). Such a relationship can be defined adequately by the following Michaelis–Menten or Monod equation:

$$u = U \frac{C}{(C + C_s)} \tag{13.1}$$

where

    u = rate (per time step in model simulation)
    U = maximum rate (mass or concentration per time step)
    C = solute concentration (mg/L), as defined by equation (13.2)
    C_s = half-saturation constant (mg/L)

$$C = \frac{m}{V} \tag{13.2}$$

where m is mass of solute and V is volume of solution.

In plant N uptake, the mass of DIN solute (m) can be taken as the DIN storage in soil, and V is the volume of soil moisture. The maximum uptake rate, U, is considered to be a constant in a specific time or season that is assigned/interpolated in the user input for a specific plant and soil condition. The calculated DIN uptake rate, u, varies at each computational time step solely due to changes in DIN concentration (C).

Equation (13.1) and Figure 13.4 show the relationship between the uptake rate and DIN concentration, but do not explicitly show the relationship between uptake rate and soil moisture contents.

In equation (13.1), the unit of U is not specified. It can be in mass per time step, denoted U_m, or concentration per time step, denoted U_c, depending on the simulation. Wang and Linker (2006) pointed out that the unit of maximum uptake rate, U in equation (13.1), in mass per unit time or concentration per unit time, will determine the trend of mass uptake with soil moisture conditions, which will be summarized in section 13.2.

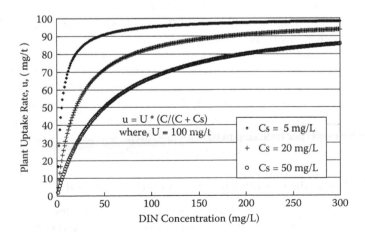

FIGURE 13.4 Hypothetical Michaelis–Menten curves showing the relationship between uptake rate and concentration.

13.2 METHOD AND APPROACH

13.2.1 BRIEF DESCRIPTION OF THE APPROACH

We used the N-uptake modules from both the SWAT and HSPF software as the basic platforms to simulate plant N uptake, but we modified the source code of the software to better reflect the response of plant N uptake to soil moisture conditions. We then compared the results of our modified modules with the initial SWAT or HSPF runs to demonstrate that the modified modules improve DIN uptake and export simulations.

We performed sensitivity analysis on the response of DIN uptake and export to soil moisture status by running various scenarios with different precipitation factors, starting with the observed precipitation and then ±10% changes from the observed precipitation time series. We then examined the changes of N uptake and export against the changes of soil moisture conditions to see whether the trend is consistent with observed plant dynamics in Figure 13.2. These analyses helped us to understand how the modified modules, implemented with the effect of soil moisture on N uptake, can improve simulations of DIN uptake and export.

13.2.2 MICHAELIS–MENTEN PLANT N-UPTAKE SIMULATION FOR FOREST IN HSPF AND OUR MODIFICATION

The HSPF Michaelis–Menten uptake module uses DIN concentration and the maximum uptake rate to calculate plant N uptake in each time step of the model simulations. The user-specified maximum uptake rates, generally monthly, are interpolated to daily values by the model. The monthly maximum uptake rates are estimated primarily based on soil nutrient storage and plant yield or biomass in the modeled land use. The values are generally higher in the summer and lower in the winter, reflecting variations of uptake rates with growing seasons and plant biomass in a northern hemispheric temperate region. The seasonal changes in the maximum uptake rate for an established forest are generally consistent with the interpolated maximum uptake rate, but the changes of maximum uptake for crop can be quite different. Therefore, the current Michaelis–Menten plant uptake module in HSPF is more applicable to a forest simulation, rather than the simulation of crops. Hence, this chapter uses forestland simulation as an example to study the effect of moisture on plant N uptake when using HSPF's Michaelis–Menten module. The related forestland sample is from the Chesapeake Bay Phase V Community Watershed Model.

As pointed out by Wang and Linker (2006), the unit of uptake rate (U), in concentration or mass (per time step in the unit land area), will exhibit different relationships between uptake rate and the soil moisture conditions.

13.2.2.1 Using Mg/L as the Unit of Maximum Uptake Rate

In the Michaelis–Menten plant N-uptake simulation in HSPF version 12, the unit of maximum uptake rate (U) is mg/L, which is denoted U_c to indicate that it is a concentration unit. The calculated mass uptake, u, in each time step is

$$u = U_c \frac{C}{(C + C_s)} V \qquad (13.3)$$

where

u = plant N-uptake rate (mass, kg, per time step)

U_c = maximum uptake rate (mg/L, per time step)

C = concentration of DIN (mg/L) in soil water, as defined by equation (13.2)

C_s = half-saturation constant (mg/L)

V = volume of soil moisture (L)

The $C/(C + C_s)$ term is unitless and approaches 1 when C is far greater than C_s. The unit of U_c is in concentration. The multiple of V on the right-hand side of equation (13.3) is to obtain a mass uptake of u. To see the relationship between mass uptake and soil moisture explicitly, we rearrange equation (13.3) by substituting m/V for C and then simplifying the equation to

$$u = \frac{U_c * m}{(m / V + C_s)} \qquad (13.4)$$

Equation (13.4) shows that the higher soil moisture (V), the smaller the term m/V, and, consequently, the greater the uptake (u). This pattern holds within the range of soil moisture being below the field capacity, and is consistent with the description in Figure 13.2. At the field capacity, DIN export begins. However, for the soil-moisture values above the field capacity, equation (13.4) will still result in greater N uptake, and therefore less DIN export potential, as the soil moisture (V) continues to increase. This is contrary to the general observations that, for moisture values above the field capacity, there is more DIN export with greater water discharge (Figure 13.3). The excessive moisture beyond the field capacity also dilutes DIN concentration and bypasses the soil–root system, causing a reduction in DIN uptake and elevating DIN export potential. Also, the excessive moisture conditions may cause anaerobiosis, resulting in possible root damage and less DIN uptake (Tinker and Nye 2000). Therefore, equation (13.3) (similar to equation 13.4) is acceptable for soil moisture below the field capacity, but is unsuitable for soil moisture above the field capacity.

13.2.2.2 Using Mass as the Unit of Maximum Uptake Rate

Wang and Linker (2001) pointed out the above problem and suggested an alternative method to simulate plant uptake at above the field capacity. In their method, the unit of maximum uptake rate is mass per area, denoted as U_m:

$$u = U_m \frac{C}{(C + C_s)} \qquad (13.5)$$

where u, C, and C_s are the same as defined in equation (13.3), and U_m is the maximum mass uptake rate (kg, for unit area per time step).

We can compute the mass of plant uptake, u, directly from equation (13.5) without a multiplication of soil moisture volume (V). To see the relationship between soil moisture and mass uptake explicitly, we rearrange equation (13.5) by substituting m/V for C, multiplying V for both the numerator and the denominator, and then simplifying it to

$$u = \frac{U_m * m}{(m + C_s V)} \qquad (13.6)$$

Equation (13.6) shows that the higher the soil moisture volume (V), the less the mass uptake (u), which increases the potential for N export. Thus, equation (13.5) (similar to equation 13.6) can reasonably simulate plant uptake and N export above the field capacity, but it is unsuitable for moisture below the field capacity.

13.2.2.3 Michaelis–Menten Plant N-Uptake Simulation Used in This Chapter

Our modified module uses the synthetic method by Wang and Linker (2006). Their method used mass as the unit of maximum uptake rate for both above and below the field capacity, and the rate is specifically defined as the "effective maximum mass uptake rate," U_e (Wang and Linker 2006), as expressed in equation (13.7).

$$u = U_e \frac{C}{(C + C_s)} \qquad (13.7)$$

The maximum mass uptake rate U_m commonly refers to the whole thickness of soil for a specific soil–plant area, which is related to the plant's growing conditions (such as the stage of plant growth and seasons), root development, soil DIN storage and physical properties, etc.

Above the field capacity, the whole soil is wet and can involve uptake processes. Therefore, the effective maximum uptake (U_e) equals the maximum uptake for the whole soil–root system (U_m), i.e., $U_e = U_m$. Under a specified DIN storage in soil, DIN concentration (C) decreases with the increase in soil water content. Equation 13.7 is equivalent to equation (13.5) and will result in less N uptake and more DIN export for a greater soil moisture, which is our expectation for model simulation of plant N uptake at soil moisture levels above the field capacity.

Below the field capacity, we can assume that only a fraction of the soil is effectively involved in N uptake. The efficiency of supplying DIN for plant uptake is reduced when soil moisture decreases. The fraction can be approximated as V/V_{fcp}, where V_{fcp} is the volume of soil moisture at the field capacity that can be assumed constant for a specific soil. Therefore, $U_e = (V/V_{fcp}) * U_m$. In low soil-moisture conditions, only a portion of soil DIN storage (m) supplies solute to DIN solution. This portion can be approximated as V/V_{fcp} (Wang and Linker 2006). Therefore, the concentration cannot be simply calculated by $C = m/V$, but can be approximated as $C = (V/V_{\text{fcp}})m/V = m/V_{\text{fcp}}$.

Substituting $U_e = (V/V_{\text{fcp}}) * U_m$ and $C = m/V_{\text{fcp}}$ into equation (13.7) and further simplifying, we obtain

$$u = \frac{U_m * m * V}{(m / V_{\text{fcp}} + C_s)} \tag{13.8}$$

Equation (13.8) shows that a greater soil moisture (V) will have more mass uptake (u). This is our expectation for a model simulation of plant N uptake at soil moisture below the field capacity.

In summary, equation (13.7) can simulate plant N uptake for both above and below the field capacity. Note again: above the field capacity, $U_e = U_m$; below the field capacity, $U_e = (V/V_{\text{fcp}}) * U_m$.

In this work, our implementation to HSPF target-based uptake and SWAT demand-based uptake is primarily based on the synthetic method proposed by Wang and Linker (2006).

13.2.3 DEMAND-BASED PLANT N-UPTAKE SIMULATION FOR CROP IN SWAT AND OUR MODIFICATION

The SWAT model simulates both plant biomass and N uptake by crops, as well as the movement and loss of nitrogen from the land. The basic N transformation processes are mineralization, immobilization, nitrification, denitrification, volatilization, plant uptake, etc., as addressed in the introduction to this chapter. Plant N uptake is one of the main DIN loss pathways and is estimated using a supply-and-demand approach. The nitrogen demand is computed on a daily basis based on the optimal N crop concentration for each growth stage. The plant N uptake considers plant growth stages within a planting–harvest cycle. In turn, the plant growth depends on water and nutrient availability and weather conditions. The plant uptake is indirectly related to soil water. To consider the effect of soil moisture and DIN concentration on plant N uptake, we modified the source code to simulate plant N uptake governed by the Michaelis–Menten kinetics proposed by equation (13.7), i.e., the synthetic method by Wang and Linker (2006). The key to the modification is to have a reasonable estimate in the maximum N-uptake rate, U_m. In this modification, the U_m is correlated to the optimal N contents in the plant biomass at the growth stages, which is simulated in the SWAT model. Therefore, plant N uptake in SWAT can also be simulated with the same Michaelis–Menten equation.

For these simulations, the input data uses the sample data set for a cornfield provided by the SWAT developer along with the source code distribution.

13.2.4 TARGET-BASED PLANT N-UPTAKE SIMULATION FOR CROP IN HSPF AND OUR MODIFICATION

The target-based plant N-uptake simulation is recommended for cropland by HSPF (Bicknell et al. 1997). Here we utilize a land use of conventional tillage from the

Chesapeake Bay Phase 5 Community Watershed Model. In the HSPF plant N-uptake simulation, the user assigns a maximum target of nitrogen uptake in a year. The annual target is split into monthly or daily targets that progressively increase from the spring planting, peak in summer, and go to nearly zero in the winter.

In the following discussion, we compare three methods of soil moisture-dependent N uptake:

A. *Plant uptake is not affected by soil moisture*: The plant takes up 100% of the daily target without considering moisture.
B. *Plant uptake is a step function of soil moisture at the wilting point*: There is no uptake in soil moisture below the wilting point; plant uptake is at 100% of the daily target as long as the moisture is above the wilting point.
C. *Plant uptake varies based on soil moisture*: There is no uptake below the wilting point; uptake increases gradually with soil moisture until the field capacity; above the field capacity, N uptake decreases slightly with soil moisture, as shown in Figure 13.2B.

Method C is our modification to the HSPF target-based N uptake. One option to modify the code is using the Michaelis–Menten simulation (equation 13.7) to calculate N uptake by the effect of soil moisture. A critical step in the modification is to have a reasonable estimate in the maximum uptake rate, U_m. We may use the same approach in the modification of SWAT demand-based uptake to estimate the maximum uptake rate from the plant's biomass simulated in HSPF. However, such a module is no longer target based, but is demand based.

To maintain the modified module's target-based status while using the Michaelis–Menten kinetics, we correlate the two parameters: the uptake target of crop and the maximum uptake rate required by the Michaelis–Menten calculation. Both parameters are user inputs in the HSPF operation. However, we found that the code modification yields only a little improvement. This is because, in a developed forest, its biomass and its maximum uptake potential vary rather regularly within a year and over the years, whereas, the variations of a crop's biomass and its maximum uptake potential with seasons are more irregular. The user input and the interpolated target does not well match the uptake demands of the crop.

Here, we use another way to modify the HSPF target-based uptake module to simulate the effect of soil moisture on plant N uptake. We use an empirical function of plant uptake versus soil moisture, consistent with the curve in Figure 13.2B. That is,

$$\text{Below the field capacity: } u = \text{target}*[(V - V_{wp})/(V_{fcp} - V_{wp})]^{0.25}$$

$$\text{Above the field capacity: } u = \text{target}*[-0.01(V - V_{fcp})^2 - 0.001(V - V_{fcp}) + 1]$$

where V_{wp} is soil moisture at the wilting point and V_{fcp} is soil moisture at the field capacity. In the initial HSPF simulations, plants always take up 100% target as long as the moisture is above the wilting point. In the modified module, the uptake will be less than the target if soil moisture is other than the field capacity.

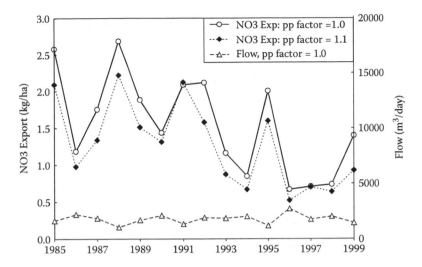

FIGURE 13.5 Response of NO_3 export to precipitation factors 1.0 and 1.1 simulated by HSPF v.12 using Michaelis–Menten kinetics in a forest land use.

13.3   RESULTS

13.3.1   COMPARING MICHAELIS–MENTEN PLANT N-UPTAKE SIMULATIONS FOR FOREST BY INITIAL AND MODIFIED HSPF

The Michaelis–Menten plant N-uptake simulation in the initial HSPF uses mg/L as the unit of the maximum uptake rate, as seen in equation (13.3), which causes more uptake at greater moisture content. It is suitable for dry conditions, but unsuitable for wet conditions, where excessive water inputs and outflow and N export occur. The model results in less N export in greater moisture conditions, as shown in Figure 13.5. There, the annual N export is less for the scenario of precipitation factor = 1.1 (the diamonds) than that by precipitation factor = 1.0 (the open circles). The solid cubes are annual outflow for precipitation factor = 1.0. The wet years, 1986, 1990, and 1996, have lower annual plant N uptakes than nearby drier years for both precipitation factors 1.0 and 1.1. This is contrary to what is observed (Figure 13.2), confirming our earlier discussion that using concentration as the unit of plant maximum uptake rate in the Michaelis–Menten equation is unsuitable for simulating plant N uptake/export above the field capacity.

Figure 13.6 compares annual N exports simulated by the modified HSPF (i.e., equation 13.7) for the scenarios of precipitation factors 1.0 and 1.1 (open circles and diamonds, respectively). The annual N export is higher for the scenario with a greater precipitation factor. The solid cubes are annual outflow for precipitation factor = 1.0. The wet years of 1986, 1990, and 1996 have lower annual plant N uptakes than their nearby drier years using either precipitation factor 1.0 or 1.1. This indicates that the modified HSPF better simulates the effects of soil moisture on plant N uptake and DIN export.

Table 13.1 provides the DIN 15-year average budget in forest simulation using precipitation factors 1.0 and 1.1 by the initial and modified modules. By the initial

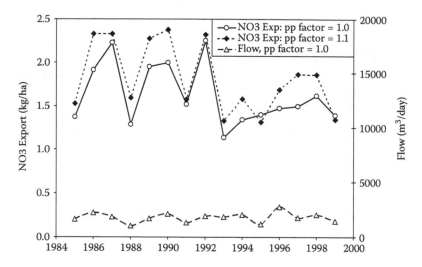

FIGURE 13.6 Response of NO_3 export to precipitation factors 1.0 and 1.1 simulated by the modified HSPF using Michaelis–Menten kinetics in a forest land use.

TABLE 13.1

DIN 15-Year Average Budget in Forest Simulation by HSPF Michaelis–Menten Plant N Uptake

	Initial HSPF		Modified HSPF	
	Precipitation Factor			
Modules	**1.0**	**1.1**	**1.0**	**1.1**
DIN Simulation (kg/ha)				
Initial storage	25.68	25.68	25.78	25.81
Air depot	8.20	8.20	8.20	8.20
Application	0.00	0.00	0.00	0.00
Mineralization–immobilization	31.86	31.68	26.14	26.36
Volatilization	0.00	0.00	0.00	0.00
Denitrification	0.00	0.00	−0.01	−0.01
Leach/runoff	−1.62	−1.34	−1.70	−1.90
Uptake	−38.36	−38.48	−32.56	−32.58
Sum (input-output)	0.08	0.06	0.07	0.06

Note: Negative means loss from the soil; positive means gain in the soil.

HSPF Michaelis–Menten module, higher precipitation yields more uptake and less export (i.e., leaching/runoff), consistent with the assessment from Figure 13.5, but contrary to what we discussed in Figure 13.2. By the modified Michaelis–Menten module, higher precipitation yields more uptake, consistent with the assessment from Figure 13.6 and what is discussed in Figure 13.2. The long-term average of uptake has no significant difference between the two precipitation conditions. This is due in part to the existence of two opposite trends in the relationship between uptake and moisture changes, and the fact that the soil moisture conditions above and below the field capacity can occur alternatively on different days.

For the modified HSPF, compared with precipitation factor = 1.0, the excessive 0.2-kg/ha DIN export by precipitation factor = 1.1 is mainly from a net gain of DIN from the mineralization–immobilization processes, specifically due to less loss of DIN by immobilization. HSPF does not have a sophisticated simulation on these processes. Nevertheless, it simulates a relatively stable forest in the sense of soil DIN storage (Table 13.1). The information of plant uptake for a 15-year average cannot be easily used to compare two simulations on the response of plant uptake to soil moisture, but the DIN export information is more useful. Table 13.1 shows that with a higher precipitation factor, the initial HSPF yields less DIN export, but the modified HSPF yields more DIN export, consistent with the above assessment from Figures 13.5 and 13.6 on yearly DIN export. This indicates that the modified HSPF simulates DIN export better. One of the important goals of the HSPF water-shed model for nutrient management is to achieve the correct relationship between soil moisture and export, which requires a good simulation in plant N uptake. Here, this goal is achieved by the implementation of N-uptake processes in the Michaelis–Menten equation.

13.3.2 Comparing Demand-Based Plant N-Uptake Simulations for Crop by Initial and Modified SWAT

Figure 13.7 shows plant N uptake simulated by the initial SWAT for precipitation factors 1.0 and 0.8. Note that the figures presented in this section only show the growing season, from day 60 to day 191. There is no significant difference in plant uptake between the two precipitation factors. Figure 13.8 presents plant N uptake simulated by the modified SWAT for precipitation factors 1.0 and 0.8. Precipitation factor 0.8 yields less uptake than the normal precipitation (i.e., factor = 1.0).

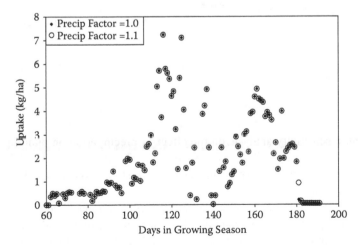

FIGURE 13.7   Response of plant N uptake to precipitation factors 1.0 and 1.1 simulated by the initial SWAT code in a corn land use.

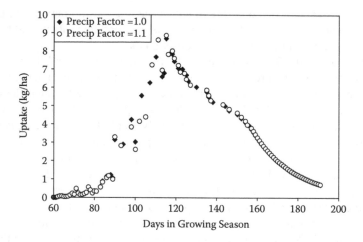

FIGURE 13.8   Response of plant N uptake to precipitation factors 1.0 and 1.1 simulated by the modified SWAT code in a corn land use.

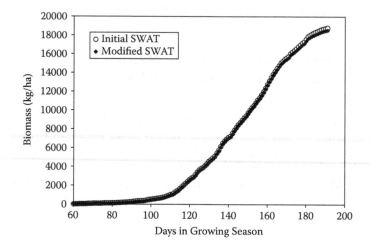

FIGURE 13.9   Plant biomass simulated by the initial and modified SWAT codes using the observed precipitation in a corn land use.

Figure 13.9 shows plant biomass simulated by the initial and modified SWAT using the observed precipitation. Both codes simulate plant biomass in a similar way.

We ran a sensitivity analysis on the effect of precipitation on plant uptake and plant biomass using precipitation factors of 0.6, 0.7, 0.8, 0.9, 1.0, 1.1, 1.2, 1.4, 1.6, 2.0, 2.5, and 3.0. Figure 13.10 presents the preharvest biomass on day 191 versus the precipitation factors.

In the simulations by the modified SWAT, plant biomass increases gradually when the precipitation factor increases from 0.8 to 1.2, then gradually decreases when the precipitation factor increases from 1.2 to 3.0. The total precipitation in the six months of growing season (March–September) in the study area is 476 mm. The simulated potential evapotranspiration is 597 mm. The observed precipitation

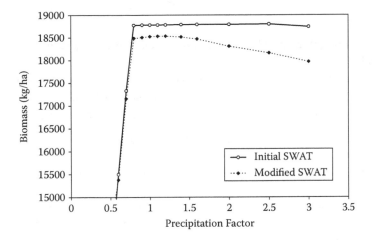

FIGURE 13.10   Response of preharvest plant biomass to precipitation factors (0.6 through 3.0) simulated by the initial SWAT and the modified SWAT codes in a corn land use.

(i.e., factor = 1.0) in this model did not meet the water demand, and was about 20% short. A higher precipitation would increase its productivity, and a lower precipitation would decrease plant productivity. When the precipitation is more than 120% of the observed precipitation, i.e., greater than the water demand by the system, the productivity decreases due to the stress of excessive water supplies that may cause anaerobiosis and root damage. When the water supply is less than 80% of the observed precipitation, i.e., less than two-thirds of the demand, plant growth is reduced significantly. The above shows that the modified SWAT reasonably simulates the relationship between plant biomass (also plant N uptake) and water supplies.

The initial SWAT also simulates the responses of plant N uptake and plant biomass to water supplies in some degrees. Plant biomass decreases when precipitation is reduced from about 0.8 to 0.7 or 0.6, and increases slightly when precipitation increases from 0.8 to 2.5.

Table 13.2 provides the nitrate budget in major processes involved in nitrate simulations. The output setting in our SWAT module provides relatively complete information on nitrate but not ammonia. The plant uptake combines ammonia and nitrate, as shown by the parentheses in Table 13.2. The portion of nitrate uptake is calculated from the DIN uptake with certain assumptions. Therefore, the nitrate budget cannot be balanced quite accurately.

Both the initial and modified SWAT modules have similar simulations according to the nitrate budget. The modified SWAT simulates less nitrate uptake than the initial SWAT. This is due to the fact that the N uptake simulations in the modified SWAT consider the effect of soil moisture and DIN concentration. Comparing precipitation factors 1.0 and 1.1, more precipitation causes slightly more DIN uptake. The values are similar to the net gain of DIN from mineralization–immobilization and nitrification. These simulations by both the initial and the modified SWAT are consistent with the common observations that increasing water supply increases productivity (i.e., plant uptake), and more water discharge causes more DIN export from the field.

TABLE 13.2

Nitrate Budget in the SWAT Demand-Based Plant N-Uptake Simulations

Modules	Initial SWAT			Modified SWAT		
	Precipitation Factor					
	1.0	1.1	2.5	1.0	1.1	2.5
	Nitrate Simulation (kg/ha)					
Initial storage	50.14	50.14	50.14	50.14	50.14	50.14
Air depot	9.16	10.08	22.90	9.16	10.08	22.90
Application	86.80	86.80	86.80	86.80	86.80	86.80
Mineralization–immobilization	89.43	90.32	94.81	83.18	83.80	79.32
Nitrification	1.06	1.16	2.68	1.06	1.16	2.68
Denitrification	0	0	0	0	0	0
Leach/runoff	−26.19	−38.37	−113.35	−27.11	−39.67	−118.17
Uptake (DIN)[a]	(−120.2)	(−120.6)	(−124.3)	(−113.4)	(−113.6)	(−107.1)
Uptake NO$_3$	−96.19	−96.50	−99.47	−85.02	−85.20	−80.32
Sum (input-output)	64.07	53.49	−5.63	68.07	56.97	−6.79

Note: Negative means loss from the soil; positive means gain in the soil.

[a] Parentheses indicate combined uptake of ammonia and nitrate.

In cases of excessive water supply, i.e., precipitation factor = 2.5, the initial SWAT will still have more uptake and productivity, which seems unreasonable. The modified SWAT, using Michaelis–Menten to control the effect of N uptake by soil moisture and DIN concentration, simulates less N uptake. Under such an excessive water supply condition, both the initial and modified SWAT simulate high DIN export and significant reduction of soil DIN storage. The modified SWAT can better simulate uptake and leaching at extreme wet conditions, whereas the initial SWAT still has high uptake to satisfy the demand of plant growth.

The above analysis shows that both the initial SWAT and the modified SWAT can simulate plant N uptake and biomass accumulation rather well, whereas the modified SWAT improves the simulation.

13.3.3 COMPARING TARGET-BASED PLANT N-UPTAKE SIMULATIONS FOR CROP BY INITIAL AND MODIFIED HSPF

13.3.3.1 Compare Plant N Uptakes Simulated by Three Methods of Moisture Effect

The three symbols—squares, triangles, and crosses—in Figure 13.11 are annual plant N uptake simulated by the three methods of soil-moisture effect as described in section 13.2.4. The symbols of pluses and dots are the observed annual N uptake by crops in two regions near the simulated area. These are calculated by multiplying the plant acreage by the nutrient content and summing over all of the crop types in a sin-

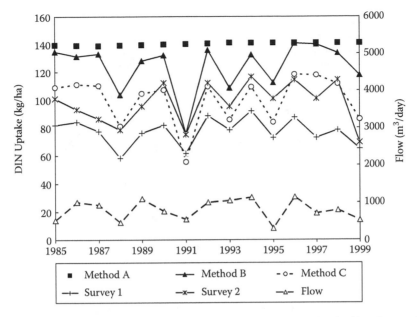

FIGURE 13.11 Annual plant N uptake simulated by three methods of soil-moisture effect using the HSPF target-based uptake, and surveyed uptake and simulated annual flows in a conventional tillage crop land use.

gle land use, and then dividing by the total number of acres to find an average uptake rate in mass per area for a particular land use. The plant acreage is found from the Ag Census (http://www.nrcs.usda.gov/technical/NRT/maps/meta/t5840.html). The nutrient content is from the USDA Plant Nutrient Content Database (http://www.nrcs.usda.gov/technical/ECS/nutrient/tbb1.html). The diamonds represent flow in the plant growing season simulated by HSPF. The variation of annual crop yield is generally in agreement with the annual water availability.

In method A at each model step, the uptake is at 100% of the target regardless of changes in soil moisture unless the nitrate and/or ammonia storage is lower than the target. Therefore, the annual uptake is almost the same. The annual plant uptake varies more for the other two methods (B and C), and the variation is in a similar pattern to the variations of annual flow and the observed annual crop yield. This indicates that the consideration of soil moisture in plant uptake is important.

A year having high water discharge usually has more DIN export, as discussed in Figure 13.2. The variation of annual NO_3 export is in agreement with the variation of annual outflows by the above three methods (Figure 13.12). This indicates that all three methods can simulate such relationships.

The plant N uptake simulated by method C (stars in Figure 13.11) is close to the observed uptake. Water discharge plays a key role in DIN export in the HSPF simulations. However, for example, the average annual NO_3 export for the land use (i.e., the conventional tillage) in the modeled area for 1985–1999 is 31 lb. per acre (Chesapeake Bay Modeling Phase 5 Community Watershed Model documentation, in preparation). The average DIN exports simulated by the three methods are 27, 28.6, and 30.8, respectively.

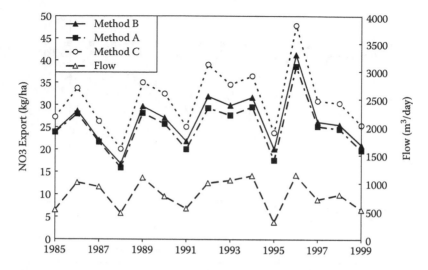

FIGURE 13.12 NO_3 export simulated by three methods of soil-moisture effect using the HSPF target-based uptake in a conventional tillage crop land use. Modeling the uptake of nitrate by a growing plant with an adjustable root nitrate uptake capacity.

13.3.3.2 Compare N Uptake and Export between Precipitation Factors 1.0 and 1.1

Figure 13.13 shows the differences of N uptake between precipitation factors 1.0 and 1.1 by the three methods of HSPF target-based uptake simulation. There is no difference in uptake between precipitation factors 1.0 and 1.1 for method A because the uptake is not affected by soil moisture. For the other two methods, N uptake is higher for precipitation factor 1.1 rather than by precipitation factor 1.0. The effect is more pronounced in the dry years (e.g., 1991 and 1995) than the wet years (e.g., 1986 and 1996). This is because 10% change in precipitation has more impact on plant uptake

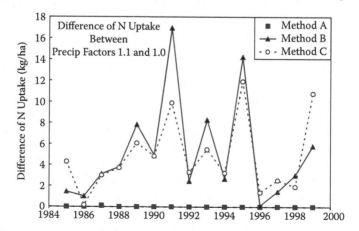

FIGURE 13.13 Difference of plant N uptake between precipitation factors 1.1 and 1.0 simulated by three methods of target-based uptake in a conventional tillage crop land use.

TABLE 13.3

DIN 15-Year Average Budget in Crop Simulation by HSPF Target-Based Plant N Uptake

Modules	Method A		Method B		Method C	
	Precipitation Factor					
	1.0	1.1	1.0	1.1	1.0	1.1
	DIN Simulation (kg/ha)					
Initial storage	58.59	58.39	58.74	57.47	58.76	58.30
Air depot	8.20	8.20	8.20	8.20	8.20	8.20
Application	133.41	132.90	133.41	132.90	133.41	132.90
Mineralization–immobilization	33.65	34.35	20.67	25.16	6.66	12.52
Volatilization	−6.97	−6.76	−7.31	−7.01	−7.64	−7.32
Denitrification	−0.12	−0.11	−0.13	−0.12	−0.15	−0.14
Leach/runoff	−27.04	−29.39	−28.59	−30.59	−30.81	−32.96
Uptake	−141.00	−140.16	−126.07	−128.47	−110.41	−112.06
Sum (input-output)	0.12	0.01	0.18	0.07	0.26	0.12

Note: Negative means loss from the soil; positive means gain in the soil.

in drier years. From Figure 13.13 we can see that method B yields a greater fluctuation in the differences of N uptake between precipitation factors 1.0 and 1.1. This is due to the fact that method B assumes plant uptake to be a step function of soil moisture crossing the wilting point, whereas method C assumes that plant uptake gradually changes with soil moisture.

All three methods simulate higher DIN export by precipitation factor 1.1 than that by precipitation factor 1.0. There is no significant difference among the three methods. This agrees with the analysis in the previous section that flow discharge plays a key role in DIN export in the HSPF simulation.

Table 13.3 provides the 15-year average DIN budget in crop simulation using precipitation factors 1.0 and 1.1 by the three methods. In this simulation, the atmospheric deposition is assumed to be unchanged with the precipitation factors. The HSPF simulates the effect of nutrient application by precipitation: If the precipitation intensity is over a certain threshold, for example 1 cm/h, the application will be delayed or eliminated. Therefore, there is a 1.5-kg/ha difference in the application between the two precipitation factors. The DIN uptake simulated by method A usually is not affected by soil-moisture conditions. The lower uptake at precipitation factor = 1.1 may be due to more DIN export, which causes the DIN storage in some time steps to be lower than the target. Methods B and C yield more uptake at the higher precipitation factor.

All three methods result in greater DIN export with the higher precipitation factor, in agreement with common observations. From the DIN budget table, however, it is difficult to evaluate which method simulates more reasonably the relationship between soil moisture and DIN uptake/export. This becomes clearer if we analyze the patterns in finer time steps.

From the overall comparison, our results showed that method C simulates N uptake and DIN export more appropriately than the other two methods. We recommend that the HSPF target-based uptake simulation consider soil-moisture effect and use method C in the watershed nutrient assessment.

13.4 DISCUSSION

13.4.1 Unit of the Maximum Uptake Rate in Michaelis–Menten Equation: Mass or Concentration?

Michaelis–Menten kinetics has been used widely to represent growth with a limiting substrate, such as bacteria growth rate in containers (Metcalf and Eddy 1979). In a fixed volume of container, there is no difference for the unit of bacteria growth rate to be expressed in bacteria concentration (mg/L) in a unit of time or bacteria mass (mg) in a unit of time.

Different from the bacteria growth in containers, watershed models are generally based on the unit area of land use, and the volume of DIN solution in a unit land area changes due to changes in hydrology. In the plant growth environment, the volume of soil moisture would increase from precipitation. The water will then dissolve DIN from the soil matrix during the infiltration, percolation, and lateral transport processes. Below the field capacity, more water generally wets more soil, and more DIN solutes actively supply the plants, consistent with observations in arid areas that more precipitation increases productivity (Law et al. 2000). The mass uptake is the product of the concentration of the DIN solution and the volume of soil moisture. Therefore, in the conditions below the field capacity, using concentration as the unit of uptake (e.g., equation 13.3) can generate a correct relationship between uptake and soil moisture. However, above the field capacity, much of the solution may bypass the plant's root system. Therefore, we can no longer use volume of soil moisture times DIN concentration to calculate uptake. Otherwise, it yields higher uptake and less export in the elevated runoff, which is contrary to our discussion of Figure 13.2.

The synthetic method (equation 13.7) uses mass as the unit of maximum uptake rate. Below the field capacity, more moisture causes more of the DIN in the soil to be involved in plant uptake, thereby increasing the uptake. Above the field capacity, more moisture will dilute DIN concentration, thereby reducing the uptake. The relationship between uptake and soil moisture is consistent with what we discussed in Figure 13.2. Therefore, the synthetic method satisfies our expectation and is recommended. The simulation results presented in Figures 13.5 and 13.6 and Table 13.1 support this analysis.

We have also observed similar results for both the synthetic method and the initial HSPF in plant uptake simulation at conditions below the field capacity. This is because both computer calculations involve the same parameter values, though in different orders of calculation. We still recommend using the synthetic method for conditions below the field capacity. This is based on the consideration of the physical meaning of the uptake and its calculation equation. The uptake is dependent on soil DIN storage, root development, plant growth stage, and plant demand. The maximum uptake is usually assessed in terms of the mass of DIN per unit area per year

(Frissel 1978; Kimmins 1987; Castrol et al. 1997), instead of the maximum level of DIN concentration that can be uptaken. Therefore, for the maximum uptake rate in the user input, it is more appropriate to use mass rather than concentration as the unit of maximum uptake.

13.4.2 Applicability of Michaelis–Menten Equation in Forest Simulations in HSPF

In the HSPF Michaelis–Menten module, the maximum uptake rate is specified in the user input. Users normally assign monthly maximum uptake rates for the first day of a month, which are then interpolated to daily or even finer time steps. The forest is relatively stable within a few weeks, and the maximum possible uptake changes rather gently day by day. The interpolated daily maximum uptake rate by the HSPF module can resemble the changes in plant development. Our observation shows that the Michaelis–Menten module does not simulate cropland as well as it simulates forest. This is primarily because the daily changes in uptake demand by crops are more irregular.

To overcome the problems with the user-defined maximum uptake, it is better to link the maximum uptake rate directly to the biomass that is simulated within the model, as we have done in the SWAT modification. However, we did not follow the same procedure because (a) HSPF does not simulate plant growth using as sophisticated a procedure as SWAT, and (b) it requires much more code modification on the HSPF Michaelis–Menten module to link plant biomass to uptake, as well as a change of input format. Our modified HSPF Michaelis–Menten module proposed in this chapter does not change the format of model input. There is no noticeable change in the user operation. Because the unit of the maximum mass uptake in the input is different from concentration, the numbers are of course different, and new calibration is needed.

13.4.3 Improvement of N Uptake for HSPF Target-based Module

The HSPF target-based N-uptake simulation and calibration is based on the balance between the target DIN uptake and the expected DIN export. Unlike the Michaelis–Menten kinetics, it does not simulate the uptake process by the effect of DIN concentration in soil water.

In method C, we have implemented the target-based uptake simulation by incorporating soil-moisture effect. The model assumes that the uptake–moisture relation follows an empirical function: The uptake increases with soil moisture from the wilting point to the field capacity, and slightly decreases with soil moisture above the field capacity. The simulated annual N uptakes are consistent with the observed N uptakes in the years of different water availabilities. While the initial target-based module assumes plant uptake to be a step function of wilting point (i.e., method B), it simulates DIN uptake and export less accurately. Method A, which is simply based on plant demand without considering soil moisture, is the worst simulation scenario among the three methods.

In the HSPF Michaelis–Menten simulation, the monthly maximum uptake rate is in the user-defined input, which reflects approximately the monthly biomass/plant demand and the general soil–root conditions. Different from a developed forest where biomass varies rather regularly within a year and over the years, the variation of a crop's biomass with seasons is more irregular. It is difficult to obtain reasonable values of the maximum uptake rate for crops for user input in the HSPF module, and the interpolated maximum uptake in each time step is unlikely to reflect such variations. Therefore, this work does not use the Michaelis–Menten equation to implement the HSPF target-based module, but adopts the empirical function.

The better implementation is to consider plant biomass to be one of the factors affecting the maximum uptake, as we did in the modification for the SWAT demand-based uptake. In this context, the HSPF model needs to be improved for the simulation of plant biomass in each time step so that the uptake simulation of crops will be a demand-based, or a target-based, uptake, with the Michaelis–Menten kinetics as an additional step controlling the uptake process.

13.4.4 ADDITIONAL CONTROL OF UPTAKE BY MICHAELIS– MENTEN EQUATION IN SWAT MODEL

The SWAT yields reasonable simulations in plant biomass and uptake (Figures 13.7 and 13.9), as well as the response of biomass to precipitation (Figure 13.10). Its plant N uptake simulation is based on the plant's demands in terms of plant growth stages and the effect of water availability on plant growth. In this context, the availability of soil water affects the simulated plant N uptake. However, the plant uptake process itself in SWAT does not consider soil water and DIN concentration directly. By incorporating the Michaelis–Menten kinetics into the SWAT plant-uptake module, the uptake process is linked directly with soil moisture and DIN concentration. The modified SWAT, therefore, improves N-uptake simulation, e.g., in the daily uptake pattern (Figure 13.8), and has a more reasonable response of biomass to wet conditions, e.g., at precipitation factors ≥2.5. These improvements are useful in watershed simulations for nutrient management.

13.5 CONCLUSION

Plant N uptake is controlled by plant demand on nitrogen, the availability of DIN in soil, and the DIN concentration in soil water solution. The Michaelis–Menten method of plant uptake considers DIN concentration and soil moisture effects, in addition to its consideration of plant N demand. The synthetic method of plant N uptake proposed by Wang and Linker (2006) that uses mass as the unit of maximum uptake rate for both above and below field capacity can simulate plant nitrogen uptake more accurately. The plant N-uptake simulation in the current SWAT and the HSPF target-based methods considers plant demand on nitrogen, but lacks consideration of DIN concentration and soil moisture effects. The consideration of the latter improves the simulation of DIN uptake and export. Incorporating the effect of soil moisture on plant uptake, such as using the Michaelis–Menten kinetics in the demand-based or target-based method, improves the simulations of plant N uptake.

ACKNOWLEDGMENT

The authors greatly appreciate Nancy Simmons for providing portable SWAT source code, and Dr. Yackov Pachepsky for professional input.

REFERENCES

Aber, J. D., and C. A. Federer. 1992. A generalized, lumped parameter model of photosynthesis, evapotranspiration and net primary production in temperate and boreal forest ecosystems. *Ecologia* 92: 463–474.

Aber, J. D., S. V. Ollinger, and C. A. Federer. 1995. Predicting the effect of climate change on water yield and forest production in the northeastern United States. *Climate Res.* 5: 207–222.

Arnold, J. G., and N. B. Sammons. 1988. Decision support system for selecting inputs to a basin scale model. *Water Res. Bull.* 24 (4): 749–759.

Arnold, J. G., R. Srinivasan, R. S. Muttiah, and J. R. Williams. 1998. Large area hydrologic modeling and assessment, 1: Model development. *J. Am. Water Res. Assoc.* 34 (1): 73-89.

Bhuyan, S. J., J. K. Koelliker, L. J. Marzen, and J. A. Harrington. 2003. An integrated approach for water quality assessment of a Kansas watershed. *J. Environ. Modeling Software* 18 (5): 473–484.

Bicknell, B. R., J. C. Imhoff, J. L. Kittle, A. D. Donigian, and R. C. Johanson. 1997. *Hydrological simulation program—FORTRAN: User's manual for Version 11.* EPA/600/R-97/080. Athens, GA: NERL, U.S. EPA.

Bowie, G. L., W. B. Mills, D. B. Porcella, C. L. Campbell, J. R. Pagenkopf, G. L. Rupp, K. M. Johnson, P. W. H. Chan, S. A. Gherini, and C. E. Chamberlin. 1985. *Rates, constants, and kinetics formulations in surface water quality modeling.* EPA/600/3-85/040. Athens, GA: NERL, U.S. EPA.

Castrol, M. S., K. N. Eshleman, R. P. Morgan, S. W. Seagle, R. H. Gardner, and L. F. Pitelka. 1997. *Nitrogen dynamics in forested watersheds of the Chesapeake Bay.* STAC Publication 97-3. Annapolis, MD: Scientific & Technical Advisory Committee of the Chesapeake Bay Program.

Fenn, M. E., and M. A. Poth. 1999. Temporal and spatial trends in streamwater nitrate concentrations in the San Bernardino Mountains. *South. Calif. J. Environ. Qual.* 28: 822–836.

Frissel, M. J. 1978. *Cycling of mineral nutrients in agricultural ecosystems.* Symposium proceedings. North Holland: Elsevier.

Glass, A. D. M., and M. Y. Siddiqui. 1984. The control of nutrient uptake rate in relation to the inorganic composition of plants. In *Advances in plant nutrition*, ed. P. B. Tinker and A. Lauchi, 103–148. New York: Praeger.

Grant, R. F. 1991. The distribution of water and nitrogen in the soil–crop system: A simulation study with validation from a winter wheat field trial. In *Nitrogen turnover in the soil-crop system*, ed. J. J. R. Groot, P. de Willingen, and E. L. J. Verberne. Dordrecht: Kluwer Academic.

Kimmins, J. P. 1987. *Forest Ecology.* New York: Macmillan.

Law, B. E., R. H. Waring, P. M. Anthon, and J. D. Aber. 2000. Measurements of gross and net ecosystem productivity and water vapour exchange of a *Pinus ponderosa* ecosystem, and an evaluation of two generalized models. *Global Change Biology* 6: 155–168.

Likens, G. E., and F. H. Bormann. 1995. Biogeochemistry of a forested ecosystem. 2nd ed. New York: Springer-Verlag.

Metcalf and Eddy, Inc. 1979. *Wastewater engineering: Treatment, disposal, reuse.* New York: McGraw-Hill.

Neitsch, S. L, J. G. Arnold, J. K. Kintry, J. R. Williams, and K. W. King. 2002. Soil and water assessment tool theoretical documentation. TWRI Report TR-191. College Station, TX: Texas Water Resources Institute.

Ramos, C., and E. A. Carbonell. 1991. Nitrate leaching and soil moisture prediction with the LEACHM model. In *Nitrogen turnover in the soil–crop system*, ed. J. J. R. Groot, P. de Willingen, and E. L. J. Verberne, 171–180. Dordrecht: Kluwer Academic.

Timlin, D.J., Y. Pachepsky, V. A. Snyder, and R. B. Bryant. 2001. Water budget approach to quantify corn grain yields under variable rooting depths. *Soil Sci. Soc. Am. J.* 65: 1219–1226.

Tinker, P. B., and P. H. Nye. 2000. *Solute movement in the rhizosphere.* New York: Oxford University Press.

Wagenet, R. J., and J. L. Hutson. 1989. LEACHM: Leaching estimation and chemistry model: A process-based model of water and solute movement transformation, plant uptake and chemical reactions in the unsaturated zone. *Continuum.* Vol. 2, version 2. Ithaca, NY: Water Resources Inst. Cornell Univ.

Wang, P., and L. Linker. 2001. Improvement of HSPF watershed model in plant uptake and DIN export from forest. *EOS Trans. AGU* 82 (47). Fall Meet. Suppl., Abstract H12H-12, 2001, San Francisco.

Wang, P., and L. Linker. 2006. A correction of DIN uptake simulation by Michaclis–Menton saturation kinetics in HSPF watershed model to improve DIN export simulation. *J. Environ. Model. Software* 21: 45–60.

Willigen, P. De. 1991. Nitrogen turnover in the soil–crop system: Comparison of fourteen simulation models. In *Nitrogen turnover in the soil–crop system*, ed. J. J. R. Groot, P. de Willingen, and E. L. J. Verberne, 141–150. Dordrecht: Kluwer Academic.

Index

A

Aber, J. D., 73, 280
Abrahamsen, P., 9
Acock, B., 23–24
Active nitrogen uptake rates, 227
Addiscott, T. M., 136
Ahuja, L. R., 260, 263
Alanine, 113, 114
Albrecht, S. L., 21, 24
Amino acids, 75–76, 97–98, 99–100, 104–107
    auxin permease (AAAP) superfamily, 111
    plant membranes leaky to, 110
    plant uptake of, 115–116
    in root cells, 108–109, 115–116
Amino sugars, 99–100
Ammonium, 6, 55–56, 109, 222
    amino acid uptake and, 115
    assimilation, 74
    movement to root surfaces, 185–190
Annual crops and root depth, 245–246, 256–257
Antibiotic peptides, 101
Arginine, 111
Aslyng, H. C., 129
Assimilate partitioning, 176–177
Assimilation
    ammonium and nitrate, 74
    carbon, 220–240
Atmospheric nitrogen deposition, 73–74
Availability, nitrogen
    atmospheric nitrogen deposition and, 73–74
    dissolved or soluble organic nitrogen and, 75–76
    mineralization and, 71–73, 85

B

Baker, J. T., 224
Barber, S. A., 223, 239
BassiriRad, H., 5, 9, 77
Beer's law, 132
Beevers, L., 109
Benbi, D. K., 72
Bergersen, F. J., 19
Beuve, N., 50
Biomass
    accumulation and partitioning, 132–133, *140, 141*
    compartments, crop, 128–129
    leaf, 201–202
    precipitation factors and, 293–296, 298–300
    roots, 55, 213, *214*
Boon-Long, P., 20
Boote, K. J., 5, 14, 15, 18, 20, 38
Branching processes, 202, 203–204
Bray, C. M., 109
Bristow-Campbell method, 207
Brun, W. A., 34, 38
Brunt equation, 207
Bruun, S., 263
Bucci, S. J., 83
Budgets, nitrogen, 262–264, 270–272
Bulk concentration of nutrient and attendant concentration in soil solution, 78–79
Burdine, N. T., 180

C

Campbell, G. S., 180, 181
Canopy development, crop, 131–132
    leaf area index and, 175–176
Cao, W., 227
Carberry, P. S., 131
Carbon, 173–174, 199
Carbon assimilation model, 7, 9, 220–223, 237–240. *See also* Potato model
    calculations for passive and active nitrogen-uptake rates, 227
    carbon dioxide levels and, 232–235
    growth chambers, 224–225
    measured data, 232–235
    plant cultivation and environment, 226
    plant data collection, 226–227
    plant model, 231
    simulations, 235–236
Carbon dioxide, 91–93, 224, 226, 232–235
Carboxyl, 102–103
CERES models, 18, 20, 148, 164
Chapman, J. F., 56
Christou, M., 98
Clarkson, D. T., 109
Clausnitzer, V., 202
Clement, C. R., 50
Cowling, E. B., 73
CROPGRO model
    comparative analysis of temperature sensitivities in different grain legumes, 33

development of, 5, 14–16
FORTRAN code, 15, 42–46
future improvement, 39
growth and yield response to applied nitrogen
    fertilization, 36–38
nitrogen fixation and nodule growth in, 19,
    20–21
performance, 21–38
priority for assimilate in, 23–24
robustness for different grain legumes, soils,
    and climate conditions, 38
scheme and order of plant N balance and N
    fixation in, 16–18
seasonal patterns, 25–30
sensitivity analysis, 30–38
soil balance in, 14–15
source:sink relations and, 34–36
SOYGRO 4.2 model and, 15
specific legumes and, 15, 33, 38
temperature sensitivity and, 30–33
water deficit and, 24–25
Cropping System Model (CSM), 15–16
Crops. *See* Plants/crops

D

Daisy model
application, 190–191
assimilate partitioning, respiration, and net
    production, 176–177
carbon entry, 173–174
development, 170
EU-Rotate_N model and, 252
extraction of soil water by roots, 180–185
generic crop model, 173–179
leaf area index and canopy structure,
    174–176
nitrogen uptake and stress, 179
overview, 170–173
phenological development, 177–178
photosynthesis, 173–175
process descriptions in, *172–173*
root production and development, 178, 204
senscence, 177
solute movement to root surfaces, 185–190
water uptake and water stress, 178–179,
    191–193
Davies, H. V., 226
Demand-based nitrogen uptake, 4, 205–206,
    220–222, 237–240, 265, 283–285,
    289, 293–300
Deposition, atmospheric nitrogen, 73–74
Depth, root, 137
of annual crops, 245–246
crop rotations and, 252–253
deep, 246–247, 254–256
EU-rotate_N model, 252–256

experimental details and data analysis,
    247–249
important factors for nitrogen uptake and
    deep, 256
potential nitrate uptake and, 249–252
soil nitrogen and, 254–256
DeVries, J. D., 21, 34
Dewar, R. C., 220
Diffusion coefficient, soil, 186–187
Dissolved inorganic nitrogen (DIN), 278
Michaelis-Menten saturation kinetics and,
    284–289, 291–293, 300–302
plant nitrogen uptake simulations and,
    282–285
soil moisture and, 280–282, 296–297, *298*
Dissolved organic carbon (DOC), 98
Dissolved organic nitrogen (DON), 2, 4, 5–6,
    75–76
amino acids/sugars and, 97–98, 99–100,
    104–107
antibiotic peptides in soil and, 101
compared to mineral nitrogen pool, 96–98
concentrations in soil, 96–103
in drainage water, 102
energized transporters, 110–112
large-scale *vs.* small-scale processes
    involving, 116–117
ligand exchange between carboxyl and,
    102–103
microbial regulation in soil, 103–107
in nonsterile soil, 114–115
plant production of, 107–109
plant uptake of, 109–116, 113–115
soil organic matter (SOM), 97–98
study of, 95–96
watershed models, 278
Diurnal cycles of nitrogen uptake, 58–60
Drainage waters, 102
DSSAT models, 15, 16, 18
Duynisveld, W. H. M., 239
Dwyer, L. M., 131
Dynamics, nitrogen, 133–135, 141–143

E

Egli, D., 24
Enzymes
plant extracellular, 113
soil, 100–101, 103–107
EU-Rotate_N model, 252–256
Extraction of soil water by roots, 180–185

F

Fate estimation, nitrogen, 263–264, 265–266
Faure-Rabasse, S., 50, 53, 54

Federer, C. A., 280
Fertilization, nitrogen
  agricultural dependence on, 259–260,
      272–273
  Daisy model application, 191
  grain demand and use of, 260–261
  grain yield and aboveground dry matter and,
      209, *210, 211, 214–215*
  growth and yield response to, 36–38
  impact on nitrogen uptake, 61–62
  mineralization and, 71–73
  nitrogen fate estimation and, 263–264
  quantifying nitrogen budgets of agricultural
      systems and, 262–263, 270–272
  reactive nitrogen in the environment and,
      261–262
Final leaf number (FLN), 130–131
Fitter, A., 85
Flooding and nitrogen fixation, 24–25
Forde, B. G., 49, 109
French, B. K., 135

G

Galloway, J. N., 73
Gastal, F., 221
Gene expression and nitrogen deprivation, 58
Genetic parameter, rice, 160, 164
Gerendas, J., 112
Gerwitz, S., 183, 204
GLYCIM model, 228, 231
Glycine, 114, 115
Godwin, D. C., 14
Goodchild, D. J., 19
Gosse, G., 55
Grain protein. *See* Rice
Green area index (GAI), 131–132, 136, 138, *139*
Gross photosynthesis (PG), 16, *17,* 18
Growth respiration rate, 176–177

H

Hammer, G. L., 132
Hansen, J., 9, 23
Hansen, S., 129, 191
Hatfield, J. L., 260
Hayati, R., 20, 24
Heinrich, R., 79
Henry, H. A. L., 76
Herkelrath, W. N., 180
High-affinity transport systems (HATS), 5, 49,
      53–54, 66–67
  light/darkness cycle and, 58–60
  nutrient concentrations, 78–79
  ontogeny and, 60–61
  root temperature and, 60

Hirel, B., 272
Histidine, 111
Hopmans, J. W., 202
Hurley Pasture Model, 201
Hydrological Simulation Program-FORTRAN
      (HSPF), 7–8, 278–280
  first-order kinetics of plant nitrogen uptake
      in, 284
  method and approach, 286–290, *291*
  Michaelis-Menten plant nitrogen uptake
      simulation, 286–287, 300–302
  moisture effect and, 296–297, *298*
  plant nitrogen uptake simulations in,282–285
  precipitation and, 293–296, 298–300
  results, 291–293
  target-based plant nitrogen uptake in,
      289–290, *291,* 296–300
Hypoxia, 262

I

Influx rates, nitrogen fixation, 51, 54, 56–61
  nitrogen deprivation and, 58
  ontogeny and, 60–61
Integrated management practices (IMP), 1
Irrigated *vs.* rain-fed conditions, 25–30
Isoleucine, 114

J

Jakobsen, B. F., 178
Jamieson, P. D., 6, 128, 129, 131, 135, 137
Jefferies, R. L., 76
Jeuffroy, M. H., 221
Johnson, J. R., 201
Johnson model, 201
Jones, D. C., 14
Jones, D. L., 5, 106

K

Kinetics
  Michaelis-Menten saturation, 9, 284–293,
      300–302
  nitrogen influx, 56–58
  root, 77–78
Klipp, E., 79
Kristensen, H. L., 248

L

Larcher, W., 109
Lateral root biomass, 55
Lawless, C., 132
Lawn, R. J., 34, 38

Leaves
    area distribution (LAD), 176
    area index (LAI), 156–157, 171, 174–176
    biomass, 201–202
    carbon assimilation and, 231
    nitrogen dynamics, 133–135
    senescing, 65–66, 177
Legg, B. J., 135
Lemaire, G., 221
Leucine, 114
Leverage of plant attributes and soil
        environmental factors in nutrient
        acquisition and growth, 76–77
Ligand exchange, 102–103
Light/darkness cycle, 58–60
Light-use efficiency (LUE), 128, 129, 132–133
Linker, L., 285, 286, 287, 288, 289
Low-affinity transport systems (LATS), 5, 49,
        53–54, 66–67
    in induced vs. noninduced plants, 56–57
    light/darkness cycle and, 58–60
    ontogeny and, 60–61
    root temperature and, 60
Lysine, 111

M

Ma, L., 5, 264, 265, 266
Maintenance respiration, 176
Maize model, 129
    biomass accumulation and partitioning, 132
    canopy development, 131
    experimental verification, 138–141
    nitrogen dynamics, 133–135
    phenology, 130
    water use and responses to water shortage,
        135–136
Malagoli, P., 5, 50, 66, 239
Malone, R., 5, 264, 265, 266, 268
Marschner, H., 109
Martre, P., 132, 162
Mass flow, 80–83, 235, 237, 238
McGechan, M. B., 208
McKee, H. S., 109, 113
McNulty, S. G., 74
Mendham, N. J., 56
Meng, Y., 164
Michaelis-Menten function, 5, 9, 18, 53, 56–57,
        77, 82, 110, 223, 280
    roots, 205
    SWAT and HSPF models, 7–8, 284–293,
        300–302
Microbial regulation of dissolved organic
        nitrogen (DON) in soil, 103–107
Middleton, K. R., 222
Mineralization, 71–73, 85, 252–253
Mobilized nitrogen, 65–66

Model Maker software, 55
Monteith, J. L., 135
Mortality, root, 204
Muchow, R. C., 129, 131
Mycorrhizal infection of plants, 83–85, 86, 114

N

Nassar, A. H., 111
Native ecosystems
    nitrogen availability in, 71–76
    plant infections and, 83–85
    robust modeling, 85–86
    root system characteristics and, 76–83
Nitrate, 185–190, 222
    injection into deep-rooted plants, 248–249
    potential uptake at different root depths,
        249–252
Nitrogen, reactive, 261–262
Nitrogen availability
    atmospheric nitrogen deposition and, 73–74
    dissolved or soluble organic nitrogen and,
        75–76
    mineralization and, 71–73, 85
    shortages, 137, 220
Nitrogen deficiency, 137, 220
    effects on influx rates and gene expression,
        58
    nodule growth and nitrogen fixation triggered
        by, 21–23
    vegetative consequences of, 19–20
Nitrogen deposition, atmospheric, 73–74
Nitrogen fate estimation, 263–264, 265–266
Nitrogen fertilization
    agricultural dependence on, 259–260,
        272–273
    Daisy model application, 191
    grain demand and use of, 260–261
    grain yield and aboveground dry matter, 209,
        210, 211, 214–215
    growth and yield response to, 36–38
    impact on nitrogen uptake, 61–62
    mineralization and, 71–73
    nitrogen fate estimation and, 263–264
    quantifying nitrogen budgets of agricultural
        systems and, 262–263, 270–272
    reactive nitrogen in the environment and,
        261–262
Nitrogen fixation
    influx rates, 51, 54, 56–61
    kinetics, 56–58
    nodule growth and, 19, 20–21, 23–25
    in rain-fed vs. irrigated conditions, 25–30
    scheme and order, 16–18
    seasonal patterns, 25–30
    sensitivity to water deficit, 24–25
    source:sink relations effect on, 34–36

temperature sensitivities, 30–33, 60
triggered by nitrogen deficit, 21–23
Nitrogen mobilization, 65–66
Nitrogen offtakes, 210, 215
Nitrogen remobilization, 8–9, 62–66, 157–158
Nitrogen stress, 179
Nitrogen uptake
   budgets, 262–264, 270–272
   calculation of unregulated and regulated,
      54–55
   carbon assimilation rate and, 220–240
   carbon dioxide levels and, 232–235
   current status in modeling, 2–8
   by deep roots, 246–247, 254–256
   demand-based, 4, 205–206, 220–222,
      237–240, 265, 283–285, 289, 293–300
   EU-Rotate_N model, 252–256
   flow calculations, 52–53
   future research needs, 8–10
   during growth cycle, 58–62
   interactions among main processes
      contributing to, 2, 3, 4
   kinetic equations, 53–54
   lateral root biomass and, 55
   light/darkness cycle and, 58–60
   managing, 1–2
   nitrogen availability and, 71–76, 85, 220–222
   nitrogen loss in tile drainage and, 268–270
   nitrogen stress and, 179
   partitioning and remobilization to vegetative
      and reproductive tissues, 62–66
   rice, 154–157, 162–164
   root depth and, 245–257
   root system characteristics and, 76–83,
      136–137, 204–205
   root temperature and, 60
   Root Zone Water Quality Model (RZWQM)-
      estimated, 266–267, 268
   simulations, 2–4
   single-leaf plant model, 231
   soil moisture and, 280–282
   solute movement to root surfaces, 185–190
   target-based, 289–290, 291, 296–300
   tile drainage and, 268–272
   transpiration and, 2, 82
   watershed, 278–280
   winter oilseed rape, 50–54
Nitrogen use efficiency (NUE), 49
Nodule growth, 19
   applied nitrogen fertilization and, 36–38
   dynamics of, 23–24
   initialization and parameters for, 20–21
   nodule death rate and, 25
   priority for assimilate and, 23–24
   in rain-fed vs. irrigated conditions, 25–30
   rate (RGR), 19, 22, 31–33, 77–79
   seasonal pattern, 25–30

sensitivity to water deficit, 24–25
temperature sensitivities, 30–33
triggered by nitrogen deficit, 21–23
Nonsterile soil, 114–115
Noquet, C., 50
NSTRES ratio, 20
Nutrient concentration in soil solution, 78–79
Nye, P. H., 76, 77, 82

O

Offtakes, nitrogen, 210, 215
Ontogeny and nitrogen influx, 60–61
Oparka, K. J., 226
ORYZA2000 model, 164

P

Page, E. R., 183, 204
Pare, K., 264
Parkin, T. B., 263
Partitioning, 19–20, 62–66
   assimilate, 176–177
   biomass accumulation and, 132–133, 140, 141
Passive nitrogen uptake rates, 227
Penman, H. L., 135
Peptides, antibiotic, 101
Phenological development, crops, 130–131,
   177–178
Photosynthesis
   consequences of nitrogen deficiency on,
      19–20
   Daisy model, 173–175
   gross (PG), 16, 17, 18
   roots and, 201, 203–204
   temperature sensitivity and, 30, 31
Photosynthetically active radiation (PAR), 5, 132
Physiological development time (PDT), rice,
   154–156, 164, 165
Pickering, N. B., 18, 224
Plants/crops. See also Roots
   amino acid uptake, 115–116
   annual, 245–246, 256–257
   assimilate partitioning, 176
   attributes and soil environmental factors
      leveraging, 76–77
   biomass accumulation and partitioning,
      132–133, 140
   biomass and precipitation, 293–296
   biomass compartments, 128–129
   branching processes, 202
   canopy development, 131–132, 175–176
   components, 200–202
   demand-based nitrogen uptake, 4, 205–206,
      220–222, 237–240, 265, 283–285, 289
   dissolved inorganic nitrogen (DIN) export,
      280–282

dissolved organic nitrogen (DON) uptake,
    109–116, 113–115
extracellular enzymes and, 113
green area index (GAI), 131–132, 136, 138,
    139
growth and development processes, 14,
    58–62, 200–202, 220–222
growth respiration, 176–177
leaf area index (LAI), 156–157, 171, 174–176
light-use efficiency (LUE), 128, 129,
    132–133
maintenance respiration, 176
membranes leaky to amino acids, 110
membranes with dissolved organic nitrogen
    (DON) transporters that are
    energized, 110–112
mycorrhizal infections of, 83–85, 86, 114
nitrogen dynamics, 133–135, 141–143
nitrogen fate estimation and, 263–264
nitrogen losses balanced with production of,
    262
nitrogen uptake simulations in HSPF and
    SWAT, 282–285
in nonsterile soil, 114–115
ontogeny and nitrogen influx, 60–61
passive and active nitrogen uptake rates, 227
phenological development, 130–131, 177–178
precipitation and nitrogen uptake by,
    293–296, 298–300
production of dissolved organic nitrogen
    (DON), 107–109
quantifying nitrogen budgets of, 262–263
respiration, 176–177, 231
root biomass, 55
root temperature, 60
rotations, 252–253
sampling, 207
single-leaf plant model, 231
target-based nitrogen uptake, 289–290, *291*
transpiration, 2, 82, 135–136, 182, 235,
    238–239
watershed management, 278–302
water uptake, 127–128, 135–136
Polyamines, 111
Potato model, 129. *See also* Carbon assimilation
    model
biomass accumulation and partitioning, 132,
    133
canopy development, 131
experimental verification, 141
nitrogen dynamics, 133–135
phenology, 130
water use and responses to water shortage,
    135–136
Precipitation factors in nitrogen uptake, 293–296,
    298–300
Priestley, C. H. B., 135

Proteases
    plant, 113
    soil, 103–104
Protein
    grain, 158–160
    soil, 97–101
    transport, 105–106
Putrescine, 111

R

Rain-fed versus irrigated conditions, 25–30
RCSODS model, 148, 164
Reactive nitrogen in the environment, 261–262
Relative growth rate (RGR), 19, 22, 77–79
    root:shoot ratio and, 92
Remobilization, nitrogen, 8–9, 62–66
    post-anthesis, 157–158
Respiration, 176–177, 231
Rice
    field experiments, 149–152
    genetic parameter, 160, 164
    grain protein accumulation, 158–160, 165
    grain protein model development, 152–153
    grain quality, 148
    growth modeling, 148–149
    model validation, 160–162
    nitrogen uptake, 162–164
    physiological development time (PDT),
        154–156, 164, 165
    post-anthesis nitrogen remobilization,
        157–158
    post-anthesis nitrogen uptake, 156–157
    pre-anthesis nitrogen uptake, 154–156
    RiceGrow model, 148, 152–154
RiceGrow, 148, 152–154
    model description, 154–160
    model validation, 160–162
Richards equation, 200
Richter, J., 72
Ritchie, J. T., 135
RIW model, 164
Roots. *See also* Plants/crops
    amino acids and, 75–76, 108–109, 115–116
    architecture modeling, 198–199
    biomass, 55, 213, *214*
    branching, 202, 203–204
    control exerted by allocation to, 79–80
    deposited nitrogen and, 74
    depth, 137, 245–257
    dissolved or soluble organic nitrogen and,
        75–76, 112–113
    extraction of soil water by, 180–185
    field experiment modeling, 206–207
    grain yield and aboveground dry matter and
        fertilization of, 209, *210*

growth and elongation direction, 202–203, 212–213
kinetics, 77–78
and leveraging of plant attributes and soil environmental factors, 76–77
mass flow, 80–83
mineral nitrogen dynamics, 210–212
mortality, 204
mycorrhizal infection of, 83–85, 86
nitrogen demand, 205–206
nitrogen offtake, 210, 215
nitrogen uptake, 76–83, 136–137, 204–205
1-D system, 6–7, 204, 208–216
photosynthesis and, 201, 203–204
potential nitrate uptake, 249–252
production and development, Daisy model, 178
:shoot ratio optimization, 5, 91–93
solute movement to surfaces of, 185–190, 228–231
SPACYS model, 198–216
system characteristics modeling, 76–83
temperature, 60
3-D system, 6–7, 9, 202–204, 208–216
water uptake and, 136–137
Root Zone Water Quality Model (RZWQM), 7, 260
estimated nitrogen budget, 270–272
estimated nitrogen loss, 268–270
estimated nitrogen uptake under different management and years, 266–267, 268
materials and methods, 264–266
Rotations, crop, 252–253

S

Sau, F., 31
Schaeffer, S. M., 79
Schulze, E. D., 73
Seasonal pattern of nodule growth, 25–30
Seed growth
consequences of nitrogen deficiency for, 19–20
nitrogen fixation priority for assimilate over, 23–24
winter oilseed rape, 51
Semenov, M. A., 128, 129, 131, 137
Senescence, leaf, 65–66, 177
Sensitivity analysis
of different grain legumes, 33, 34
effect of source:sink relations, 34–36
growth and yield response to applied nitrogen fertilization, 36–38
impact of nitrogen fertilization, 61–62
temperature, 30–33
Sensitivity of nodule growth and nitrogen fixation to water deficit, 24–25

Sexton, P. J., 31
Shortages
nitrogen, 19–20, 21–23, 58, 137, 220
water, 128, 135–136
SIMRIW, 148
Sinclair, T. R., 129, 132, 135
Singh, U., 14
Skalar method, 55
Smith, G. S., 222
Sogbedji, J. M., 263
Soil
antibiotic peptides, 101
balance processes, 14–15
bulk concentration of nutrient and attendant concentration in, 78–79
concentrations of dissolved organic nitrogen (DON), 96–103
diffusion coefficient, 186–187
dissolved organic carbon (DOC) in, 98
environmental factors and plant attributes leveraging, 76–77
enzyme activity, 100–101, 103–107
mass flow, 80–83
microbial regulation of dissolved organic nitrogen (DON) in, 9, 103–107
mineral nitrogen, 210–212
moisture in plant nitrogen uptake and DIN export, 280–282, 296–297, 298
nitrogen, 4, 254–256
nonsterile, 114–115
physical properties, 77–78, 207–208
proteins in, 97–98
root system field experiment and, 206–207
sampling, 206–207
-vegetation-atmosphere transfer (SVAT), 171, 182
water extraction by roots, 180–185
water uptake and, 136–137, 182–185, 228–231
Soil and Water Assessment Tool (SWAT), 7–8, 278–280
demand-based nitrogen uptake in, 283–285, 289, 293–296
initial versus modified, 293–296
method and approach, 286–290, 291
plant nitrogen uptake simulations, 282–285
SOIL model, 200
Soil organic matter (SOM)
amino acids and amino sugars in, 99–100
concentration, 36–38, 97–98
Soluble organic nitrogen (SON), 75–76, 95
in drainage waters, 102
Solute movement to root surfaces, 185–190, 228–231
Somma, F., 205
Source:sink relations effect on nitrogen fixation, 34–36

SOYGRO 4.2 model, 15
SPACSYS model, 198–199
    accuracy, 213–216
    description, 199–205
    field experiment, 206–207
    inputs and parameterization, 207–208
    results, 208–213
Specific nitrogenase activity (SNA), 19, 21–23
    seasonal patterns, 28–30
    temperature sensitivity and, 31–33
Spiers, J. A., 82
Stanton, M. A., 24
Stewart, D. W., 131
Strebel, O., 239
Stress
    nitrogen, 179
    water, 178–179, 191–193

T

Tanner, C. B., 135
Target-based nitrogen uptake, 289–290, *291,* 296–300
Taylor, R. J., 135
Temperature and nodule growth/nitrogen fixation
    in different grain legumes, 33
    root, 60
    sensitivity analysis, 30–33
Thornton, B., 116
Thorup-Kristensen, K., 248
Tibbitts, T. W., 227
Tietema, A., 74
Tile drainage, 268–272
Tinker, P. B., 76, 77, 280
Titus, J. H., 79
Total dissolved nitrogen (TDN), 98–99
Transpiration, 2, 82, 135–136, 182, 235, 238–239
Transporters, dissolved organic nitrogen (DON), 110–112
TURFAC, 24, 30
2DSOIL model, 228–231

U

Urea, 112, 113–114

V

Valine, 114
Van Genuchten model, 200
Vegetative tissues, consequences of nitrogen deficiency on, 19–20
Verkroost, A. W. M., 221, 222

W

Wang, P., 285, 286, 287, 288, 289
Wassen, M. J., 221, 222
Water
    deficit and nitrogen fixation, 24–25
    demand, 128, 135–136
    drainage, 102
    extraction of soil, 180–185
    interactions between shortages of nitrogen and, 137
    rain-fed *vs.* irrigated conditions, 25–30
    shortages, 128, 135–136
    uptake and root:shoot ratio, 91–93
    uptake and stress, 178–179, 191–193
    uptake by plants, 127–128, 135–136
    uptake by soil, 136–137, 182–185, 228–231
Watershed management, 7–8, 9–10, 278–280
    demand-based nitrogen uptake in, 283–285, 289, 293–296
    first-order kinetics of plant nitrogen uptake in, 284
    Michaelis-Menten plant nitrogen uptake simulation, 286–287, 300–302
    modeling method and approach, 286–290, *291*
    moisture effect and, 296–297, *298*
    plant nitrogen uptake simulations and, 282–285
    precipitation and, 293–296, 298–300
    results, 291–293
    soil moisture in, 280–282
    target-based plant nitrogen uptake in, 289–290, *291,* 296–300
Weather data, 207
Wheat model, Sirius, 129
    biomass accumulation and partitioning, 132
    canopy development, 131–132
    experimental verification, 137–141
    nitrogen dynamics, 133–135
    phenology, 130–131
    water use and responses to water shortage, 135–136
Whitmore, A. P., 136
Wilkerson, G. G., 15
Wilson, D. R., 129
Winter oilseed rape
    calculation of unregulated and regulated uptake by, 54–55
    cycling in cereal crop rotations, 48
    developmental stage effect experiment, 51
    effects of nitrogen deprivation on influx rates and gene expression by, 58
    experimental treatment, labeling, and harvest for field, 52
    experimental treatments for nitrogen uptake measurements, 51
    field conditions, 51

growth conditions, 50–51
growth cycle, 58–62
induced versus noninduced plants, 56–57
input variables, 55
light/darkness cycle and, 58–60
low- and high-affinity transport systems, 5,
        49–50, 53–54, 66–67
modeling method, 53–56
nitrogen deprivation experiment, 50–51
nitrogen flow calculations, 52–53
nitrogen remobilization in, 62–66
nitrogen use efficiency, 48–50

ontogeny, 60–61
partitioning of nitrogen uptake in, 62–66
root temperature, 60
senescing leaves, 65–66
total nitrogen and isotopic analyses, 52
        yields, 48
Wright, G. C., 132

Y

Yanai, R. D., 76, 83